Titanic Day by Day

Titanic Day by Day

366 Days with the *Titanic*

Simon Medhurst

Great-grandson of Robert Hichens, quartermaster and helmsman of RMS *Titanic*

PEN & SWORD
HISTORY

First published in Great Britain in 2022 and reprinted in 2026 by
Pen & Sword History
An imprint of
Pen & Sword Books Ltd
Yorkshire – Philadelphia

ISBN 978 1 39901 143 3

A CIP catalogue record for this book is available from the British Library.

Typeset by Mac Style
Printed in the UK by CPI Group (UK) Ltd, Croydon, CR0 4YY.

MIX
Paper | Supporting responsible forestry
FSC® C013604

The Publisher's authorised representative in the EU for product safety is Authorised Rep Compliance Ltd., Ground Floor, 71 Lower Baggot Street, Dublin D02 P593, Ireland.
www.arccompliance.com

For a complete list of Pen & Sword titles please contact:

PEN & SWORD BOOKS LIMITED
47 Church Street, Barnsley, South Yorkshire, S70 2AS, England
E-mail: enquiries@pen-and-sword.co.uk
Website: www.pen-and-sword.co.uk
or
PEN AND SWORD BOOKS
1950 Lawrence Rd, Havertown, PA 19083, USA
E-mail: Uspen-and-sword@casematepublishers.com
Website: www.penandswordbooks.com

Contents

Foreword

by Helen Benziger
Great-granddaughter of Margaret (Molly) Brown,
rescued on lifeboat 6

By a happy accident of birth (as she puts it), Helen was born the great-granddaughter of Margaret and James Joseph Brown (more widely known as Molly and J. J.). She grew up as the daughter of G. Peter Benziger, a New York newspaper publisher and grandson of Margaret Brown, and Edith, an actress/model in Chicago.

To take Margaret's side, there was certainly enough room in the boat for many more. I can understand that her heart was breaking for those left behind and her instinct was most certainly to return for them. That she made her feelings known quite forcefully to Hichens is undeniable. I would have done the same thing without question.

She was not wrong. Was Quartermaster Hichens wrong? No. When Hichens was given command of lifeboat 6, he was given charge of the safety of the lives in that boat. They, and only they, were his sole responsibility.

Lifeboat 6 left early and there were not many in the water at that time. I am sure that had there been people near the lifeboat, as anyone would, he would have rescued them. They were too far out for that to be the case. Had he gone back to the ship, would they have been able to save more lives? Perhaps, but he would most certainly have been putting those in his charge at risk.

Many people that night had to make very hard decisions. I would ask that we all understand that none of us really know how we would have reacted in the same horrific situation.

Simon has written a book that is, in my opinion, of epic importance. When I speak of my great-grandmother, I always remind people that everyone has people in their family worth remembering. The truth is that if we don't tell their stories, they will fade from our hearts and minds. They will cease to exist. Only through telling their stories will they go on. Simon has, through this book, made sure that those on *Titanic* will live on and on.

To all: fair winds and following seas.

Foreword

by Rory Golden

Rory Golden was privileged to take part in one dive on the Titanic, *becoming the first Irish diver to do so. He left a memorial plaque from Ireland on the wreck. During this 12-hour dive, the ship's wheel was spotted by Rory and recovered. He was the first person to touch the wheel of the* Titanic *since it sank in 1912.*

> **'I repeated the order: "Hard a starboard. The helm is hard over, sir. Hard a starboard."'**

With these words Quartermaster Robert Hichens, in the wheelhouse of RMS *Titanic*, confirmed that he had spun the main helm in a desperate attempt to steer the ship away from an iceberg, looming out of the darkness at 11.40 pm, on 14 April 1912. This was part of his testimony given at the United States Senate Court of Inquiry into the sinking the following month in May.

88 years later, in August 2000, I was peering through the viewport of the Russian submersible, *MIR 2*, as we moved along the *Titanic*'s starboard officers' deck for one last look after a 9-hour dive. The time was 7.00 pm. As the ruptured expansion joint loomed into view, a semi-circular shape caught my eye up against the steel wall of the officers' smoking room. My friend, the late Ralph White, on his thirtieth dive to *Titanic*, was viewing the camera monitor. Despite all the previous expeditions that had also travelled along this same deck, I couldn't believe what I was looking at. 'Do you see what I see, Ralph?' I asked. His quick answer affirmed my thoughts about what the object in front of us was: 'I think we've just redeemed ourselves,' he said. As the robotic arm lifted the piece of wreckage, astonishingly, in front of us were the remains of that main helm from the wheelhouse. We now looked upon the bronze centre boss, which had three wooden stumps of the Brown Brothers-manufactured wheel spokes defiantly protruding. The twisted angle-steel leg supports and the bent drive shaft were a testament to the ships' destructive plunge to the bottom, nearly 4 kilometres deep.

Three hours later, we were back on the deck of the mother ship, the *Akademik Mstislav Keldysh.* As we climbed out of the sub, Ralph stood back and said to me, 'You spotted it, you touch it first.'

The significance of that moment took on an even greater meaning, when, in the spring of 2017, I first made contact with Simon Medhurst, great-grandson of Robert Hichens. At the time of writing, we have yet to meet in person, but there is no doubt that we have a special connection.

Titanic does that to you. It connects people. *Titanic Day by Day: 366 Days with the Titanic* will connect even more of them. This simple and brilliant concept, in day, month and year sequence, will give you an insight into the lives of survivors and victims. Perhaps you will find a new connection.

One day, Simon and I will stand beside the helm that his great-grandfather stood at. Another connection will be complete then. Now that is something to look forward to.

Foreword

by Timothy Trower

Timothy Trower is a United States-based historian with a special emphasis on the RMS Titanic, *letterpress printing, and the Hotels Statler Company. Timothy has written extensively for the* Titanic *Historical Society in the past by reviewing books and working as a contributing writer, and often assists other historians with their research.*

As a maritime historian with an emphasis on my first love, the *Titanic*, I have long advocated the publication of books and articles written by those who have the closest of ties to the disaster that changed the future for 2,208 men, women, and children. It is one thing for an observer like myself to spend time researching and writing about something that happened in the distant past, but another for someone with a direct family connection to draw on that kinship, their family history, and their love for the subject to write a book about the sinking of the now-legendary White Star liner.

As a reviewer of numerous *Titanic* and maritime-related books, I have always loved to open a parcel and find a new work written by someone with a familiar name, and it was with great anticipation that I received a copy of the manuscript for *Titanic Day by Day: 366 Days with the Titanic*. The author, Simon Medhurst, is one who I have grown to love and respect as I've watched his devotion to the life of his great-grandfather unfold through the years; that his great-grandfather was Quartermaster Robert Hichens has only served to focus Simon's quest to find out every possible detail about the sometimes-troubled life of the man who stood at the helm of the *Titanic* when the ship met its destiny. But where one might expect a book about this seaman, gone but not forgotten, we are instead treated to a broad look at just one facet of the overall story.

Although one might expect that a descendant of the sinking would be concerned only with the part that his or her ancestor played, I have enjoyed talking to these men and women about a myriad of interests we share, from ships to children to even cats. It is a well-rounded researcher who can draw on a lifetime of learning that can also craft the finest of books. Simon Medhurst has done just that with his look at the lives of every person who was on board the *Titanic* the night of 14–15 April. Written in a chronological format, each day of the calendar adds detail to the lives of the great and small alike – from Dr Max Stähelin-Maeglin (who, it is noted, was a lawyer and not a physician) to Louis Kinsella, who worked as a fireman in the boiler rooms and, aged 30, died with the ship.

The importance of remembering that average people were aboard the ship is often forgotten, as survivors are often thought of as royalty, and certain victims almost regarded as deity. There are times when it is almost inevitable that Bruce Ismay or Lady Duff-Gordon will receive more attention than Francis Rice, just 2 years old when he died, or Amelia Lemore, who lived another 38 years past the sinking, but both Francis and Amelia had lives just as important as John Jacob Astor in the sight of God. Too often I have picked up a new book and found that the same tired stories are just presented in a different typeface or format; these are the books that either do not find a place in my collection or that are relegated to a box in the garage. The strength of this book is that instead of concentrating on any one person or class (not even his famous great-grandfather receives special treatment), all aboard are given the same respect, and it cannot be overemphasised that too often this is not the case with books written about a certain topic.

When my list of the top ten books that any *Titanic* enthusiast should have is again updated, I will be including *Titanic Day by Day* for its completeness, accuracy, organisation, and devotion to detail. Family members have that special fire within them to tell the story of this part of their past, and Simon Medhurst has not only paid proper homage to Robert Hichens, but to the memories of all.

Introduction

My great-grandfather was Robert Hichens, one of the six quartermasters of the *Titanic.* He was at the helm of the ship on that fateful night when it struck an iceberg in the North Atlantic Ocean. Robert Hichens survived on lifeboat 6.

This book has been a labour of love for me over a number of years. It is unique in that no other work covers the facts and information on the *Titanic*, its passengers and crew in such daily detail. It is a book that can be picked up for casual reading, to check a date, learn a fact or be used every day of the week and enjoyed for its historical content. More importantly, it serves as an easily accessible record of all those who were lost and survived in the disaster, so that they may never be forgotten.

In *Titanic Day by Day: 366 Days with the Titanic*, each page relates to a day in the year and gives historical information relevant to that day, allowing the reader to discover facts relating to the *Titanic.* Each day also lists the births and deaths of *Titanic* passengers and crew, who either survived or were tragically lost; this is as near a complete list as possible.

Other interesting and revealing historical details fill the pages, including the ages of passengers and crew members when they were on the *Titanic*, the occupation of each crew member, and information about tickets.

The *Titanic* was all about luxury and the tickets were expensive. The prices were very individual and depended a great deal on which accommodation you wanted, in which class, or even where you booked. A third-class ticket would have cost approximately £7 in 1912, which is nearly £800 in today's money. Second class cost around £13 (nearly £1,500 today) and a first-class ticket was a minimum of £30, or more than £3,300 today. Ticket prices in this book show the cost of combined family or group tickets, not individual ones, and give an idea of the lifestyle of the opulent, contrasting with those who had set out for a new life with just the bags they carried.

There are also newspaper articles from the time relating to different passengers or crew members. *Titanic* survivor quotes; *Titanic* facts; and snippets from the British and American inquiries are also amongst the wealth of information in this book.

One of my motivations behind writing this book was to shine a light on the personalities of the crew, passengers, historians and explorers. The introduction and three forewords reflect this aim. As great-grandson of Robert Hichens, I represent the heroic crew that he was a part of: the crew members who did everything in their power to rescue and protect all aboard the *Titanic*, from the captain and officers on the bridge to the engineers, firemen/stokers, trimmers and greasers, rarely if ever seen, down amongst the twenty-four double-ended boilers and five single-ended ones.

The first foreword is by Helen Benziger, great-granddaughter of Molly Brown, who represents the passengers that night. Then there are forewords by Timothy Trower, representative of the many historians who have made this book possible; and by Rory Golden, diver, presenter and explorer, who went down to the wreck site, so that today we can actually view the mysterious ship of dreams, two and a half miles down on the seabed.

The late Phil Gowan, my great friend and renowned *Titanic* historian, was very much the inspiration for this project. During the years I knew him, we would keep in regular contact and the information I received from him became the very building blocks of the book. This quote from Phil speaks volumes: 'There are a million different stories that played out that night and each one of them is a different drama. Each one of them has been interesting to me and it's those stories that I think keep most people interested in what happened that night. It's not so much what kind of engines that the boat had, it's not so much how many rivets it had, it's not the fact that it was the largest thing on the water in those days, it is what happened to the people on that individual night.' This book captures and details facts from across time, allowing the reader to immerse themselves in those very stories and that of the *Titanic*.

Credits

With many thanks and credit to:

www.encyclopedia-titanica.org
www.titanicinquiry.org
www.britishtitanicsociety.com
Artist Nady Gubanova for the plate illustrations
Titanic explorer Rory Golden for the *Titanic* wreck cover images

January

Day 1

1 January | On this day in *Titanic* history

The following people were born on this day:

Hipkins, Mr William Edward – 1 January 1857 – aged 55 when on *Titanic*; first-class passenger (ticket price: £50); *Titanic* victim

Hakkarainen, Mr Pekka Pietari – 1 January 1884 – aged 28 when on *Titanic*; third-class passenger (ticket price: £15 17s); *Titanic* victim

Phillips, Miss Kate Florence – 1 January 1893 – aged 19 when on *Titanic*; second-class passenger (ticket price: £26); rescued in lifeboat 11

The following people died on this day:

Potter, Mrs Lily Alexenia – 1 January 1954 – aged 56 when on *Titanic*; first-class passenger (ticket price: £83 3s 2d); rescued in lifeboat 7

Olsen, Master Artur Karl – 1 January 1975 – aged 9 when on *Titanic*; third-class passenger (ticket price: £3 3s 5d); rescued in lifeboat 13

Titanic survivor quote

Mrs Lily Alexenia Potter (née Wilson)

> 'The men took to the oars. The sea was absolutely calm and the stars were out. We kept rowing and suddenly someone cried out, "I feel water on my feet!" We checked and found that the drainage plug was not in. It was quickly put back. I asked, "Are there any provisions aboard?" The men looked and could find none whatsoever. After rowing from an eighth to a quarter of a mile, we stood off and watched the mammoth ship. About fifteen minutes after we left the *Titanic*, we were drifting in water filled with cakes of floating ice with our eyes on the great vessel we had deserted. Within a short time, we saw the *Titanic* begin to settle and then we knew that we had been wise to take to the small boats. On the *Titanic*, the crew kept sending up the distress signals. The rockets would roar upward and light the water for miles around. The orchestra kept playing and their music helped to calm us.'

Titanic facts

There were 840 staterooms in all: 416 in first class, 162 in second class and 262 in third class.

The breed of dog that John Jacob Astor owned was an Airedale named Kitty.

Day 2

2 January | On this day in *Titanic* history

The following people were born on this day:
Canner, Mr John – 2 January 1872 – aged 40 when on *Titanic*; fireman; *Titanic* victim

Frost, Mr Anthony Wood – 2 January 1874 – aged 38 when on *Titanic*; fitter; *Titanic* victim

Brobeck, Mr Karl Rudolf – 2 January 1890 – aged 22 when on *Titanic*; third-class passenger (ticket price: £7 15s 11d); *Titanic* victim

Hirvonen, Mrs Helga Elisabeth Lindqvist – 2 January 1890 – aged 22 when on *Titanic*; third-class passenger (ticket price: £12 5s 9d); rescued in lifeboat 15

The following people died on this day:
Roberts, Mrs Mary Kezia – 2 January 1932 – aged 41 when on *Titanic*; stewardess; rescued in lifeboat 16

Dillon, Mr Thomas Patrick – 2 January 1939 – aged 33 when on *Titanic*; trimmer; rescued in lifeboat 4

Hakkarainen, Mrs Elin Matilda – 2 January 1957 – aged 24 when on *Titanic*; third-class passenger (ticket price: £15 17s); rescued in lifeboat 15

Andrews, Mr Charles Edward – 2 January 1961 – aged 19 when on *Titanic*; saloon steward in second class; rescued in lifeboat 16

Titanic survivor quote

Mr Charles Williams

> 'The racket player of Harrow, who was on his way to New York to defend his title of world's champion said he left the squash court in the *Titanic* at 10.30. He was in the smoking room when he first felt the shock. He rushed out and saw an iceberg which seemed to loom over a hundred feet above the deck. It broke up amidships and floated away. Eventually he jumped from the boat deck on the starboard side into the sea, getting as far away from the steamer as possible. He was nine hours in a small boat standing with the water up to his knees before he was picked up. Mr Williams said that the sailors conducted themselves admirably.'

Titanic fact

Harold S. Bride, *Titanic*'s assistant wireless telegraph operator, was paid a wage of £4 a month.

Day 3

3 January | On this day in *Titanic* history

The following people were born on this day:
Dinenage, Mr James Richard – 3 January 1863 – aged 49 when on *Titanic*; saloon steward in first class; *Titanic* victim

Gatti, Sig. Gaspare Antonio Pietro – 3 January 1875 – aged 37 when on *Titanic*; à la carte restaurant manager (restaurant staff); *Titanic* victim

Montvila, Fr Juozas – 3 January 1885 – aged 27 when on *Titanic*; minister; second-class passenger (ticket price: £13); *Titanic* victim

Widener, Mr Harry Elkins – 3 January 1885 – aged 27 when on *Titanic*; first-class passenger (ticket price: £211 10s); *Titanic* victim

Coffey, Mr John – 3 January 1889 – aged 23 when on *Titanic*; fireman (engineering crew); Mr Coffey left the *Titanic* at Queenstown and joined the *Mauretania* shortly afterwards

Pålsson, Master Gösta Leonard – 3 January 1910 – aged 2 when on *Titanic*; third-class passenger (ticket price: £21 1s 6d); *Titanic* victim

Mr Harry Elkins Widener helped his mother into boat 4 and then stood back with the other men to await his fate. He was joined by William Ernest Carter who told him to try for a boat, but Harry's reply was, 'I think I'll stick to the big ship, Billy, and take a chance.'

On 3 January 1935, it was announced that the *Olympic*, *Titanic*'s younger and almost identical sister ship, was to be withdrawn from service. She stood abandoned for a little over six months. In September that year, she was sold for £100,000 to Sir John Jarvis, who then sold her to the Jarrow ship-breakers of Thomas Ward & Sons. *Olympic* had been in service for 24 years. Her interiors were stripped and sold off by the London auctioneers of Knight, Frank & Rutley.

Titanic survivor quote

Mr Jack Thayer

> 'It sounded like locusts on a midsummer night in the woods. This terrible cry lasted for twenty or thirty minutes, gradually dying away, as one after another could no longer withstand the cold and exposure.'

Titanic fact

The official number of the ship was 131,428.

Day 4

4 January | On this day in *Titanic* history

The following people were born on this day:
Paice, Mr Richard Charles John – 4 January 1880 – aged 32 when on *Titanic*; fireman; *Titanic* victim

Stanley, Miss Amy Zillah Elsie – 4 January 1888 – aged 24 when on *Titanic*; third-class passenger (ticket price: £7 11s); rescued in collapsible lifeboat C

The following person died on this day:
Mellinger, Mrs Elizabeth Anne – 4 January 1962 – aged 41 when on *Titanic*; second-class passenger (ticket price: £19 10s); rescued in lifeboat 14

Belfast Newsletter, 1 June 1911

> 'So thorough are the precautions which have been taken to prevent the ship from sinking in the event of a serious accident that any two compartments may be flooded without endangering the safety of the vessel.'

Titanic survivor quotes

Miss Amy Zillah Elsie Stanley, speaking about meeting Rosa Abbott, the only female passenger who went down with the *Titanic* and survived.

> 'We were very close since we were on the *Titanic* together. And her stateroom had been near mine. I was the only one that she could talk to about her sons because I knew them myself. She told me that she would get in the lifeboat if there hadn't been so many people around. So she and her sons kept together. She was thankful that [the] three of them had stayed with her on that piece of wreckage. The youngest went first then the other son went. She grew numb and cold and couldn't remember when she got on the *Carpathia*. There was a piece of cork in her hair and I managed to get a comb and it took a long time but finally we got it out.'

Mr Harry Senior

> 'I swam around for about half an hour, and was swimming on my back when the *Titanic* went down. I tried to get aboard a boat, but some chap hit me over the head with an oar. There were too many in her. I got around to the other side of the boat and climbed in.'

Titanic fact

The approximate number of lamp bulbs used on the ship was 10,000.

Day 5

5 January | On this day in *Titanic* history

The following people were born on this day:
Young, Miss Marie Grice – 5 January 1876 – aged 36 when on *Titanic*; first-class passenger (ticket price £135 12s 8d); rescued in lifeboat 8

Senior, Mr Harry – 5 January 1881– aged 31 when on *Titanic*; engineering crew; rescued in collapsible lifeboat B

MacKay, Mr George William – 5 January 1892 – aged 20 when on *Titanic*; third-class passenger (ticket price: £7 11s); *Titanic* victim

The following person died on this day:
Jussila, Mr Eiriik – 5 January 1944 – aged 32 when on *Titanic*; third-class passenger (ticket price: £7 18s 6d); rescued in lifeboat 15

Titanic survivor quote

Mr Joseph George Scarrott

> 'We were amongst hundreds of dead bodies floating in lifebelts. We could only see four alive. The first one we picked up was a male passenger. He died shortly after we got him in the boat. After a hard struggle we managed to get the other three. One of these we saw kneeling as if in prayer upon what appeared to be a part of a staircase. He was only about twenty yards away from us but it took us half an hour to push our boat through the wreckage and bodies to get to him; even then we could not get very close so we put out an oar for him to get hold of and so pulled him to the boat... All the bodies we saw seemed as if they had perished with the cold as their limbs were all cramped up. As we left that awful scene we gave way to tears. It was enough to break the stoutest heart. Just then we sighted the lights of a steamer, which proved to be the steamship *Carpathia* of the Cunard line. What a relief that was... We then made sail and went back to our other boats. By this time day was just beginning to dawn. We then saw we were surrounded with icebergs and field ice. Some of the fields of ice were from sixteen to twenty miles long. On our way back we saw one of our collapsible boats waterlogged; there were about eighteen persons on it, so we went and took them off. We left two dead bodies on it, and we were told two others had died and had fallen off.'

Titanic fact

There were ten decks in total. From top to bottom they were the Boat Deck, the Promenade Deck (deck A), passenger decks B to G, Orlop Deck and the Tank Top.

Day 6

6 January | On this day in *Titanic* history

The following people were born on this day:
Rheims, Mr George Alexander Lucien – 6 January 1879 – aged 33 when on *Titanic*; first-class passenger (ticket price: £39 12s); rescued in collapsible lifeboat A

McQuillan, Mr William – 6 January 1886 – aged 26 when on *Titanic*; fireman; *Titanic* victim

The following person died on this day:
Duquemin, Mr Joseph Pierre – 6 January 1950 – aged 19 when on *Titanic*; third-class passenger (ticket price: £7 11s); rescued in collapsible lifeboat D

Mr Joseph Pierre Duquemin, mentioned in *Guernsey Evening Press*, Thursday, 2 May 1912

> 'Joseph was doing his best to help the women and children. At one point he took off his overcoat and wrapped it round a shivering seven-year-old girl (Eva Hart). Both of them, Denbuoy and Duquemin, worked together until they were waist deep in water. By that time all the boats had left. Finally Joseph Duquemin turned to his friend and said, "I'm off." He swam away from the deck and headed for the last lifeboat but Bert Denbuoy hesitated. When he started to follow, the suction from the sinking ship dragged him under. For months afterwards Joseph had nightmares. Time and again he reheard his friend crying out as the current pulled him to his death.'

Titanic survivor quote

Miss Marie Grace Young, quoted in *The Buffalo Courier*, 20 April 1912

> 'Then he stepped to the deck of the steamer and the boat I was in was lowered to the water. It was the last boat to leave the ship; of this I am perfectly certain. And I know that I was the last of those who were saved to whom 'Archie' Butt spoke. As our boat was lowered and left the side of the steamer, 'Archie' still was standing at the rail, looking down at me. His hat was raised and the same old, genial, brave smile was on his face. The picture he made as he stood there, hat in hand, brave and smiling was one that will always linger in my memory.'

Titanic fact

There were twenty lifeboats: two emergency cutters and fourteen standard lifeboats (both types made of wood); and four Engelhardt collapsible lifeboats.

Day 7

7 January | On this day in *Titanic* history

The following people were born on this day:
Spencer, Mr William Augustus – 7 January 1855 – aged 57 when on *Titanic*; first-class passenger (ticket price: £146 10s 5d); *Titanic* victim

Gosling, Mr Frank Henry – 7 January 1887 – aged 25 when on *Titanic*; trimmer; *Titanic* victim

McCrae, Mr Arthur Gordon – 7 January 1880 – aged 32 when on *Titanic*; second-class passenger (ticket price: £13 10s); *Titanic* victim

The following people died on this day:
White, Mr Alfred – 7 January 1922 – aged 32 when on *Titanic*; greaser; rescued in lifeboat 4

Tannūs, Mrs Thamīn – 7 January 1974 – aged 16 when on *Titanic*; third-class passenger (ticket price: £8 10s 4d); rescued in lifeboat 14

Barkworth, Mr Algernon Henry – 7 January 1945 – aged 47 when on *Titanic*; first-class passenger (ticket price: £30); rescued in collapsible lifeboat B

Titanic survivor quotea

Mr Alfred White, a greaser in the engine room

> 'I went down into the engine room at 12.40 am. We even made coffee, so there was not much thought of danger. An hour later I was still working at the light engines. I heard the chief engineer tell one of his subordinates that number six bulkhead had given way. At that time things began to look bad... I was told to go up and see how things were, and made my way up a dummy funnel to the bridge deck. By that time all the boats had left the ship, yet everyone in the engine room was at his post. I was near the captain and heard him say, "Well boys, it's every man for himself now."'

Dr Washington Dodge

> 'As the excitement began, I saw an officer of the *Titanic* shoot down two steerage passengers who were endeavouring to rush the lifeboats. I have learned since that twelve of the steerage passengers were shot altogether, one officer shooting down six. The first cabin men and women behaved with great heroism.'

Titanic fact

Buglers called first-class passengers to dinner by playing 'The Roast Beef of Old England'.

Day 8

8 January | On this day in *Titanic* history

The following people were born on this day:

Stanton, Mr Samuel Ward – 8 January 1870 – aged 42 when on *Titanic*; second-class passenger (ticket price: £15 11d); *Titanic* victim

Morley, Mr William – 8 January 1878 – aged 34 when on *Titanic*; third-class passenger (ticket price: £8 1s); *Titanic* victim

Collett, Mr Sidney Clarence Stuart – 8 January 1887 – aged 25 when on *Titanic*; second-class passenger (ticket price: £10 10s); rescued in lifeboat 9

Snyder, Mr John Pillsbury – 8 January 1888 – aged 24 when on *Titanic*; first-class passenger (ticket price: £82 5s 4d); rescued in lifeboat 7

Ahier, Mr Percy Snowden – 8 January 1892 – aged 20 when on *Titanic*; saloon steward; *Titanic* victim

The following person died on this day:

Knowles, Mr Thomas – 8 January 1951 – aged 42 when on *Titanic*; fireman's messman

The Daytona Daily News, 8 January 1910

> 'The White Star Company is to have two of these leviathans, and they will be in commission early in 1911. Their launching will signalize a most important era in marine achievement, for they will be by all odds the largest vessels in the world. The *Titanic* and *Olympic* will each be 860 feet in length, 92 feet beam, 45,000 tons register and 60,000 tons displacement. These enormous dimensions are best expressed in comparisons. The Washington monument is 555 feet high and the Metropolitan tower in New York city rises 700 feet. The combined tonnage of the *Olympic* and *Titanic* far exceeds the tonnage of all the vessels engaged on both sides in the historic conflict of the Spanish armada.'

Titanic survivor quote

Mr Sidney Clarence Stuart Collett

> 'The ladies stepped into the boat, then the officer, with drawn revolver, said to me, "Well, what of you, where are you going?" To which I replied I [had] these young ladies in my charge and felt it my duty to take care of them. "Get in," said the officer and a moment later the boat was lowered.'

Titanic fact

First-class passengers were given copies of *The White Star Music Book*, comprising 352 songs, so they could make requests. The musicians had to know them all.

Day 9

9 January | On this day in *Titanic* history

The following people were born on this day:
Myles, Mr Thomas Francis – 9 January 1849 – aged 63 when on *Titanic*; second-class passenger (ticket price: £9 13s 9d); *Titanic* victim

Spedden, Mr Frederic Oakley – 9 January 1867 – aged 45 when on *Titanic*; first-class passenger (ticket price: £134 10s); rescued in lifeboat 3

Small, Mr William McMillan – 9 January 1870 – aged 42 when on *Titanic*; leading fireman; *Titanic* victim

Minahan, Miss Daisy E. – 9 January 1879 – aged 33 when on *Titanic*; first-class passenger (ticket price: £90); rescued in lifeboat 14

Lundin, Miss Olga Elida – 9 January 1889 – aged 23 when on *Titanic*; second-class passenger (ticket price: £7 17s 1d); rescued in lifeboat 10

Zanetti, Minio – 9 January 1892 – aged 20 when on *Titanic*; assistant waiter (restaurant staff); *Titanic* victim

The following people died on this day:
Maynard, Mr Isaac Hiram – 9 January 1948 – aged 31 when on *Titanic*; entrée cook; rescued in collapsible lifeboat B

Taussig, Miss Ruth – 9 January 1925 – aged 18 when on *Titanic*; first-class passenger (ticket price: £79 13s); rescued in lifeboat 8

Richards, Master William Rowe – 9 January 1988 – aged 3 when on *Titanic*; second-class passenger (ticket price: £18 15s); rescued in lifeboat 4

Titanic survivor quote

Mrs Tillie Taussig, quoted in *New York Times*, Monday, 22 April 1912

> 'Only twenty women were near the boat, and these were put in. My daughter Ruth was among the first, but I said that I wouldn't go if my husband did not accompany me. There was room for fourteen more after the last woman had found her place, and they all pleaded to let the men take the empty seats. But the Captain said that he would not allow it. I was frantic. There was that boat, ready to be lowered into the water and only half full. Then the order came to lower. The men were pleading for permission to step in, and one came forward to take a place next to his wife. I heard a shot and I am sure it was he that went down.'

Titanic fact

At 11.40 pm, Frederick Fleet was the first person on the *Titanic* to see the iceberg, describing it as something 'even darker than the darkness'.

Day 10

10 January | On this day in *Titanic* history

The following people were born on this day:
Smith, Miss Marion Elsie – 10 January 1863 – aged 39 when on *Titanic*; second-class passenger (ticket price: £13); rescued in lifeboat 9

Mackay, Mr Charles Donald – 10 January 1878 – aged 34 when on *Titanic*; saloon steward; rescued in lifeboat 11

Barrett, Mr Frederick William – 10 January 1879 – aged 33 when on *Titanic*; fireman; *Titanic* victim

Blaney, Mr James – 10 January 1880 – aged 32 when on *Titanic*; fireman; *Titanic* victim

Peters, Mr William Chapman – 10 January 1886 – aged 26 when on *Titanic*; able seaman; rescued in lifeboat 9

Bradley, Miss Bridget Delia – 10 January 1890 – aged 22 when on *Titanic*; third-class passenger (ticket price: £7 14s 6d); rescued in lifeboat 13

The following people died on this day:
Threlfall, Mr Thomas – 10 January 1934 – aged 44 when on *Titanic*; leading fireman (engineering crew); rescued in lifeboat 14

Blair, Mr David – 10 January 1955 – aged 3 when on *Titanic*; second-class passenger (ticket price: £18 15s); rescued in lifeboat 14

Fleet, Mr Frederick – 10 January 1965 – aged 24 when on *Titanic*; able seaman and lookout; rescued in lifeboat 6

Laroche, Mme Juliette Marie Louise – 10 January 1980 – aged 22 when on *Titanic*; second-class passenger (ticket price: £41 11s 7d); rescued in lifeboat 14

Titanic survivor quote

Mr Thomas Threlfall

> 'From the wreckage we picked up four men. Then Mr Lowe called out, "There's a boat sinking over there." Although we were then towing a collapsible boat with about eighty people in her we reached the sinking boat just as the water was up to her gunwale and took twenty-six men and one woman, a Mrs Abbott off her. I held the woman in my arms till we reached the *Carpathia*.'

Titanic fact

On board, there were 40 tons of potatoes, 36,000 apples, 1,000 pounds of grapes and 36,000 oranges.

Day 11

11 January | On this day in *Titanic* history

The following people were born on this day:
Cullen, Mr Charles James – 11 January 1863 – aged 49 when on *Titanic*; bedroom steward in first class; rescued in lifeboat 11

Petty, Mr Edwin Henry – 11 January 1887 – aged 25 when on *Titanic*; bedroom steward in second class; *Titanic* victim

Self, Mr Albert Charles Edward – 11 January 1888 – aged 24 when on *Titanic*; fireman; rescued in lifeboat unknown

Bride, Mr Harold Sydney – 11 January 1890 – aged 22 when on *Titanic*; assistant telegraphist; rescued in collapsible lifeboat B

The following people died on this day:
Ridsdale, Miss Lucy – 11 January 1946 – aged 58 when on *Titanic*; second-class passenger (ticket price: £10 10s); rescued in lifeboat 13

Endres, Miss Caroline Louise – 11 January 1964 – aged 39 when on *Titanic*; first-class passenger (ticket price: £247 10s 6d); rescued in lifeboat 4

Titanic survivor quote

Mr Harold Sydney Bride (assistant wireless telegraph operator)

> 'We could hear a terrible confusion outside, but there was not the least thing to indicate that there was any trouble. The wireless was working perfectly. "Send the call for assistance," said the captain, barely putting his head in the door. "What call should I send?" Phillips asked. "The regulation international call for help. Just that." Then the captain was gone. Phillips began to send CQD. He flashed away at it and we were joking while he did so. All of us made light of the disaster. We joked that way while he flashed signals for about five minutes. Then the captain came back. "What are you sending?" he asked. "CQD," Phillips replied. The humour of the situation appealed to me. I cut in with a little remark that made us all laugh, including the captain. "Send SOS," I said. "It's the new call, and it may be your last chance to send it." Phillips with a laugh changed the signal to SOS.'

Titanic fact

Edward John Smith, captain of *Titanic*, gained his master mariner certificate at the age of 24.

Day 12

12 January | On this day in *Titanic* history

The following people were born on this day:
Ryerson, Mr Arthur Larned – 12 January 1851 – aged 61 when on *Titanic*; first-class passenger (ticket price: £262 7s 6d); *Titanic* victim

Mousselmani, Mrs Fatima – 12 January 1890 – aged 22 when on *Titanic*; third-class passenger (ticket price: £7 4s 6d); rescued in collapsible lifeboat C

Mr Arthur Larned Ryerson, mentioned in *New York Times*, Thursday, 5 December 1912

> 'About the time we were ordered to take the boats I passed through the A deck, going from the stern toward the bow. I saw four gentlemen in the smoking room, three of whom I recognized as Mr Millet, Mr Moore, and Mr Butt. The fourth gentleman I did not know, but afterward ascertained he was Mr Ryerson. They seemed to be absolutely intent upon doing what they were doing and disregarding anything that was going on on the decks outside.'

Titanic survivor quote

Miss Anna Katherine Kelly

> 'Miss Kelly is a nervous wreck', said Dr Thomas J. O'Malley, who is attending her. 'I doubt if she ever will completely recover her normal condition. Her life is in jeopardy now. Unless she can overcome her awful fear and terror at every sound, I fear for her life.' Despite her condition the young woman gave a graphic account of the wreck and her escape in one of the last lifeboats to leave the ship. She blames the stewards for not awakening the steerage passengers in time, declaring that the first steerage passengers to become alarmed were ordered to go back to bed 'because there was no danger.'

Charles Joughin (chief baker)

> 'The electric lights were burning right to the very last. I saw the time by my watch at a quarter-past two.'

Titanic fact

In the film *A Night to Remember* (1958), Arthur Ryerson was portrayed by the actor Stuart Nicholl. He appeared during the scene of loading of the lifeboats, where Arthur's son John was denied entry to the lifeboat by Second Officer Lightoller because he felt John was old enough to stay with the men. But Arthur refused to be stopped and said, "Of course he can go, he's only thirteen." Lightoller finally relented and allowed John into the lifeboat.

Day 13

13 January | On this day in *Titanic* history

The following people were born on this day:
Hansen, Mr Claus Peter – 13 January 1871 – aged 41 when on *Titanic*; third-class passenger (ticket price: £14 2s 2d); *Titanic* victim

Meyer, Mr Edgar Joseph – 13 January 1884 – aged 28 when on *Titanic*; first-class passenger (ticket price: £82 3s 5d); *Titanic* victim

The following people died on this day:
Ryerson, Miss Susan Parker 'Suzette' – 13 January 1921 – aged 21 when on *Titanic*; first-class passenger (ticket price: £262 7s 6d); rescued in lifeboat 4

Osman, Mrs Mara – 13 January 1930 – aged 31 when on *Titanic*; third-class passenger (ticket price: £8 13s 8d); rescued in lifeboat unknown

Haggan, Mr John – 13 January 1952 – aged 35 when on *Titanic*; fireman/stoker; rescued in lifeboat 3

Minahan, Mrs Lillian E. 'Daisy' – 13 January 1962 – aged 37 when on *Titanic*; first-class passenger (ticket price: £90); rescued in lifeboat 14

Mrs Mara Osman, mentioned in *Marysville Journal-Tribune*, Wednesday, 24 April 1912

> 'Mrs Mary Osman, a *Titanic* wreck survivor, arrived here and will go to Columbus to see her husband, Mele Osman, who is doing time in the penitentiary for murder. Mrs Osman has not seen her husband since he left the old country to come to America.'

Titanic survivor quote

Mrs Meyer (recalling her parting with her beloved husband Edgar Meyer)

> 'I tried and tried to get Edgar to come into the lifeboat with me, and pleaded to be allowed to stay behind and wait until he could leave, he not caring to leave before all the women had been saved. Mr Meyer finally persuaded me to leave, reminding me of our one-year-old child at home. I entered the lifeboat and watched until the *Titanic* sank, but only for a short time did I see my husband standing beside the rail and assisting other women into boats in which he might have been saved.'

Titanic fact

The number of the slipway on which *Titanic* was built was 3.

Day 14

14 January | On this day in *Titanic* history

The following people were born on this day:
Abbott, Mrs Rhoda Mary 'Rosa' – 14 January 1873 – aged 39 when on *Titanic*; third-class passenger (ticket price: £20 5s); rescued in collapsible lifeboat A

Backström, Mr Karl Alfred – 14 January 1880 – aged 32 when on *Titanic*; third-class passenger (ticket price: £15 17s); *Titanic* victim

Mellors, Mr William John – 14 January 1893 – aged 19 when on *Titanic*; second-class passenger (ticket price: £10 10s); rescued in collapsible lifeboat A

The following people died on this day:
McCarthy, Mr William – 14 January 1932 – aged 48 when on *Titanic*; able seaman; rescued in lifeboat 4

Thorneycroft, Mrs Florence Kate – 14 January 1950 – aged 32 when on *Titanic*; third-class passenger (ticket price: £16 2s); rescued in lifeboat 10

McCoy, Miss Agnes – 14 January 1957 – aged 29 when on *Titanic*; third-class passenger (ticket price: £23 5s); rescued in lifeboat 16

Caldwell, Mrs Sylvia Mae – 14 January 1965 – aged 28 when on *Titanic*; second-class passenger (ticket price: £29); rescued in lifeboat 13

Mr William John Mellors, mentioned in *Evening Banner*, Friday, 26 April 1912

> 'Mellor had a terrible experience while the ship was sinking in being hit by a wave that rushed over the forward deck and swept him against a stanchion. He had a foot frozen and his other ankle was seriously injured. He used to be a valet for Sir Frederick Schuster and was coming over here to better himself, having saved up a little money, and Mr Barkworth said yesterday he had been looking after him and hoped to see him with a job before he himself went back to England.'

Titanic survivor quote

Mrs Washington Dodge

> '…the awful scene of the great steamer as we drew away. From the upper rails heroic husbands and fathers were waving and throwing kisses to their womenfolk in the receding lifeboats.'

Titanic fact

The managers of the Harland & Wolff shipyard were expected to be on site by 6.00 am.

Day 15

15 January | On this day in *Titanic* history

The following people were born on this day:
Snyder, Mrs Nelle Stevenson– 15 January 1889 – aged 23 when on *Titanic*; first-class passenger (ticket price: £82 5s 4d); rescued in lifeboat 7

Donohoe, Miss Bridget – 15 January 1891 – aged 21 when on *Titanic*; third-class passenger (ticket price: £7 15s); *Titanic* victim

The following people died on this day:
Robert, Mrs Elisabeth Walton McMillan – 15 January 1956 – aged 43 when on *Titanic*; first-class passenger (ticket price: £211 6s 9d); rescued in lifeboat 2

Ray, Mr Frederick Dent – 15 January 1977 – aged 32 when on *Titanic*; saloon steward; rescued in lifeboat 13

Hart, Mr John Edward – 15 January 1954 – aged 31 when on *Titanic*; steward (victualling crew); rescued in lifeboat 15

Mr John Edward Hart, mentioned in *The Courier News*, 17 May 1912

> 'Third class steward on the *Titanic*, examined when the *Titanic* inquiry was resumed yesterday, said there were 58 passengers in his section, and he served lifebelts to all who would put them on. Later, when the order came for women and children to be passed to the boat deck, all those willing to go were shown the way. Some were not willing to go, and some who went and found it cold returned to their cabins.'
>
> 'When witness left the third-class compartment for the last time, there were still passengers there who would not go to the boat deck.
>
> The Solicitor-General – By this time you realised it was a serious matter?
> Witness – Yes, but they would not be convinced.
> The Solicitor-General – Did you do your best to convince them?
> Witness – Yes. Everybody did their best. [I] saw no one trying to keep back third-class passengers.'

Mr Joseph G. Boxhall, witness at United States Senate Inquiry (Day 3)

> 'Mr Lightoller was on watch on the bridge when I went on watch at 8 o'clock with Sixth Officer Moody. Mr Lightoller was relieved at 10 o'clock by Mr Murdoch. Mr Murdoch was on watch until the accident happened.'

Titanic fact

The only surviving senior officer of *Titanic* was Second Officer Charles H. Lightoller.

Day 16

16 January | On this day in *Titanic* history

The following person was born on this day:
Dyer, Mr William Henry – 16 January 1879 – aged 33 when on *Titanic*; saloon steward in first class; rescued in lifeboat 9

The following people died on this day:
Herman, Mrs Jane – 16 January 1937 – aged 48 when on *Titanic*; second-class passenger (ticket price: £65); rescued in lifeboat 9

Guy, Mr Elgar John – 16 January 1922 – aged 28 when on *Titanic*; boots steward (victualling crew); rescued in lifeboat 5

Cavendish, Mrs Julia Florence – 16 January 1963 – aged 25 when on *Titanic*; first-class passenger (ticket price: £78 17s); rescued in lifeboat 6

Maugé, Mr Paul Achille Maurice Germain – 16 January 1971 – aged 31 when on *Titanic*; kitchen clerk (restaurant staff); rescued in lifeboat 13

On 10 April 1912, the day *Titanic* departed from Southampton, the following was received by Captain Benjamin Steele, White Star's marine superintendent:

> 'I herewith report this ship loaded and ready for sea. The engines and boilers are in good order for the voyage, and all charts and sailing directions up-to-date.

Your obedient servant, Edward J. Smith.'

The St. Louis Star, **15 April 1912**

> 'Captain and crew of 860 men refuse to leave vessel after passengers are placed in lifeboats and believe ship will reach harbor – Anxious relatives crowd offices of company in New York to learn latest news from headquarters – many millionaires aboard – from four St. Louisiana passengers.'

Asbury Park Evening Press, **15 April 1912**

> 'Saved by bulkheads – Many notables included in passenger list of 1,360 – $5,000,000 in bonds and diamonds aboard. – Huge ice fields stretch for miles.'

Titanic survivor quote

Mrs Jane Herman (née Laver), mentioned in *Newark Evening News*, Saturday, 20 April 1912

> 'Mrs Herman and her daughters were in great distress of mind over Mr Herman's fate and their own experience. They were at a loss to understand how the disaster occurred on such a clear night. They said they were near enough to the big ship when she sank to feel the suction. Members of the crew who manned their boat thought that the boat was full. Mrs Herman said that a boy named Sweet, fifteen years old, who accompanied Mr Herman and whose first name they did not know, was also lost in the wreck.'

Day 17

17 January | On this day in *Titanic* history

The following people were born on this day:
Gracie IV, Colonel Archibald – 17 January 1859 – aged 53 when on *Titanic*; first-class passenger (ticket price: £28 10s); rescued in collapsible lifeboat B

Wells, Mrs 'Addie' Dart – 17 January 1883 – aged 29 when on *Titanic*; second-class passenger (ticket price: £23); rescued in lifeboat 14

Jeffery, Mr William Alfred – 17 January 1884 – aged 28 when on *Titanic*; à la carte restaurant controller; *Titanic* victim

Moen, Mr Sigurd Hansen – 17 January 1885 – aged 27 when on *Titanic*; third-class passenger (ticket price: £7 13s); *Titanic* victim

Driscoll, Miss Bridget – 17 January 1885 – aged 27 when on *Titanic*; third-class passenger (ticket price: £7 15s); rescued in collapsible lifeboat D

The following person died on this day:
Maioni, Miss Roberta Elizabeth Mary 'Cissy' – 17 January 1963 – aged 19 when on *Titanic*; personal maid to Lucy Noël Martha, Countess of Rothes; first-class passenger (ticket price: £86 10s); rescued in lifeboat 8

Titanic survivor quote

Mr Colonel Archibald Gracie

> 'What impressed me at the time that my eyes beheld the horrible scene was a thin light-gray smoky vapour that hung like a pall a few feet above the broad expanse of sea that was covered with a mass of tangled wreckage. That it was a tangible vapour, and not a product of my imagination, I feel well assured. It may have been caused by smoke or steam rising to the surface around the area where the ship had sunk. Add to this, within the area described, which was as far as my eyes could reach, there arose to the sky the most horrible sounds ever heard by mortal man except by those of us who survived this terrible tragedy. The agonising cries of death from over a thousand throats, the wails and groans of the suffering, the shrieks of the terror-stricken and the awful gaspings for breath of those in the last throes of drowning, none of us will ever forget to our dying day.'

Titanic fact

After *Titanic* hit the iceberg, she began taking on a lot of water. After 20 minutes, the ship had almost 2 million gallons on board. To put this into perspective, this amount of water would fill four Olympic-sized swimming pools.

Day 18

18 January | On this day in *Titanic* history

The following people were born on this day:
Kemp, Mr Thomas Hulman – 18 January 1869 – aged 43 when on *Titanic*; extra assistant fourth engineer (refrigeration); *Titanic* victim

Gallagher, Mr Martin – 18 January 1883 – aged 27 when on *Titanic*; third-class passenger (ticket price: £7 14s 10d); *Titanic* victim

Maréchal, Mr Pierre – 18 January 1883 – aged 29 when on *Titanic*; first-class passenger (ticket price: £29 14s); rescued in lifeboat 7

Barton, Mr David John – 18 January 1890 – aged 22 when on *Titanic*; third-class passenger (ticket price: £8 1s); *Titanic* victim

Andrews, Mr Charles Edward – 18 January 1893 – aged 19 when on *Titanic*; saloon steward in second class; rescued in lifeboat 16

The following people died on this day:
Romaine, Mr Charles 'Hallace' – 18 January 1922 – aged 45 when on *Titanic*; first-class passenger (ticket price: £26 11s); rescued in lifeboat 9

Lewis, Mr Arthur Ernest Read – 18 January 1973 – aged 27 when on *Titanic*; steward; rescued in lifeboat 15

Herman, Miss Kate – 18 January 1983 – aged 23 when on *Titanic*; second-class passenger (ticket price: £65); rescued in lifeboat 9

Abelson, Mrs Anna – 18 January 1972 – aged 28 when on *Titanic*; second-class passenger (ticket price: £24); rescued in lifeboat 10

Sir Archibald Denny, unveiling the *Titanic* Engineers' Memorial on 22 April 1914

> 'They must have known that pumping could do no more than delay the final catastrophe, yet they stuck pluckily to their duty.'

Titanic survivor quote

Mr Charles Edward Andrews, witness at United States Senate Inquiry (Day 7)

> 'Besides these six men, I should think there were about fifty passengers. There was no effort on the part of the steerage men to get into our boat. I was told by the officer to allow none in it. When the officer started to fill the boat with passengers and the men to man it, there were no individuals who tried to get in, or that he permitted to get in. There was no confusion whatever. The officer asked me if I could take an oar. I said I could…'

Titanic fact

It took twenty horses to transport the main anchor of *Titanic* from Netherton Ironworks to Dudley railway station on 30 April 1911. It was then loaded on a cargo steamer for the trip to *Titanic*'s shipyard in Belfast.

Day 19

19 January | On this day in *Titanic* history

The following people were born on this day:
Collins, Mr John (Samuel) – 19 January 1874 – aged 38 when on *Titanic*; fireman (engineering crew); rescued in lifeboat 1

Carlsson, Mr Carl Robert – 19 January 1888 – aged 24 when on *Titanic*; third-class passenger (ticket price: £7 17s 1d); *Titanic* victim

Badman, Miss Emily Louisa – 19 January 1894 – aged 18 when on *Titanic*; third-class passenger (ticket price: £8 1s); rescued in collapsible lifeboat C

Andersson, Miss Ellis Anna Maria – 19 January 1910 – aged 2 when on *Titanic*; third-class passenger (ticket price: £31 5s 6d); *Titanic* victim

The entire Andersson family was lost, including little Ellis Anna Maria.

Titanic survivor quotes

Miss Emily Louisa Badman, quoted in *Syracuse Herald*, Tuesday, 23 April 1912

> 'When we reached the water we could see that the *Titanic* was fast sinking and the six men who were rowing pulled us away as fast as they could...We had gone about a mile when there were two explosions and the *Titanic* split in two. The front end went down at once and the back stood up so that it was almost straight and then went out of sight. Our boat had about thirty women in it and it was filled so that I could not move. Cramped in there in light clothing I suffered horribly from the cold and I caught a severe cold. We were in there until about 7 o'clock the next morning, when the *Carpathia* came along and took us off.'

Dr Washington Dodge

> 'When the boats were drawing away from the ship they could hear the orchestra playing 'Lead, Kindly Light', and rockets were going up from the *Titanic* in the wonderfully clear night. We could see from the distance that two boats were being made ready to be lowered. The panic was in the steerage, and it was [in] that portion of the ship that the shooting was made necessary.'

Titanic fact

Captain Herbert James Haddock was the first captain of the *Titanic*. He commanded her in Belfast before her delivery to the White Star Line. He was relieved on 31 March by Captain Edward J. Smith at Belfast; Haddock then returned to Southampton to take command of Smith's previous ship, RMS *Olympic*.

Day 20

20 January | On this day in *Titanic* history

The following people were born on this day:
Lindblom, Miss Augusta Charlotta – 20 January 1867 – aged 45 when on *Titanic*; third-class passenger (ticket price: £7 15s); *Titanic* victim

McGarvey, Mr Edward Joseph – 20 January 1878 – aged 34 when on *Titanic*; fireman; *Titanic* victim

Broom, Mr Herbert George – 20 January 1879 – aged 33 when on *Titanic*; bath steward; *Titanic* victim

Peacock, Mrs Edith – 20 January 1886 – aged 26 when on *Titanic*; third-class passenger (ticket price: £13 15s 6d); *Titanic* victim

The following people died on this day:
Naṣr Allāh, Mrs Adāl – 20 January 1970 – aged 14 when on *Titanic*; second-class passenger (ticket price: £30 1s 5d); rescued in lifeboat unknown

Brown, Miss Edith Eileen – 20 January 1997 – aged 15 when on *Titanic*; second-class passenger (ticket price: £39); rescued in lifeboat 14

***Great Falls Daily Tribune*, 15 April 1912**

> 'The *Titanic* strikes an iceberg, and last heard from her she was sinking and women were being taken off on lifeboats – wireless messages break off abruptly.'

Titanic survivor quotes

Miss Edith Eileen Brown

> 'Father appeared a few minutes later. He told us, "You'd better put on your life jackets and something warm, it's cold on deck. It's just a precaution. We've struck an iceberg, it's nothing much. The steward in the corridor says it's nothing to worry about." We waited for ages on the boat deck for someone to tell us what to do. The ship's band was playing ragtime. They played to keep our spirits up. Everybody kept saying, "She's unsinkable. She won't go down." Father kissed us and saw us into lifeboat 14. Up to fifty people got in as it swung perilously over the side. One man jumped into the boat dressed as a woman. As we rowed away from the ship, we could still hear the band playing, but now it was hymns. We were almost six hours in the lifeboat and during that time we had no water and nothing to eat. I kept wondering if my father had gotten off the ship, that's all I could think of.'

Mr Samuel Hemming (lamp trimmer)

> 'The boatswain told us to turn out; the ship had only half an hour to live, from Mr Andrews, but not to tell anyone. The boatswain heard it from Mr Andrews, and he told us.'

Day 21

21 January | On this day in *Titanic* history

The following people were born on this day:

Andersson, Mr Anders Johan – 21 January 1873 – aged 39 when on *Titanic*; third-class passenger (ticket price: £31 5s 6d); *Titanic* victim

Aronsson, Mr Ernst Axel Algot – 21 January 1887 – aged 25 when on *Titanic*; third-class passenger (ticket price: £7 15s 6d); *Titanic* victim

Hunt, Mr George Henry – 21 January 1878 – aged 34 when on *Titanic*; second-class passenger (ticket price: £12 5s 6d); *Titanic* victim

The following people died on this day:

Ryerson, Master John Borie – 21 January 1986 – aged 13 when on *Titanic*; first-class passenger (ticket price: £262 7s 6d); rescued in lifeboat 4

Lang, Mr Fang – 21 January 1986 – aged 16 when on *Titanic*; third-class passenger (ticket price: £7); rescued in lifeboat 14

John Ryerson almost didn't make it onto lifeboat 4 but was reluctantly allowed by Officer Lightoller as his father stated he was only 13. John Ryerson helped Walter Lord do research for his book *A Night to Remember*, but was reluctant to speak of the tragedy.

Titanic survivor quote

Miss Anna Sofia Nysten

> 'Of course the women and children were put into the boats first, but several men tried to get in them too. It was sad, the way families were separated…I'm not sure the band was playing – as the stories say – but I think it was.'

Miss Elizabeth Weed Shutes

> 'The first wish on the part of all was to stay near the *Titanic*. We all felt so much safer near the ship. Surely such a vessel could not sink. I thought the danger must be exaggerated, and we could all be taken aboard again. But surely the outline of that great, good ship was growing less. The bow of the boat was getting black. Light after light was disappearing, and now those rough seamen put to their oars and we were told to hunt under seats, any place, anywhere, for a lantern, a light of any kind. Every place was empty. There was no water no stimulant of any kind. Not a biscuit, nothing to keep us alive had we drifted long…'

Titanic fact

The length of *Titanic* was 882' 9" (269.1 metres).

Day 22

22 January | On this day in *Titanic* history

The following people were born on this day:
Jones, Mr Charles Cresson – 22 January 1866 – aged 46 when on *Titanic*; first-class passenger (ticket price: £26); *Titanic* victim

Nysten, Miss Anna Sofia – 22 January 1890 – aged 22 when on *Titanic*; third-class passenger (ticket price: £7 15s); rescued in lifeboat 13

Cecil, Mr Charles Thomas – 22 January 1891– aged 21 when on *Titanic*; steward; *Titanic* victim

The following person died on this day:
Stengel, Mrs Annie May Morris – 22 January 1956 – aged 44 when on *Titanic*; first-class passenger (ticket price: £55 8s 10d); rescued in lifeboat 5

Miss Anna Sofia Nysten, mentioned in *Jersey Journal*, Monday, 22 April 1912

> 'One of the fortunate survivors of the ill-fated steamship *Titanic* was Miss Anna Nysten, a Swedish girl who sailed from England to accept a position as mother's helper in the home of Mr and Mrs Nielsen, in Central Avenue. At the present time Miss Nysten is staying with friends in Passaic, until she has fully recovered from the shock of her tragic experience, but was able to report at the Nielsen home, Saturday and reassure them of her safety. At the time of the crash, Miss Nysten, was one of a party of ten, disregarding companions' entreaty to not go up on deck to see the cause of the shock. Miss Nysten insisted and was the only one of the party who was saved from the wreck. Both Mr and Mrs Nielsen were greatly relieved when hearing of the young woman's safety, as they were instrumental in having the maid come to America.'

Mr Robert Hichens (quartermaster), witness at United States Senate Inquiry (Day 5)

> 'Capt. Smith [said]...to Mr Murdoch; "Close the emergency doors." Mr Murdoch replied, "The doors are already closed." The captain sent then for the carpenter to sound the ship. He also came back to the wheelhouse and looked at the commutator in front of the compass, which is a little instrument like a clock to tell you how the ship is listing. The ship had a list of 5' to the starboard.'

Titanic fact

There was one fully equipped photographic darkroom on board *Titanic*.

Day 23

23 January | On this day in *Titanic* history

The following people were born on this day:
Daly, Mr Eugene Patrick – 23 January 1883 – aged 29 when on *Titanic*; third-class passenger (ticket price: £7 15s); rescued in collapsible lifeboat B

Smyth, Mr Thomas – 23 January 1886 – aged 26 when on *Titanic*; third-class passenger (ticket price: £7 15s); *Titanic* victim

Pearcey, Mr Albert Victor – 23 January 1887 – aged 25 when on *Titanic*; pantry steward in third class; rescued in collapsible lifeboat C

Noss, Mr Bertram Arthur – 23 January 1892 – aged 20 when on *Titanic*; fireman; *Titanic* victim

Quick, Miss Winifred Vera – 23 January 1904 – aged 8 when on *Titanic*; second-class passenger (ticket price: £26); rescued in lifeboat 11

The following people died on this day:
Nichols, Mr Walter Henry – 23 January 1960 – aged 35 when on *Titanic*; assistant saloon steward in second class; rescued in lifeboat 15

Lee, Mr Henry Reginald – 23 January 1965 – aged 29 when on *Titanic*; scullion; rescued in lifeboat 13

Madsen, Mr Fridtjof Arne – 23 January 1972 – aged 24 when on *Titanic*; third-class passenger (ticket price: £7 2s 10d); rescued in lifeboat 13

Nakid, Mrs Waika 'Mary' – 23 January 1963 – aged 19 when on *Titanic*; third-class passenger (ticket price: £15 14s 10d); rescued in collapsible lifeboat C

Titanic survivor quotes

Mr Eugene Patrick Daly

> '...an officer pointed a revolver and said if any man tried to get in he would shoot him on the spot. I saw the officer shoot two men dead because they tried to get into the boat. Afterwards there was another shot, and I saw the officer himself lying on the deck. They told me he shot himself, but I did not see him.'

Mr Alfred Crawford, witness at United States Senate Inquiry (Day 9)

> 'Capt. Smith could see the light quite plain, as he pointed in the direction that we were to make for. We pulled toward the light, and we could not reach it.'

Mr John Hart (steward in third class)

> 'Those that were willing to go to the boat deck were shown the way. Some were not willing to go to the boat deck, and stayed behind. Some of them went to the boat deck, and found it rather cold, and saw the boats being lowered away, and thought themselves more secure on the ship, and consequently returned to their cabin.'

Day 24

24 January | On this day in *Titanic* history

The following people were born on this day:
Hold, Mr Stephen – 24 January 1868 – aged 44 when on *Titanic*; second-class passenger (ticket price: £26); *Titanic* victim

Hoyt, Mr William Fisher – 24 January 1869 – aged 43 when on *Titanic*; first-class passenger (ticket price: £30 13s 11d); *Titanic* victim

Pennal, Mr Thomas Frederick Cohen – 24 January 1878 – aged 34 when on *Titanic*; bathroom steward in first class; *Titanic* victim

Berglund, Mr Karl Ivar Sven – 24 January 1890 – aged 22 when on *Titanic*; third-class passenger (ticket price: £9 7s); *Titanic* victim

Kiernan, Mr Philip – 24 January 1890 – aged 22 when on *Titanic*; third-class passenger (ticket price: £7 15s); *Titanic* victim

The following people died on this day:
Bradley, Miss Bridget Delia – 24 January 1956 – aged 22 when on *Titanic*; third-class passenger (ticket price: £7 14s 6d); rescued in lifeboat 13

Lord, Captain Stanley – 24 January 1962 – aged 34 at the time of the disaster; captain of the *Californian*

New York Times, Wednesday, 24 January 1912

> 'Nine Hundred First Cabin Passengers on *Olympic*, *Cedric*, and *Lapland*
>
> More than 1,200 cabin passengers, of whom nearly 900 will be in the first cabin, will sail from New York for Europe today. This big crowd of travellers, in numbers one of the largest that has sailed from New York in a single day this winter, is booked on the White Star liners *Olympic* and *Cedric*, and the *Lapland* of the Red Star Line. Of the 900 who are going in the saloon apartments the *Olympic* will have about 500, the *Cedric* 300, and the *Lapland* the rest.
>
> 'Among the *Olympic*'s passengers will be Col. and Mrs John Jacob Astor, who go abroad for the first time since their marriage; Mr and Mrs Reginald Brooks, J. Bruce Ismay, President of the International Mercantile Marine; Prince and Princess del Drago, Lewis Einstein, the American Minister to Costa Rica, and the Countess Camilla Hoyos.'

Titanic fact

50,000 items of bone china crockery were supplied to the *Titanic* by Stonier and Co. of Liverpool.

Day 25

25 January | On this day in *Titanic* history

The following people were born on this day:

Isham, Miss Ann Elizabeth – 25 January 1862 – aged 50 when on *Titanic*; first-class passenger (ticket price: £28 14s 3d); *Titanic* victim

Meo (Martino), Mr Alfonzo – 25 January 1864 – aged 48 when on *Titanic*; third-class passenger (ticket price: £8 1s); *Titanic* victim

White, Mr Alfred – 25 January 1880 – aged 32 when on *Titanic*; greaser; rescued in lifeboat 4 after being pulled from the water

Hogan, Mr Albert John – 25 January 1893 – aged 19 when on *Titanic*; scullion; *Titanic* victim

Heikkinen, Miss Laina – 25 January 1896 – aged 26 when on *Titanic*; third-class passenger (ticket price: £7 18s 6d); rescued (probably) in lifeboat 14

The following people died on this day:

Midtsjø, Mr Karl Albert – 25 January 1939 – aged 21 when on *Titanic*; third-class passenger (ticket price: £7 15s 6d); rescued in lifeboat 15

Laroche, Miss Louise – 25 January 1998 – aged 1 when on *Titanic*; second-class passenger (ticket price: £41 11s 7d); rescued in lifeboat 14

Titanic survivor quote

Mr Karl Albert Midtsjø

> 'I have lost everything I had of clothes and money, and all I have are the clothes I stand in. Well, it was no fun to be in a little boat out on the Atlantic. It was pretty well in the middle of the Atlantic. We sat in the lifeboats and thought now it is their turn and soon it will be ours. It is no joke when such a big ship is sinking. It was a serious situation when the rockets went off. It was a clear, starry night and the stars shone just as they do at home, even though it was a four-day voyage by steamer to reach land. It would take a whole day to tell it all, but I must be content with a few words.
>
> 'We are treated like royalty and looked up to almost as if we were unnatural and there are plenty of people who want to help us. I am well and fit and the disaster has not affected me. However, you can be quite solemn when you have seen something as terrible as this. I can still hear the cries for help. And someone was shot when they tried to push their way into the boats.'

Titanic fact

Two bathtubs were available to the 700 third-class passengers.

Day 26

26 January | On this day in *Titanic* history

The following people were born on this day:
Douglas, Mrs Mahala Dutton – 26 January 1863 – aged 49 when on *Titanic*; first-class passenger (ticket price: £106 8s 6d); rescued in lifeboat 2

Thorne, Miss Gertrude Maybelle – 26 January 1875 – aged 37 when on *Titanic*; first-class passenger (ticket price: £80 4d); rescued in collapsible lifeboat D

Moore, Mr George Alfred – 26 January 1880 – aged 32 when on *Titanic*; able seaman; rescued in lifeboat 3

Olsson, Miss Elina – 26 January 1881 – aged 31 when on *Titanic*; third-class passenger (ticket price: £7 17s 1d); *Titanic* victim

Scheerlinck, Mr Jean – 26 January 1883 – aged 29 when on *Titanic*; third-class passenger (ticket price: £9 10s); rescued in lifeboat 11

Phillips, Miss Alice Frances Louisa – 26 January 1891 – aged 21 when on *Titanic*; second-class passenger (ticket price: £21); rescued in lifeboat 12

The following person died on this day:
Street, Mr Thomas Albert – 26 January 1964 – aged 25 when on *Titanic*; fireman; rescued in lifeboat 9

Titanic survivor quotes

Miss Alice Frances Louisa Phillips

> 'I met the cabin steward, and asked him what had happened, but he assured me there was nothing wrong. Everything was all right, he said, and advised me to go back to the cabin. I could not understand it, and felt there must be something amiss, but I listened to his advice, and, with many doubts, went back to the cabin. Then I heard shouts and the sounds of general confusion on the deck, and determined to at least see what was being done for myself. Without a moment's further hesitation I rushed to the upper deck, and no sooner had I got there than someone picked me up and put me into one of the lifeboats. That was the last I saw of the '*Titanic*', and I shall never see my poor father again.'

Mr Paul Chevré

> '…We did nothing but row, row, row to escape from the death cries. In our little boat we were frozen with cold, having left the ship without overcoats or rugs. We shouted from time to time to attract attention, but obtained no reply.'

Day 27

27 January | On this day in *Titanic* history

The following people were born on this day:
Smith, Captain Edward John – 27 January 1850 – aged 62 when on *Titanic*; Master; *Titanic* victim

Goldsmith, Mr Frank John – 27 January 1879 – aged 33 when on *Titanic*; third-class passenger (ticket price: £20 10s 6d); *Titanic* victim

Gilardino, Sig. Vincenzo Pio – 27 January 1881 – aged 31 when on *Titanic*; waiter (restaurant staff); *Titanic* victim

Corben, Mr Ernest Theodore – 27 January 1885 – aged 27 when on *Titanic*; assistant printer steward (victualling crew); *Titanic* victim

Cottam, Harold Thomas – 27 January 1891 – aged 21 when on *Carpathia*; telegraphist (victualling crew)

Orešković, Mr Luka – 27 January 1892 – aged 20 when on *Titanic*; third-class passenger (ticket price: £8 13s 3d); *Titanic* victim

Riordan, Miss Hannah – 27 January 1894 – aged 18 when on *Titanic*; third-class passenger (ticket price: £7 14s 5d); rescued in lifeboat 13

The following person died on this day:
Goldsmith, Master Frank John William – 27 January 1982 – aged 9 when on *Titanic*; third-class passenger (ticket price: £20 10s 6d); rescued in collapsible lifeboat C

Captain Smith, commander of *Titanic*

> 'When anyone asks how I can best describe my experience in nearly 40 years at sea, I merely say, uneventful. Of course there have been winter gales, and storms and fog and the like, but in all my experience, I have never been in any accident of any sort worth speaking about...I never saw a wreck and never have been wrecked, nor was I ever in any predicament that threatened to end in disaster of any sort. You see, I am not very good material for a story. I cannot imagine any condition which would cause a ship to founder. I cannot conceive of any vital disaster happening to this vessel.'

Titanic survivor quote

Mr George A. Harder

> 'About a quarter to 11 I went down to my stateroom with Mrs Harder and retired for the night; and at 20 minutes to 12 we were not asleep yet, and I heard this thump. It was not a loud thump; just a dull thump. Then I could feel the boat quiver and...a sort of rumbling, scraping noise along the side of the boat. When I went to the porthole I saw this iceberg go by. The porthole was closed. The iceberg was...about 50 to 100 feet away.... it was about as high as the top deck of the boat. I just got a glimpse of it, and it is hard to tell how high it was.'

Day 28

28 January | On this day in *Titanic* history

The following people were born on this day:
Olsen, Mr Karl Siegwart Andreas – 28 January 1870 – aged 42 when on *Titanic*; third-class passenger (ticket price: £8 8s 1d); *Titanic* victim

Pelham, Mr George – 28 January 1873 – aged 39 when on *Titanic*; trimmer; rescued in lifeboat 16

Kieran, Mr Edgar Michael – 28 January 1878 – aged 34 when on *Titanic*; storekeeper; *Titanic* victim

Hesketh, Mr John Henry – 28 January 1879 – aged 33 when on *Titanic*; second engineer (refrigeration); *Titanic* victim

Ryan, Mr Edward – 28 January 1888 – aged 24 when on *Titanic*; third-class passenger (ticket price: £7 15s); rescued in lifeboat 14

Henry, Miss Bridget – 28 January 1890 – aged 22 when on *Titanic*; third-class passenger (ticket price: £7 15s); *Titanic* victim

Collyer, Miss Marjorie Charlotte – 28 January 1904 – aged 8 when on *Titanic*; second-class passenger (ticket price: £26 5s); rescued in lifeboat 14

Johnson, Master Harold Theodor – 28 January 1908 – aged 4 when on *Titanic*; third-class passenger (ticket price: £11 2s 8d); rescued in lifeboat 15

The following person died on this day:
Carter, Master William Thornton II – 28 January 1985 – aged 11 when on *Titanic*; first-class passenger (ticket price: £120); rescued in lifeboat 4

Titanic survivor quote

Mr Edward Ryan

> 'I stood on the *Titanic* and kept cool, although she was sinking fast. She had gone down about forty feet by now. The last boat was about being rowed away when I thought in a second if I could only pass out [i.e. get into the boat] I'd be all right. I had a towel round my neck. I just threw this over my head and left it hang in the back. I wore my waterproof overcoat. I then walked very stiff past the officers, who had declared they'd shoot the first man that dare pass out. They didn't notice me. They thought I was a woman. I grasped a girl who was standing by in despair, and jumped with her thirty feet into the boat.'

Titanic fact

One drink we know for sure that was served on the *Titanic* is Punch à la Romaine, a shaved-ice concoction popularised by famed French chef Georges Auguste Escoffier. Made with rum and Champagne, it was served as a palate cleanser between courses.

Day 29

29 January | On this day in *Titanic* history

The following people were born on this day:
Stone, Mrs Martha Evelyn Stevens – 29 January 1850 – aged 62 when on *Titanic*; first-class passenger (ticket price: £80); rescued in lifeboat 6

Storey, Mr Thomas – 29 January 1853 – aged 59 when on *Titanic*; third-class passenger (ticket price: £7 5s); *Titanic* victim

Locke, Mr Albert George – 29 January 1879 – aged 33 when on *Titanic*; scullion; *Titanic* victim

King, Mr Ernest Waldron – 29 January 1884 – aged 28 when on *Titanic*; clerk in first class; *Titanic* victim

Williams, Mr Richard Norris II – 29 January 1891 – aged 21 when on *Titanic*; first-class passenger (ticket price: £61); rescued in collapsible lifeboat A

The following people died on this day:
Prior, Mr Harold John Arnold – 29 January 1949 – aged 21 when on *Titanic*; steward in third class; rescued (perhaps) in lifeboat 11 or 13

Webber, Miss Susan – 29 January 1952 – aged 37 when on *Titanic*; second-class passenger (ticket price: £13); rescued in lifeboat 12

Titanic survivor quotes

Mr Richard Norris II Williams, quoted in *Washington Times*, 22 April 1912

> 'I was not under water very long, and as soon as I came to the top I threw off the big fur coat. I also threw off my shoes. About twenty yards away I saw something floating. I swam to it and found it to be a collapsible boat. I hung on to it and after a while got aboard and stood up in the middle of it. The water was up to my waist. About thirty of us clung to it. When Officer Lowe's boat picked us up eleven of us were still alive; all the rest were dead from cold. My fur coat was found attached to this Engelhardt boat 'A' by the *Oceanic*, and also a cane marked 'C. Williams.' This gave rise to the story that my father's body was in this boat, but this as you see, is not so. How the cane got there I do not know.'

Miss Susan Webber, quoted in *Western Times*, Monday, 13 May 1912

> 'I was facing the *Titanic* and could see her going down. I saw the lights go out deck after deck. When the water got into the engine room there was an explosion, and then I saw the leviathan part in the middle. The stern rose high in the air; the bow less high. Then she went down slowly amid heartrending cries for help of hundreds of doomed men and women.'

Day 30

30 January | On this day in *Titanic* history

The following people were born on this day:
Newell, Mr Arthur Webster – 30 January 1854 – aged 58 when on *Titanic*; first-class passenger (ticket price: £113 5s 6d); *Titanic* victim

Hawksford, Mr Walter James – 30 January 1867 – aged 45 when on *Titanic*; first-class passenger (ticket price: £30); rescued in lifeboat 3

Marvin, Mrs Mary Graham Carmichael Farquharson – 30 January 1894 – aged 18 when on *Titanic*; first-class passenger (ticket price: £53 2s); rescued in lifeboat 10

The following people died on this day:
Wick, Mrs Mary 'Mollie' – 30 January 1920 – aged 45 when on *Titanic*; first-class passenger (ticket price: £164 17s 4d); rescued in lifeboat 8

Moor, Mrs Beila – 30 January 1958 – aged 29 when on *Titanic*; third-class passenger; (ticket price: £12 9s 6d); rescued in lifeboat 14

McGowan, Miss Anna 'Annie' – 30 January 1990 – aged 17 when on *Titanic*; third-class passenger (ticket price: £8 7d); rescued (possibly) in lifeboat 13

Navratil, Master Michel Marcel – 30 January 2001 – aged 3 when on *Titanic*; second-class passenger (ticket price: £26); rescued in collapsible lifeboat D

Mr Joseph G. Boxhall, witness at United States Senate Inquiry (Day 3)
'I went right down below, in the lowest steerage, as far as I could possibly get without going into the cargo portion of the ship, and inspected all the decks as I came up, in the vicinity of where I thought she had struck.'

Titanic survivor quote

Mrs Mary Graham Carmichael Marvin, mentioned in *Daily Sketch*, Wednesday, 17 April 1912

> 'One of the saddest of the many sad stories which has to be told concerns the fate of Mr Daniel W. Marvin, a young American who was returning with his wife from the honeymoon trip spent in Europe. Mr Marvin is missing, but Mrs Marvin is reported to have been saved. The husband was only 19 years of age, and the wife is a year younger. The Marvins are well known in New York society, and the young man himself was a friend of Edison.'

Titanic fact

Around 15,000 men were employed by Harland & Wolff shipyard at the peak of construction.

Day 31

31 January | On this day in *Titanic* history

The following people were born on this day:
Latimer, Mr Andrew L. – 31 January 1857 – aged 55 when on *Titanic*; chief steward; *Titanic* victim

Fleming, Miss Margaret – 31 January 1870 – aged 42 when on *Titanic*; first-class passenger (ticket price: £110 17s 8d); rescued in lifeboat 4

Lavington, Miss Elizabeth – 31 January 1872 – aged 40 when on *Titanic*; stewardess; rescued in lifeboat 11

Lindell, Mr Edvard Bengtsson – 31 January 1876 – aged 36 when on *Titanic*; third-class passenger (ticket price: £15 11s); *Titanic* victim

Lucas, Mr William Watson – 31 January 1881 – aged 31 when on *Titanic*; saloon steward; rescued in collapsible lifeboat A

Asplund, Mr Johan Charles – 31 January 1899 – aged 23 when on *Titanic*; third-class passenger (ticket price: £7 15s 11d); rescued in lifeboat 13

Hart, Miss Eva Miriam – 31 January 1905 – aged 7 when on *Titanic*; second-class passenger (ticket price: £26 5s); rescued (possibly) in lifeboat 14

The following people died on this day:
Lurette, Miss Elise Eugénie – 31 January 1940 – aged 59 when on *Titanic*; first-class passenger (ticket price: £146 10s 5d); rescued in lifeboat 6

White, Mrs Ella Holmes – 31 January 1942 – aged 55 when on *Titanic*; first-class passenger (ticket price: £135 12s 8d); rescued in lifeboat 8

Titanic survivor quotes

Mr Dickinson H. Bishop, witness at United States Senate Inquiry (Day 11)

> 'One of the officers in charge of the lowering; which one, I could not tell. There was some confusion there at the time, and I did not pay much attention. There was an officer stationed at the side of the lifeboat, and as my wife got in I followed immediately, and he helped me into the boat, or rather indicated, and I fell into the boat.'

Mr Albert E. J. Horswell (able seaman)

> 'I was ordered to the port emergency boat. I was ordered to put the lantern in the boat and a chronometer. I did that and came out again, and I was sent to the starboard boats. Then I was ordered into No. 1 boat.'

Titanic fact

The height from the top of the funnels to the keel was 175' (53.3 metres).

February

Day 32

1 February | On this day in *Titanic* history

The following people were born on this day:
Goodwin, Mr Frederick Joseph – 1 February 1870 – aged 42 when on *Titanic*; first-class passenger (ticket price: £46 18s); *Titanic* victim

Rigozzi, Mr Abele – 1 February 1890 – aged 22 when on *Titanic*; assistant waiter; *Titanic* victim

The following people died on this day:
Sundman, Mr Johan Julian – 1 February 1920 – aged 45 when on *Titanic*; third-class passenger (ticket price: £7 18s 6d); rescued in lifeboat 15

White, Mr William George – 1 February 1927 – aged 42 when on *Titanic*; trimmer; rescued in lifeboat 15

Hays, Mrs Clara Jennings Gregg– 1 February 1955 – aged 52 when on *Titanic*; first-class passenger (ticket price: £93 10s); rescued in lifeboat 3

Crow, Mr George Frederick – 1 February 1960 – aged 26 when on *Titanic*; saloon steward; rescued in lifeboat 14

Evans, Mr Alfred Frank – 1 February 1964 – aged 25 when on *Titanic*; lookout; rescued in lifeboat 15 (Evans and Hogg were on lookout duty from 6.00 pm–8.00 pm; Evans was in bed at the time of the collision. Hogg was rescued in lifeboat 7.)

Mrs Clara Jennings Hays, mentioned in *Chicago Examiner*, Friday, 19 April 1912

> 'WIDOW AND DAUGHTERS OF HAYS SPEED HOME
> New York, April 18. Mrs Charles M. Hays and her two daughters, Miss Margaret Hays and Mrs Thornton Davidson, survivors of the *Titanic* wreck, were met at the pier by a large party of friends. Mr Hays, president of the Grand Trunk Railway, went down with the *Titanic*. In taxicabs the party went at once to the Grand Central Station, where they boarded a special train and proceeded to Montreal. None of them would discuss the affair.'

1 February 1904
CQD becomes the regulation distress signal call for radio Marconi installations.

Titanic fact

There were four men in each riveting crew at the Harland & Wolff shipyard.

Day 33

2 February | On this day in *Titanic* history

The following people were born on this day:
Toomey, Miss Ellen Mary – 2 February 1864 – aged 48 when on *Titanic*; second-class passenger (ticket price: £10 10s); rescued in lifeboat 9

Goldenberg, Mrs Nella Carlynne Wiggins – 2 February 1869 – aged 43 when on *Titanic*; first-class passenger (ticket price: £89 2s 1d); rescued in lifeboat 5

Coutts, Mrs Winnie 'Minnie' – 2 February 1876 – aged 36 when on *Titanic*; third-class passenger (ticket price: £15 18s); rescued in lifeboat 2

Diaper, Mr John Joseph – 2 February 1885 – aged 27 when on *Titanic*; fireman (engineering crew); rescued in lifeboat unknown

Street, Mr Thomas Albert – 2 February 1887 – aged 25 when on *Titanic*; fireman, rescued in lifeboat 9

Dean, Miss Elizabeth Gladys 'Millvina' – 2 February 1912 – aged 2 months when on *Titanic*; third-class passenger (ticket price: £20 11s 6d); rescued in lifeboat 10

The following people died on this day:
Flaherty, Mr Edward – 2 February 1940 – aged 43 when on *Titanic*; fireman; rescued in lifeboat unknown

Cameron, Miss Clear Annie – 2 February 1962 – aged 35 when on *Titanic*; second-class passenger (ticket price: £21); rescued in lifeboat 14

Hannah, Mr Borak ('Hannah Assi Borah') – 2 February 1952 – aged 27 when on *Titanic*; third-class passenger (ticket price: £7 4s 7d); rescued in lifeboat 15

Mr Marconi, witness at United States Senate Inquiry (Day 2)

> 'C.Q. is the call for all stations. If you call C.Q. on a ship it means, "All other stations stand at attention and reply." I did not make the signal originally. I presume the object was to indicate, in a certain way, to all stations, the danger, or peril that existed.'

Titanic survivor quote

Miss Ellen Mary Toomey

> 'There were thirty-five in our boat, which was the second or third boat to put off…The *Titanic* was gradually sinking and when we were about two miles away, we could hear the awful shrieks of the drowning people…we were in the small boats about seven hours…'

Day 34

3 February | On this day in *Titanic* history

The following people were born on this day:
Nannini, Sig. Francesco Luigi Arcangelo – 3 February 1870 – aged 42 when on *Titanic*; head waiter (restaurant); *Titanic* victim

Wilkinson, Mrs Elizabeth Anne – 3 February 1882 – aged 30 when on *Titanic*; second-class passenger (ticket price: £26); rescued in lifeboat 16

Evans, Mr Alfred Frank – 3 February 1887 – aged 25 when on *Titanic*; lookout; rescued in lifeboat 15

Fletcher, Mr Percy William – 3 February 1887 – aged 30 when on *Titanic*; bugler steward; *Titanic* victim

The following people died on this day:
Spedden, Mr Frederic Oakley – 3 February 1947 – aged 45 when on *Titanic*; first-class passenger (ticket price: £134 10s); rescued in lifeboat 3

Windebank, Mr Alfred Edgar – 3 February 1961 – aged 38 when on *Titanic*; cook; rescued in lifeboat 13

Mr Frederic Oakley Spedden, mentioned in *New York Times*, Thursday, 30 May 1912

> 'When the Cunarder *Carpathia* arrived at her pier yesterday from Naples, her first return to this port since she landed the survivors from the *Titanic*, she was boarded by a committee of those survivors bearing a silver loving cup, and gold, silver, and bronze medals for Capt. Arthur Henry Rostron and his officers and crew. The committee was headed by Frederick K. Seward, and included Mrs J. J. Brown of Denver, Lieut. H. B. Steffanson of the Swedish Army, F. O. Spedden, Karl Behr, the tennis player; I. G. Frauenthal, and George A. Harder. At the head of the table, beside cases of medals, was the silver loving cup, standing fifteen inches high, on an ebony base and bearing the following inscription:
>
> > 'Presented to Capt. A. H. Rostron, R. N. R., commander of the RMS *Carpathia*. In grateful recognition and appreciation of his heroism and efficient service in the rescue of the survivors of the *Titanic* on April 15, 1912, and of the generous and sympathetic treatment he accorded us on his ship.'

Titanic fact

Construction workers at the Harland & Wolff shipyard toiled for an average of 49 hours per week.

Day 35

4 February | On this day in *Titanic* history

The following people were born on this day:
Rintamäki, Mr Matti – 4 February 1877 – aged 35 when on *Titanic*; third-class passenger (ticket price: £7 2s 6d); *Titanic* victim

Harvey, Mr Herbert Gifford – 4 February 1878 – aged 34 when on *Titanic*; junior assistant second engineer; *Titanic* victim

Kelly, Miss Anna Katherine – 4 February 1892 – aged 20 when on *Titanic*; third-class passenger (ticket price: £7 15s); rescued in lifeboat 16

A tribute to the engineering staff
All the engineers were under the control of Joseph Bell, the chief engineer officer. His staff consisted of twenty-four engineers, all lost.

It was her engineers who kept the lights burning, and in the list of heroes who went down with the vessel the names of the men of the engineering force will have a high place. Not one of them was saved, although many of them were off duty, and these had some chance of climbing to the deck. While it will never be known just what happened, it is believed that everyone went back to his post instead of to the decks.

The New York Times; New York, Tuesday, 23 April 1914

> 'To the heroic marine engineers of *Titanic*
>
> 'Sir Archibald Denny Bt, LL.D; unveiling ceremony for the Engineers' Memorial, Southampton, 22 April 1914
>
> 'They must have known that pumping could do no more than delay the final catastrophe, yet they stuck pluckily to their duty. Driven back from boiler-room to boiler-room, fighting for every inch of draught to give time for the launching of the boats, not one of those brave officers was saved.' [It was estimated that over 100,000 people attended the unveiling ceremony.]

Mr Joseph G. Boxhall, witness at United States Senate Inquiry (Day 3)

> 'When I was on duty on board ship whilst at Southampton during the daytime there was always a senior officer along with me; so that any questions that were to be asked could be answered by him, or if anything was to be found out I would always refer to him, to the senior officer. At night-time the two junior officers were in charge of the ship, with men on watch with them.'

Titanic fact

The yard number given to the *Titanic* when her keel was laid down was known as 'Harland & Wolff Yard No. 401'.

Day 36

5 February | On this day in *Titanic* history

The following people were born on this day:
McCaffry, Mr Thomas Francis – 5 February 1866 – aged 46 when on *Titanic*; first-class passenger (ticket price: £75 4s 10d); *Titanic* victim

Jönsson, Mr Nils Hilding – 5 February 1885 – aged 27 when on *Titanic*; third-class passenger (ticket price: £7 17s 1d); *Titanic* victim

Dahlberg, Miss Gerda Ulrika – 5 February 1890 – aged 22 when on *Titanic*; third-class passenger (ticket price: £10 10s 4d); *Titanic* victim

The following person died on this day:
Silvén, Miss Lyyli Karoliina – 5 February 1974 – aged 17 when on *Titanic*; second-class passenger (ticket price: £13); rescued in lifeboat 16

Mr Robert Hichens (quartermaster), witness at United States Senate Inquiry (Day 5)

> 'I left the wheel at 23 minutes past 12, sir. I was relieved by Quartermaster Perkis. He relieved me at 23 minutes past 12. I think the first officer, or one of the officers said, "That will do with the wheel; get the boats out." I went out to get the boats out on the port side. I think I got in No. 6 boat, sir; put in charge of her by the second officer, Mr Lightoller. We lowered away from the ship, sir, and were told to "Pull toward that light," which we started to do, to pull for that light. I had 38 women in the boat, sir, 1 seaman [Fleet] and myself, with 2 male passengers, 1 Italian boy and a Canadian Major Peuchen who testified here yesterday.'

Titanic survivor quote

Miss Lyyli Karoliina Silvén mentioned in *Karlshamns Allehanda*, Wednesday, 15 May 1912

> 'Having found her relatives she was reluctant to enter even the last boats, but Rev. Lahtinen had forced her to do so. Rev. Lahtinen had been looking calm and cheerful. Mrs Lahtinen was supposed to be helped into the boat as well, but she clung to her husband and said she would not be parted from him, but rather die with him. In the boat where Miss Silvén was there were 50–60 passengers. The last thing she saw was some first class men, who fetched some deck chairs, placed them on the deck and started conversing [with] each other, whereas the bow kept sinking. When her boat had come some way away from the liner and taken a capsized boat's passengers aboard, there was a terrible noise from the *Titanic*, hollering and shouting. According to Miss Silvén, they had been rowing for half an hour, when the *Titanic*'s stern rose and the gigantic ship sank. The thousands of lights were still glowing…'

Day 37

6 February | On this day in *Titanic* history

The following people were born on this day:
Straus, Mr Isidor – 6 February 1845 – aged 67 when on *Titanic*; first-class passenger (ticket price: £221 15s 7d); *Titanic* victim

Straus, Mrs Rosalie Ida Blun – 6 February 1849 – aged 63 when on *Titanic*; first-class passenger (ticket price: £221 15s 7d); *Titanic* victim

Bott, Mr William Thomas – 6 February 1866 – aged 46 when on *Titanic*; greaser; *Titanic* victim

Randall, Mr Frank Henry – 6 February 1883 – aged 29 when on *Titanic*; saloon steward in second class; *Titanic* victim

Blumet, Mr Jean Baptiste – 6 February 1886 – aged 26 when on *Titanic*; pantryman (restaurant staff); *Titanic* victim

Olsen, Master Artur Karl – 6 February 1903 – aged 9 when on *Titanic*; third-class passenger (ticket price: £3 3s 5d); rescued in lifeboat 13

The following people died on this day:
Leitch, Miss Jessie Wills – 6 February 1963 – aged 31 when on *Titanic*; second-class passenger (ticket price: £33); rescued in lifeboat 11 (travelling on the same ticket as Rev John Harper and Miss Annie Jessie Harper)

Tenglin, Mr Gunnar Isidor – 6 February 1974 – aged 25 when on *Titanic*; third-class passenger (ticket price: £7 15s 7d); rescued in lifeboat 13

Kemish, Mr George – 6 February 1966 – aged 22 when on *Titanic*; fireman; rescued in lifeboat 9

Collins, Mr John – 6 February 1941 – aged 17 when on *Titanic*; scullion; rescued in collapsible lifeboat B

If you notice, Ida and Isidor Straus shared the same birthday and they both perished on the same day.

Mrs Rosalie Ida Straus

> 'We have lived together for many years. Where you go, I go.'

Mr Hugh Woolner, witness at United States Senate Inquiry (Day 10)

> 'I saw a faint line, what looked like a faint line along the horizon; but when we got on the *Carpathia*, we saw it was a huge floe which stretched out, I do not know how far but we were several hours steaming along it.'

Titanic fact

The *Titanic* held no passenger lifeboat drills during its voyage.

Day 38

7 February | On this day in *Titanic* history

The following people were born on this day:
Andrews, Mr Thomas – 7 February 1873 – aged 39 when on *Titanic*; first-class passenger (complimentary ticket); *Titanic* victim

Minahan, Mrs Lillian Mae Thorpe – 7 February 1875 – aged 37 when on *Titanic*; first-class passenger (ticket price: £90); rescued in lifeboat 14

Vanderplancke, Mr Julius – 7 February 1881 – aged 31 when on *Titanic*; third-class passenger (ticket price: £18); *Titanic* victim

Maguire, Mr John Edward – 7 February 1882 – aged 30 when on *Titanic*; first-class passenger (ticket price: £26); *Titanic* victim

Keane, Mr Andrew – 7 February 1889 – aged 23 when on *Titanic*; third-class passenger (ticket price: £7 15s); *Titanic* victim

Olsvigen, Mr Thor Anderson – 7 February 1892 – aged 20 when on *Titanic*; third-class passenger (ticket price: £9 4s 6d); *Titanic* victim

Hiltunen, Miss Martta Maria – 7 February 1894 – aged 18 when on *Titanic*; second-class passenger (ticket price: £13); *Titanic* victim

Mr Thomas Andrews (managing director of Harland & Wolff Shipyards)
Andrews boarded with a complimentary ticket No. 112050.

> 'You weren't there at my first meeting with Ismay. To see the little red marks all over the blueprints. First thing I thought was: "Now here's a man who wants me to build him a ship that's gonna be sunk." We're sending gilded egg shells out to sea.
>
> 'Let the Truth be known, no ship is unsinkable. The bigger the ship, the easier it is to sink her. I learned long ago that if you design how a ship'll sink, you can keep her afloat. I proposed all the watertight compartments and the double hull to slow these ships from sinking. In that way, you get everyone off. There's time for help to arrive, and the ship's less likely to break apart and kill someone while she's going down…The *Titanic* is now about complete and will I think do the old Firm credit tomorrow when we sail.'

Mr Robert Hichens (quartermaster), witness at United States Senate Inquiry (Day 5)
'The cries I heard lasted about two minutes, and some of them were saying, "It is one boat aiding the other." There was another boat aside of me, the boat the master-at-arms was in, full right up.'

Titanic fact

Tennis players Dick Williams and Karl Behr survived and eventually became teammates on the U.S. Davis Cup Team.

Day 39

8 February | On this day in *Titanic* history

The following people were born on this day:
Hagland, Mr Ingvald Olai Olsen – 8 February 1884 – aged 28 when on *Titanic*; third-class passenger (ticket price: £6 19s 4d); *Titanic* victim

Panula, Master Jaako Arnold – 8 February 1897 – aged 15 when on *Titanic*; third-class passenger (ticket price: £39 13s 9d); *Titanic* victim (15-year-old Jaako perished, along with his mother, four siblings and their neighbour's daughter, who was travelling with them.)

The following people died on this day:
Najib Kiamie, Miss Adele 'Jane' – 8 February 1924 – aged 15 when on *Titanic*; third-class passenger (ticket price: £7 4s 6d); rescued in collapsible lifeboat C

Clarke, Mrs Ada Maria – 8 February 1953 – aged 28 when on *Titanic*; second-class passenger (ticket price: £27); rescued in lifeboat 14

Mr Robert Hichens (quartermaster), witness at United States Senate Inquiry (Day 5)

> 'Everybody seemed in a very bad condition in the boat, sir. Everybody was quite upset, and I told them somebody would have to pull; there was no use stopping there alongside of the ship, and the ship gradually going by the head. We were in a dangerous place, so I told them to man the oars, ladies and all, "All of you do your best." We got away about a mile, I suppose, from the ship, going after this light, which we expected to be a "cod banker", a schooner that comes out on the Banks.'

Titanic survivor quote

Mrs Ada M Clarke, quoted in *Cleveland Plain Dealer*, Saturday, 20 April 1912

> 'I was in one of three boats lashed together,' said Mrs Clarke, 'when one of the officers unlashed one of these and, taking all but two of the seamen, rowed off to rescue another boat in the distance which was crowded to the sinking point. "Be brave, women!" the officer said as they rowed away. And they were brave. The women and the two men manned the oars until we were rescued later.
>
> 'He made me leave him,' she said brokenly. 'I shouldn't have done it otherwise. Oh, they were brave and splendid, all the men. They died like brave men. At the last, all the men were kneeling and there floated out across the water the strains of "Nearer, My God, to Thee". I could hear it and saw the band men kneeling, too.'

Day 40

9 February | On this day in *Titanic* history

The following people were born on this day:

Baumann, Mr John David – 9 February 1864 – aged 48 when on *Titanic*; first-class passenger (ticket price: £25 18s 6d); *Titanic* victim

Van Billiard, Mr Austin Blyler – 9 February 1877 – aged 35 when on *Titanic*; third-class passenger (ticket price: £14 10s); *Titanic* victim

Lamb, Mr John J. – 9 February 1881 – aged 31 when on *Titanic*; second-class passenger (ticket price: £10 14s 2d); *Titanic* victim

Tu'mah, Master Jirjis Yūsuf– 9 February 1904 – aged 8 when on *Titanic*; third-class passenger (ticket price: £15 4s 11d); rescued in collapsible lifeboat C

The following people died on this day:

Mayzes, Mr Thomas Jubilee – 9 February 1928 – aged 25 when on *Titanic*; fireman; rescued in lifeboat 3

Crawford, Mr Alfred George – 9 February 1938 – aged 43 when on *Titanic*; bedroom steward; rescued in lifeboat 8

Chambers, Mr Norman Campbell – 9 February 1966 – aged 27 when on *Titanic*; first-class passenger (ticket price: £53 2s); rescued in lifeboat 5

Mr Joseph G. Boxhall, witness at United States Senate Inquiry (Day 3)

> 'After the boats are cleared the chocks are knocked down, or dropped down by patent levers, and the boat is hanging free. Then the davits are screwed out and the boat is suspended over the ship's side all ready for lowering away.'

Titanic survivor quote

Mr Norman Campbell Chambers

> 'I personally felt no sense of danger, as this water was forward of the bulkhead.' Three officers came down and reported that the ship was not taking any more water. Norman and Bertha returned to their stateroom and their steward came by and told them they could go back to bed. He finished dressing and Bertha went out. She 'came rushing back to me, saying that she had seen another passenger who informed her that the call had been given out for lifebelts and on the boat deck.' He went out and found their steward, who verified the order.

Titanic fact

The last dinner served in the first-class saloon consisted of eleven courses.

Day 41

10 February | On this day in *Titanic* history

The following people were born on this day:
Rothschild, Mrs Elizabeth Jane Anne Barrett – 10 February 1858 – aged 54 when on *Titanic*; first-class passenger (ticket price: £59 8s); rescued in lifeboat 6

Harper, Mrs Myra Raymond Haxtun – 10 February 1863– aged 49 when on *Titanic*; first-class passenger (ticket price: £76 14s 7d); rescued in lifeboat 3

Mintram, Mr William – 10 February 1866 – aged 46 when on *Titanic*; fireman; *Titanic* victim

Allen, Mr Ernest Frederick – 10 February 1888 – aged 24 when on *Titanic*; trimmer; rescued in collapsible lifeboat B

Augustsson, Mr Albert – 10 February 1889 – aged 23 when on *Titanic*; third-class passenger (ticket price: £7 17s 1d); *Titanic* victim

Māmā, Mr Hannā Mikā'īl – 10 February 1892 – aged 20 when on *Titanic*; third-class passenger (ticket price: £7 4s 7d); rescued in lifeboat 15

Čačić, Mr Jego Grga – 10 February 1894 – aged 18 when on *Titanic*; third-class passenger (ticket price: £8 13s 3d); *Titanic* victim

Ibrāhīm, Mrs Sāfiyah – 10 February 1894 – aged 18 when on *Titanic*; third-class passenger (ticket price: £7 4s 7d); rescued in collapsible lifeboat C

Blades, Mr Frederick Charles – 10 February 1895 – aged 17 when on *Titanic*; lift steward; *Titanic* victim

The following people died on this day:
Dyer-Edwardes, Mr Thomas – 10 February 1926 – aged 65 when on *Titanic*; first-class passenger (ticket price: £3); disembarked at Cherbourg

Homer, Mr Harry Haven 'Harry' – 10 February 1939 – aged 40 when on *Titanic*; first-class passenger (ticket price: £26 11s); rescued in lifeboat 15

Spedden, Mrs Margaretta Corning Stone 'Daisy'– 10 February 1950 – aged 39 when on *Titanic*; first-class passenger (ticket price: £34); rescued in lifeboat 3

Titanic survivor quote

Mrs Sophie Abrahim, quoted in *Greensburg Herald Tribune*, Monday, 22 April 1912

> '"Oh, I feel bad, ver' bad, los' money, los' friend. When I stand up I see ship come again, when I shut my eyes. I see people go down, down, down – Oh ver' bad time." And, with the remembrance of it, Mrs Sophie Abrahim, 20 years old, shuddered and buried her face in the dark hair of her little niece who leaned against her knee.'

Day 42

11 February | On this day in *Titanic* history

The following people were born on this day:
Rommetvedt, Mr Knud Paust – 11 February 1863 – aged 49 when on *Titanic*; third-class passenger (ticket price: £7 15s 6d); *Titanic* victim

Carr, Miss Jane – 11 February 1867 – aged 45 when on *Titanic*; third-class passenger (ticket price: £7 15s); *Titanic* victim

The following people died on this day:
Beckwith, Mrs Sallie Monypeny – 11 February 1955 – aged 46 when on *Titanic*; first-class passenger (ticket price: £52 11s 1d); rescued in lifeboat 5

Crosby, Miss Harriette Rebecca – 11 February 1941 – aged 39 when on *Titanic*; first-class passenger (ticket price: £26 11s); rescued in lifeboat 7

Williams, Mr Walter John – 11 February 1971 – aged 28 when on *Titanic*; saloon steward in second class; rescued in lifeboat 13

Titanic survivor quotes

Mr Walter John Williams

> 'During the night we were passed by what looked like a sailing ship. As we got closer we realised it was a small iceberg.' On 14 April 1912, Walter remembered a work mate joking to him, 'We've only hit an iceberg. It's another job for Belfast.' The colleague believed the ship would simply need repairing in the shipyard at Belfast. Walter made a jump for lifeboat 13 and he later said that many lifeboats were far from full.

Mr G Hogg (lookout)

> 'On a very nice night, with the stars shining, sometimes you might think it was a ship when it was a star on the horizon. If you had glasses, you could soon find out whether it was a ship or not.'

William Ward, witness at United States Senate Inquiry (Day 7)

> 'Everybody was moving around and in a most orderly manner. There did not seem to be any excitement. In fact, there was a lot of ladies and gentlemen there that were just treating it as a kind of a joke.'

Titanic fact

Titanic's lifeboat drill was cancelled on the morning of the day it hit the iceberg.

Day 43

12 February | On this day in *Titanic* history

The following people were born on this day:
Crumplin, Mr Charles George Chandler – 12 February 1877 – aged 35 when on *Titanic*; bedroom steward; *Titanic* victim

Caton, Miss Annie – 12 February 1879 – aged 33 when on *Titanic*; Turkish bath stewardess (victualling crew); rescued in lifeboat 11

Wilson, Miss Helen Alice – 12 February 1881 – aged 31 when on *Titanic*; first-class passenger (ticket price: £134 10s); rescued in lifeboat 3

Carlsson, Mr August Sigfrid – 12 February 1884 – aged 28 when on *Titanic*; third-class passenger (ticket price: £7 15s 11d); *Titanic* victim

Newell, Miss Marjorie Anne – 12 February 1889 – aged 23 when on *Titanic*; first-class passenger (ticket price: £113 5s 6d); rescued in lifeboat 6

Marvin, Mr Daniel Warner – 12 February 1894 – aged 18 when on *Titanic*; first-class passenger (ticket price: £53 2s); *Titanic* victim

The following person died on this day:
Shiers, Mr Alfred Charles – 12 February 1946 – aged 25 when on *Titanic*; fireman; rescued in lifeboat 5

Mr Arthur J. Bright (quartermaster), witness at United States Senate Inquiry (Day 9)

> 'After we had finished firing the distress signals there were two boats left. I went and assisted to get out the starboard one; that is, the starboard collapsible boat.'

Titanic survivor quote

Miss Annie Caton

> '...as we stood waiting for our boat we felt the ship shiver under our feet. After what seemed hours of waiting we were lowered in the boat with great difficulty as the ropes were stiff and we nearly turned turtle.
>
> 'Then the pull from the ship – oh! How those boys worked to get clear of the suction. Just imagine us adrift on the ocean in a small boat in the dead of the night, no lamp, no water, no food and ice all around us, lumps knocking against us all the time. It was a nightmare, expecting every minute to meet death.
>
> 'The last we saw and heard of the *Titanic* was our brave men standing with set faces waiting for the last. One officer called out "Now boys, remember you are English."
>
> 'The bandsmen were kneeling and playing "Nearer my God to Thee" and then the ship turned and hurled all those brave fellows to their death. The shrieks I heard will forever ring in my ears.'

Day 44

13 February | On this day in *Titanic* history

The following people were born on this day:
Jensen, Mr Niels Rasmus – 13 February 1864 – aged 48 when on *Titanic*; third-class passenger (ticket price: £7 1s 1d); *Titanic* victim

Ross, Mr Horace Leopold – 13 February 1876 – aged 38 when on *Titanic*; scullion; rescued in lifeboat 13

Thomas, Mr Albert Charles – 13 February 1889 – aged 23 when on *Titanic*; saloon steward; rescued in lifeboat 15

Lindqvist, Mr Eino William – 13 February 1892 – aged 20 when on *Titanic*, third-class passenger (ticket price: £7 18s 6d); rescued in lifeboat 15

Andersen-Jensen, Miss Carla Christine Nielsine – 13 February 1893 – aged 19 when on *Titanic*; third-class passenger (ticket price: £7 17s 1d); rescued in lifeboat 16

The following people died on this day:
Woolner, Mr Hugh – 13 February 1925 – aged 23 when on *Titanic*; first-class passenger (ticket price: £113 5s 6d); rescued in lifeboat 6

Dahl, Mr Charles Edward – 13 February 1933 – aged 23 when on *Titanic*; third-class passenger (ticket price: £113 5s 6d); rescued in lifeboat 6

Titanic survivor quote

Miss Carla Christine Nielsine Andersen-Jensen

> 'I shared the cabin with three young girls, two English and a Swede and we had turned in early because the passengers in 3rd class had instructions to be below deck before 22 hours. We felt a bump a half hour later, but after the first shock, we went back to sleep again. We had confidence in *Titanic*, the other three girls would never wake again!
>
> 'At 12.30 I heard a knock and my uncle said: "You better put a coat on and come up on the deck." So I put a coat over my night dress.
>
> 'We were now up on the deck and there was not much commotion, we had hit an iceberg, but everyone felt the ship would stay afloat. The ship was fully lit and there was music in the 1st class saloon. There were no panic even when the lifeboats were launched, no one seemed to push to get into them and the women and children went into the boats first.'

Titanic fact

The weekly wage of a Harland & Wolff construction worker was £2.

Day 45

14 February | On this day in *Titanic* history

The following people were born on this day:

Clarke, Mr Charles Valentine – 14 February 1883 – aged 29 when on *Titanic*; second-class passenger (ticket price: £26); *Titanic* victim

Basilico, Sig. Giovanni – 14 February 1885 – aged 27 when on *Titanic*; waiter (restaurant staff); *Titanic* victim

Haas, Miss Aloisia – 14 February 1888 – aged 24 when on *Titanic*; third-class passenger (ticket price: £8 17s); *Titanic* victim

The following people died on this day:

Beesley, Mr Lawrence – 14 February 1967 – aged 34 when on *Titanic*; second-class passenger (ticket price: £13); rescued in lifeboat 13

Rowe, Mr George Thomas – 14 February 1974 – aged 32 when on *Titanic*; quartermaster; rescued in collapsible lifeboat C

Madill, Miss Georgette Alexandra – 14 February 1974 – aged 16 when on *Titanic*; first-class passenger (ticket price: £211 6s 9d); rescued in lifeboat 2

Hart, Miss Eva Miriam – 14 February 1996 – aged 7 when on *Titanic*; second-class passenger (ticket price: £26 5s); rescued in lifeboat 14

***Harrisburg Telegraph*, 15 April 1912**

> 'Ship sends out distress signals declaring that immediate relief is necessary; Majority of 1,500 passengers are American; All transferred in safety – Women reported to have been rescued and lowered to lifeboats; Officials of steamship company say "Greyhound will not sink"; Prominent men and women on trip.'

Titanic survivor quote

Mrs Eva Hart

> 'I was only seven but I remember thinking that everything in the world was standing still. And it wasn't until we were in the lifeboat and rowing away, it wasn't until then I realised that ship's going to sink. It hits me there.
>
> 'The sounds of people drowning are something that I cannot describe to you, and neither can anyone else. It's the most dreadful sound and there is a terrible silence that follows it. I never closed my eyes. I didn't sleep at all. I saw it, I heard it, and nobody could possibly forget it.'

Titanic fact

3,500 bags of mail were lost when the ship sank.

Day 46

15 February | On this day in *Titanic* history

The following people were born on this day:
Frölicher-Stehli, Mrs Margaretha Emerentia (née Stehli) – 15 February 1864 – aged 48 when on *Titanic*; first-class passenger (ticket price: £79 4s); rescued in lifeboat 5

Hyman, Mr Joseph Abraham – 15 February 1881 – aged 31 when on *Titanic*; third-class passenger (ticket price: £7 17s 9d); rescued in collapsible lifeboat C

Hegarty, Miss Hanora – 15 February 1892 – aged 20 when on *Titanic*; third-class passenger (ticket price: £6 15s); *Titanic* victim

Hirvonen, Miss Hildur Elisabeth – 15 February 1910 – aged 2 when on *Titanic*; third-class passenger (ticket price: £12 5s 9d); rescued in lifeboat 15

The following people died on this day:
Whilems, Mr Charles – 15 February 1940 – aged 31 when on *Titanic*; second-class passenger (ticket price: £13); rescued in lifeboat 9

Becker, Miss Marion Louise – 15 February 1944 – aged 4 when on *Titanic*; second-class passenger (ticket price: £39); rescued in lifeboat 11

Becker, Mrs Nellie E. (née Baumgardner) – 15 February 1961 – aged 35 when on *Titanic*; second-class passenger (ticket price: £39); rescued in lifeboat 11

Mrs Nellie E Becker, mentioned in *Gettysburg Complier*, Wednesday, 15 May 1912

> 'Dr L. B. Wolf Secretary of Foreign Missionary Board of The General Synod Lutheran Church, tells that he met Mrs Becker, a Lutheran missionary from India, returning on the ill-fated *Titanic*, at New York when the *Carpathia* docked with the *Titanic*'s survivors. Mrs Becker quietly told of [her] escape as follows: When the shock came she went into the gangway and inquired of the steward the cause of confusion. She was told what happened and being assured of safety was told to retire. The three children were in bed. Twenty minutes later the steward called her and shouted all hands on deck. She awakened the children and wanted to dress them but was told there was no time. She grasped shoes and stockings and in their night clothes rushed them on deck, where the two smaller children were put in to lifeboat No.18. She appealed to be placed in with them, and as the boat was being lowered, she was thrown in with the two small children. [Her daughter] Ruth was left behind and placed into the next boat lowered and she found her on board the *Carpathia*. She says she saw the *Titanic* slowly sink bow first, and after the explosion disappear.'

Day 47

16 February | On this day in *Titanic* history

The following people were born on this day:
Spencer, Mrs Marie Eugenie DeMougeot – 16 February 1864 – aged 48 when on *Titanic*; first-class passenger (ticket price: £146 10s 5d); rescued in lifeboat 6

Graham, Miss Margaret Edith – 16 February 1893 – aged 19 when on *Titanic*; first-class passenger (ticket price: £153 9s 3d); rescued in lifeboat 3

The following people died on this day:
Crimmins, Mr James – 16 February 1956 – aged 21 when on *Titanic*; third-class passenger (ticket price: £12 5s 9d); rescued in lifeboat 15

Bishop, Mr Dickinson H. – 16 February 1961 – aged 25 when on *Titanic*; fireman; rescued in lifeboat 13

Mousselmani, Mrs Fatima – 16 February 1971 – aged 22 when on *Titanic*; third-class passenger (ticket price: £7 4s 6d); rescued in collapsible lifeboat C

Mr Guglielmo Marconi, witness at United States Senate Inquiry (Day 1)

> 'The wireless equipment on the *Titanic* was a fairly powerful set, capable, I should say, of communicating four or five hundred miles during the daytime and much further during the night-time.'

Miss Fatima Muselmanie mentioned in *The South Bend Tribune*, Wednesday, 8 May 1912

> 'Mohammed Mustafa Ajamy and Miss Fatima Muselmanie, the latter a *Titanic* survivor and the woman who brought word of the drowning of four young Syrians bound for South Bend, were married here yesterday by Justice Schwein. Mrs Ajamy is a native of Syria as is also Mr Ajamy, and although she survived the terrible disaster at sea and is today a bride, she is not as happy as one might suppose. When the great ship went down her two older brothers gave up their lives that she and others of the women passengers might safely escape in the lifeboats. The young woman was prostrated for several days after the disaster and when she was able to travel came here from New York. She relates in her native tongue a thrilling account of the great disaster and her personal experiences.'

Titanic fact

Titanic was launched in Belfast, Northern Ireland, into the Victoria Channel in Belfast Lough from a slipway at the Queen's Yard of the Harland & Wolff shipyard.

Day 48

17 February | On this day in *Titanic* history

The following people were born on this day:

Carter, Reverend Ernest Courtenay – 17 February 1858 – aged 54 when on *Titanic*; second-class passenger (ticket price: £26); *Titanic* victim

Weikman, Mr Augustus Henry – 17 February 1860 – aged 52 when on *Titanic*; barber (victualling crew); rescued in collapsible lifeboat A

Connaughton, Mr Michael – 17 February 1872– aged 40 when on *Titanic*; third-class passenger (ticket price: £7 15s); *Titanic* victim

Smith, Mr James Muil – 17 February 1873 – aged 39 when on *Titanic*; junior fourth engineer; *Titanic* victim

Lahtinen, Mrs Anna Amelia – 17 February 1878 – aged 34 when on *Titanic*; second-class passenger (ticket price: £26); *Titanic* victim

West, Mrs Ada Mary – 17 February 1879 – aged 33 when on *Titanic*; second-class passenger (ticket price: £27 15s); rescued in lifeboat 11

Olsen, Mr Ole Martin – 17 February 1885 – aged 27 when on *Titanic*; third-class passenger (ticket price: £7 6s 3d); *Titanic* victim

Prentice, Mr Frank Winnold – 17 February 1889 – aged 23 when on *Titanic*; storekeeper; rescued in lifeboat 4

Crimmins, Mr James – 17 February 1891 – aged 21 when on *Titanic*; fireman; rescued in lifeboat 13

McMahon, Mr Martin – 17 February 1893 – aged 19 when on *Titanic*; third-class passenger (ticket price: £59 8s); *Titanic* victim

The following person died on this day:

Gibson, Miss Dorothy Winifred Brown – 17 February 1946 – aged 22 when on *Titanic*; first-class passenger (ticket price: £59 8s); rescued in lifeboat 7

Titanic survivor quotes

Miss Dorothy Winifred Gibson

> 'I will never forget the terrible cry that rang out from people who were thrown into the sea and others who were afraid for their loved ones.'

Mr John Hart (steward in third class)

> 'After the collision, on being aroused first by a man coming from forward, a steward, he said there had been an accident, and I closed my eyes and went to sleep again; I did not believe it.'

Day 49

18 February | On this day in *Titanic* history

The following people were born on this day:
Crosby, Captain Edward Gifford – 18 February 1842 – aged 70 when on *Titanic*; first-class passenger (ticket price: £71); *Titanic* victim

Stähelin-Maeglin, Dr Max – 18 February 1880 – aged 32 when on *Titanic*; first-class passenger (ticket price: £30 10s); rescued in lifeboat 3 (while his title was doctor, he was actually a lawyer)

Taylor, Mr William John – 18 February 1881 – aged 31 when on *Titanic*; saloon steward in first class; *Titanic* victim

The following person died on this day:
Abbott, Mrs Rhoda Mary 'Rosa' – 18 February 1946 – aged 54 when on *Titanic*; second-class passenger (ticket price: £20 5s); rescued in collapsible lifeboat A

Titanic survivor quotes

Rhoda 'Rosa' Abbott

> 'Soon the raft tilted and all slid off into the water. Many of them managed to get back on it and some did not. I managed somehow to get on it, but I don't know how. Had it not been for Officer Lowe, I would have been drowned. I was nearly exhausted when he lifted me into his lifeboat. It would have been impossible for an officer to show more courtesy and many of the criticisms that have been made against this man are very unjust.'

Mr Frederick Barrett

> 'I was standing talking to the second engineer. The bell rang, the red light showed. We sang out, shut the doors [indicating the ash doors to the furnaces] and there was a crash just as we sung out. The water came through the ship's side. The engineer and I jumped to the next section. The next section to the forward section is No. 5.'

Titanic fact

The weight of the freight and passengers' baggage was 900 tonnes.

Day 50

19 February | On this day in *Titanic* history

The following people were born on this day:
Kent, Mr Edward Austin – 19 February 1854 – aged 58 when on *Titanic*; first-class passenger (ticket price: £29 14s); *Titanic* victim

Davison, Mrs Mary E. – 19 February 1878 – aged 34 when on *Titanic*; third-class passenger (ticket price: £16 2s); rescued in lifeboat 16

Wheeler, Mr Edwin Charles – 19 February 1886 – aged 26 when on *Titanic*; second-class passenger (ticket price: £12 17s 6d); *Titanic* victim

Nilsson, Miss Helmina Josefina – 19 February 1886 – aged 26 when on *Titanic*; third-class passenger (ticket price: £7 17s 1d); rescued in lifeboat 13

Johnston, Miss Catherine Nellie – 19 February 1905 – aged 7 when on *Titanic*; third-class passenger (ticket price: £23 9s); *Titanic* victim

Laroche, Miss Simonne Marie Anne Andrée – 19 February 1909 – aged 3 when on *Titanic*; second-class passenger (ticket price: £41 11s 7d); rescued in lifeboat 14

The following people died on this day:
Dyker, Mrs Anna Elisabeth Judith – 19 February 1961 – aged 54 when on *Titanic*; second-class passenger (ticket price: £20 5s); rescued in collapsible lifeboat A

Fortune, Miss Mabel Helen – 19 February 1968 – aged 23 when on *Titanic*; first-class passenger (ticket price: £263); rescued in lifeboat 10

***Oakland Tribune*, 15 April 1912**

> '*Titanic*'s passengers saved; Liner sinking.
>
> Ill-fated vessel begins to founder while limping toward Halifax after all aboard are rescued.
>
> Monster of ocean badly damaged by collision with iceberg; Many notables on list.'

Titanic survivor quote

Mrs Elisabeth Judith Dyker (née Andersson)

> 'I lost my husband and everything I owned, I was wounded when I was brought from *Titanic*, that together with sorrow and worries has completely broke me down. My health has not recovered, I have tried to do some work to earn my living, as I'm now living in my parents' home and my father is old and cannot earn too much, I cannot fully be a burden to him, but my strength doesn't allow me to do further work.'

Day 51

20 February | On this day in *Titanic* history

The following people were born on this day:
Hodgkinson, Mr Leonard – 20 February 1866 – aged 46 when on *Titanic*; senior fourth engineer; *Titanic* victim

Holm, Mr John Fredrik Alexander – 20 February 1869 – aged 43 when on *Titanic*; third-class passenger (ticket price: £6 9s); *Titanic* victim

Ashe, Mr Henry Wellesley – 20 February 1871 – aged 41 when on *Titanic*; glory hole steward in third class; *Titanic* victim

Teuton, Mr Thomas Moore – 20 February 1877 – aged 35 when on *Titanic*; saloon steward; *Titanic* victim

Zimmermann, Mr Leo – 20 February 1883 – aged 29 when on *Titanic*; third-class passenger (ticket price: £7 17s 6d); *Titanic* victim

The following people died on this day:
Chevré, Mr Paul Romaine Marie Léonce – 20 February 1914 – aged 26 when on *Titanic*; third-class passenger (ticket price: £7 17s 1d); rescued in lifeboat 13

Bonnell, Miss Elizabeth – 20 February 1936 – aged 61 when on *Titanic*; first-class passenger (ticket price: £26 11s); rescued in lifeboat 8

Titanic survivor quote

Mr Paul Chevré

> 'Suddenly the lights went out and an immense clamor filled the air in one supreme cry for help. Little by little the *Titanic* settled down, and for three hours cries of anguish were heard. At moments the cries of terror were lulled and we thought it was all over, but the next instant they were renewed in still keener accents. As for us we did nothing but row, row, row to escape from the death cries. In our little boat we were frozen with cold, having left the ship without overcoats or rugs. We shouted from time to time to attract attention, but obtained no reply. A German baron who was with us fired off all the cartridges in his revolver. This agonizing suspense lasted for many hours until at last the *Carpathia* appeared. We shouted "Hurrah!" and all the boats scattered on the sea made towards her.'

Titanic fact

It took *Titanic* 62 seconds to travel the length of the slipway and into the water.

Day 52

21 February | On this day in *Titanic* history

The following people were born on this day:
Dyer-Edwardes, Mr Thomas – 21 February 1847 – aged 65 when on *Titanic*; first-class passenger (ticket price: £3); father of the Countess of Rothes, disembarked at Cherbourg

Greenfield, Mrs Blanche Strouse – 21 February 1867 – aged 45 when on *Titanic*; first-class passenger (ticket price: £63 7s 2d); rescued in lifeboat 7

Oviés Y Rodríguez, Mr Servando José Florentino – 21 February 1876 – aged 36 when on *Titanic*; first-class passenger (ticket price: £27 14s 5d); *Titanic* victim

Abbott, Mr Rossmore Edward – 21 February 1896 – aged 16 when on *Titanic*; third-class passenger (ticket price: £20 5s); *Titanic* victim

(Edward and his 13-year-old brother Eugene did not survive, but their mother did.)

The following people died on this day:
Baggott, Mr Allen Marden – 21 February 1921 – aged 28 when on *Titanic*; saloon steward (victualling crew); rescued in lifeboat 9

McLaren, Mrs Hypatia – 21 February 1959 – aged 42 when on *Titanic*; stewardess; rescued in lifeboat 5

Salkjelsvik, Miss Anna Kristine – 21 February 1977 – aged 21 when on *Titanic*; third-class passenger (ticket price: £7 13s); rescued in collapsible lifeboat C

New York Times, 21 February 1913

> 'New York Times reports that Sir William Arrol the designer of *Titanic*'s huge Arrol Gantries has passed away.'

Titanic survivor quote

Mr Joseph Scarrott (able seaman)

> 'We started taking the women and children. There would be 20 women got into the boat, I should say, when some men tried to rush the boats, foreigners they were, because they could not understand the order which I gave them, and I had to use a bit of persuasion. The only thing I could use was the boat's tiller.'

Titanic fact

The *New York Times* paid assistant wireless telegraph operator Harold Bride $1000 for his account of the disaster.

Day 53

22 February | On this day in *Titanic* history

The following people were born on this day:
Crow, Mr George Frederick – 22 February 1886 – aged 26 when on *Titanic*; saloon steward; rescued in lifeboat 14

Nilsson, Miss Berta Olivia – 22 February 1894 – aged 18 when on *Titanic*; third-class passenger (ticket price: £7 15s 6d); rescued in collapsible lifeboat D

Mellinger, Miss Madeleine Violet – 22 February 1899 – aged 13 when on *Titanic*; second-class passenger (ticket price: £19 10s); rescued in lifeboat 14

The following person died on this day:
Maréchal, Mr Pierre – 22 February 1942 – aged 29 when on *Titanic*; first-class passenger (ticket price: £29 14s); rescued in lifeboat 7

Head of Harland & Wolff, Lord Pirrie, successfully undergoes a serious operation today in London in 1912, and soon recovers.

Mr Robert Hichens (quartermaster), witness at United States Senate Inquiry (Day 5)

> 'Before the *Titanic* sank we were all pulling for a light which we thought was to be a cod banker. We all made for this light.'

Titanic survivor quotes

Mr Pierre Maréchal

> 'When three-quarters of a mile away we stopped, the spectacle before our eyes was in its way magnificent. In a very calm sea, beneath a sky moonless but sown with millions of stars, the enormous *Titanic* lay on the water, illuminated from the waterline to the boat deck. The bow was slowly sinking into the black water.'

Miss Madeleine Violet Mellinger

> 'I could see the lights of the ship starting to go under water, then soundlessly, perhaps a mile away, it just went down. It was gone. Oh yes, the sky was very black and the stars were very bright. They told me the people in the water were singing, but I knew they were screaming.'

Titanic fact

Titanic's 'home port', her Port of Registry, was Liverpool.

Day 54

23 February | One this day in *Titanic* history

The following people were born on this day:
Mayzes, Mr Thomas Jubilee – 23 February 1887 – aged 25 when on *Titanic*; fireman; rescued in lifeboat 3

Symons, Mr George Thomas Macdonald – 23 February 1888 – aged 24 when on *Titanic*; lookout; rescued in lifeboat 1

Ball, Mr Percy – 23 February 1893 – aged 19 when on *Titanic*; plate steward (victualling crew); rescued (possibly) in lifeboat 13

The following people died on this day:
Gregson, Miss Mary – 23 February 1923 – aged 45 when on *Titanic*; stewardess; rescued in lifeboat 16

Wright, Mr William – 23 February 1940 – aged 47 when on *Titanic*; glory hole steward; rescued in lifeboat 13

Mr Olaus Abelseth, witness at United States Senate Inquiry (Day 13)

> 'I went to bed about 10 o'clock Sunday night, and I think it was about 15 minutes to 12 when I woke up; and there was another man in the same room – two of us in the same room – and he said to me, "What is that?" I said, "I don't know, but we had better get up." So we did get up and put our clothes on, and we two went up on deck in the forward part of the ship.
>
> 'We walked to the hind part of the ship and got two Norwegian girls up. One was in my charge and one was in charge of the man who was in the same room with me. He was from the same town that I came from. The other one was just 16 years old, and her father told me to take care of her until we got to Minneapolis. The two girls were in a room in the hind part of the ship, in the steerage. We all went up on deck and stayed there. We walked over to the port side of the ship, and there were five of us standing, looking, and we thought we saw a light. Then there was quite a lot of ice on the starboard part of the ship. They wanted us to go down again, and I saw one of the officers, and I said to him: "Is there any danger?" He said, "No." I was not satisfied with that, however, so I went down and told my brother-in-law and my cousin, who were in the same compartment there. They were not in the same room, but they were just a little ways from where I was. I told them about what was happening, and I said they had better get up. Both of them got up and dressed, and we took our overcoats and put them on. We did not take any lifebelts with us. There was no water on the deck at that time.'

Day 55

24 February | On this day in *Titanic* history

The following people were born on this day:
Rowe, Mr Alfred G. – 24 February 1853 – aged 59 when on *Titanic*; first-class passenger (ticket price: £26 11s); *Titanic* victim

Hart, Mr John Edward – 24 February 1881 – aged 31 when on *Titanic*; steward; rescued in lifeboat unknown

Lulić, Mr Nikola – 24 February 1883 – aged 29 when on *Titanic*; third-class passenger (ticket price: £8 13s 3d); rescued in lifeboat 15

Jefferys, Mr Ernest Wilfred – 24 February 1891 – aged 21 when on *Titanic*; second-class passenger (ticket price: £31 10s); *Titanic* victim

The following person died on this day:
Quick, Mrs Jane – 24 February 1965 – aged 33 when on *Titanic*; second-class passenger (ticket price: £26); rescued in lifeboat 11

Mr Nikola Lulić
Jan Jalševac, Nikola's friend, spoke of the circumstances of his rescue:

> 'As to my friend Nikola Lulić, I have to say, that he was rescued by fetching the cap of a sailor, which he put on his head. So he could make his way to a boat.'

Titanic survivor quotes

Mr William Ward

> 'She gave a kind of sudden lurch forward, and I heard a couple of reports, reports more like a volley of musketry than anything else. You would not exactly call them a heavy explosion. It did not seem to me like an explosion at all.'

Mr Hugh Woolner

> 'An officer came down and said he wanted to empty some of the people out of his boat, because he wanted to go and rescue some people who were on what he called a raft; and they put some people out of the boat, as many as our boat would hold. Probably 5 or 6 were put in our boat, which brought us down very close to the water.'

Titanic fact

The typical range of the *Titanic*'s wireless at night was 1,200 miles.

Day 56

25 February | On this day in *Titanic* history

The following people were born on this day:
Hocking, Mr Samuel James Metcalfe – 25 February 1876 – aged 36 when on *Titanic*; second-class passenger (ticket price: £13); *Titanic* victim

Larsson, Mr August Viktor – 25 February 1883 – aged 29 when on *Titanic*; third-class passenger (ticket price: £9 9s 8d); *Titanic* victim

Sedunary, Mr Samuel Francis – 25 February 1887 – aged 25 when on *Titanic*; second steward in third class; *Titanic* victim

Hunt, Mr Albert Sylvanus – 25 February 1889 – aged 23 when on *Titanic*; trimmer, known as 'Ginger'; rescued in collapsible lifeboat C

Moore, Mr Leonard Charles – 25 February 1893 – aged 19 when on *Titanic*; third-class passenger (ticket price: £8 1s); *Titanic* victim

Rosblom, Miss Salli Helena – 25 February 1910 – aged 2 when on *Titanic*; third-class passenger (ticket price: £20 4s 3d); *Titanic* victim

The following person died on this day:
Pfropper, Mr Richard Paul Jozef – 25 February 1964 – aged 30 when on *Titanic*; saloon steward; rescued in lifeboat unknown

Titanic survivor quotes

Mrs Addie Wells

> 'I thought even then it was some sort of a drill or something, except that just as we went down I saw a revolver in an officer's hand. A Mrs Davis and a little boy were in the boat with us, and she asked me what it was all about.
>
> 'As soon as the boat struck water, the seaman began pulling away with all their might. As we got away, we saw a lot of wild eyed men come rushing up from steerage, but they were met by a man with a gun who pushed them back into a crowd of men and said, "Stand back there now, the first word out of you and I'll – " I didn't catch the rest.'

Mr Harold Godfrey Lowe (fifth officer)

> 'It was not the launching of the boats that took the time. We got the whole boat out and in the water in less than ten minutes. It was getting the people together that took the time.'

Titanic fact

There were only two women aboard lifeboat 1.

Day 57

26 February | On this day in *Titanic* history

The following people were born on this day:

Byles, Father Thomas Roussel Davids – 26 February 1870 – aged 42 when on *Titanic*; second-class passenger (ticket price: £13); *Titanic* victim

Burton, Mr Edward John – 26 February 1877 – aged 35 when on *Titanic*; fireman; *Titanic* victim

Major, Mr William James – 26 February 1880 – aged 32 when on *Titanic*; fireman; rescued in lifeboat 13

Renouf, Mrs Lillian – 26 February 1882 – aged 30 when on *Titanic*; second-class passenger (ticket price: £21); rescued in lifeboat 12

Nourney, Mr Alfred – 26 February 1892 – aged 20 when on *Titanic*; first-class passenger (ticket price: £13 17s 3d); rescued in lifeboat 7

Wells, Miss Joan – 26 February 1908 – aged 4 when on *Titanic*; second-class passenger (ticket price: £23); rescued in lifeboat 14

The following people died on this day:

Sanderson, Mr Harold Arthur – 26 February 1932 – aged 52 when on *Titanic*; the sole passenger on the delivery trip from Belfast to Southampton; shipowner

Tucker, Mr Gilbert Milligan Jr – 26 February 1968 – aged 31 when on *Titanic*; first-class passenger (ticket price: £28 10s 9d); rescued in lifeboat 7

Glynn, Miss Mary Agatha – 26 February 1955 – aged 19 when on *Titanic*; third-class passenger (£7 15s); rescued in lifeboat 13

Collyer, Miss Marjorie Charlotte – 26 February 1965 – aged 8 when on *Titanic*; second-class passenger (ticket price £26 5s); rescued in lifeboat 14

Titanic survivor quote

Miss Marjorie Charlotte Collyer

> 'The stars were shining, and it was just like day. Some sailor put a rug around my mother to keep her warm. There were so many in our boat that we had to sit up all the time. Nobody could lie down…Only, when we left, all the people on the decks were kneeling down praying, while the band played, "Nearer My God to Thee".'

Titanic fact

41° 43′ 32″ N, 49° 56′ 49″ W are the coordinates where the boilers lie.

Day 58

27 February | On this day in *Titanic* history

The following people were born on this day:
Chibnall, Mrs Edith Martha Bowerman – 27 February 1864 – aged 48 when on *Titanic*; first-class passenger (ticket price: £55); rescued in lifeboat 6

Robinson, Mrs Annie – 27 February 1865 – aged 47 when on *Titanic*; stewardess; rescued in lifeboat 11

Crawford, Mr Alfred George – 27 February 1869 – aged 43 when on *Titanic*; bedroom steward; rescued in lifeboat 8

Davison, Mr Thomas Henry – 27 February 1880 – aged 32 when on *Titanic*; third-class passenger (ticket price: £16 2s); *Titanic* victim

Sheath, Mr Frederick – 27 February 1892 – aged 20 when on *Titanic*; trimmer; rescued in lifeboat 1

The following people died on this day:
Diaper, Mr John Joseph – 27 February 1955 – aged 29 when on *Titanic*; third-class passenger (ticket price: £8 13s 3d); rescued in lifeboat 15

Gibbons, Mr Jacob William – 27 February 1965 – aged 36 when on *Titanic*; saloon steward (victualling crew); rescued in lifeboat 11

Titanic survivor quotes

Mr Jacob William Gibbons

On 20 April 1912 he sent a telegram to his family; it read simply:

> 'Saved, well, Daddy.'

Mr Walter Wynn (quartermaster), witness at British Wreck Commissioner's Inquiry (Day 11)

> 'What are the duties of a quartermaster?'
>
> 'To take the wheel and attend to the Officers as a stand-by man, and when you have got the last trick of the wheel to go aft on the poop and stand by there during the whole watch, and in case of an accident, if a passenger falls overboard or attempts to commit suicide, to throw a lifebelt or a lifebuoy, and warn them on the bridge at once.'

Titanic fact

41° 43′ 35″ N, 49° 56′ 54″ W are the coordinates where the stern lies.

Day 59

28 February | On this day in *Titanic* history

The following people were born on this day:
Moraweck, Dr Ernest – 28 February 1858 – aged 54 when on *Titanic*; second-class passenger (ticket price: £14); *Titanic* victim

Puzey, Mr John Edward – 28 February 1868 – aged 44 when on *Titanic*; saloon steward in first class; *Titanic* victim

Murdoch, Mr William McMaster – 28 February 1873 – aged 39 when on *Titanic*; first officer; *Titanic* victim

Ryan, Mr Patrick – 28 February 1880 – aged 32 when on *Titanic*; third-class passenger (ticket price: £24 3s); *Titanic* victim

Hoyt, Mrs Jane Anne Forby – 28 February 1879 – aged 33 when on *Titanic*; first-class passenger (ticket price: £100); rescued in collapsible lifeboat D

Villvarlange, Mr Pierre Léon Gabriel – 28 February 1893 – aged 19 when on *Titanic*; assistant soup cook (restaurant staff); *Titanic* victim

Van Billiard, Master Walter John – 28 February 1903 – aged 9 when on *Titanic*; third-class passenger (ticket price: £14 10s); *Titanic* victim

The following person died on this day:
Sloan, Miss Mary Jane – 28 February 1953 – aged 45 when on *Titanic*; stewardess; rescued in lifeboat unknown

Mr Robert Hichens (quartermaster), witness at United States Senate Inquiry (Day 5)

> 'I was standing to attention, exposed, steering the boat all night, which is a very cold billet. I would rather be pulling the boat than be steering. But I seen no one there to steer, so I thought, being in charge of the boat, it was the best way to steer myself, especially when I seen the ladies get very nervous with the nasty tumble on.'

Mr William McMaster Murdoch, mentioned in *Washington Times*, Friday, 19 April 1912

> 'Mrs Widener kissed her husband goodbye. He told her not to worry, as it was possible that all would be saved, and the danger did not seem great. He stood by Colonel Astor and Major Butt, and was one of those who upheld Chief Officer Murdoch, when the latter at the point of a pistol drove back the foreigners who started the panic and stampede. Daniels, himself, had a narrow escape. He jumped into the water, and was picked up by a lifeboat that was in the vicinity of the wreck.'

Titanic fact

41° 43′ 57″ N, 49° 56′ 49″ W are the coordinates where the bow lies.

Day 60

29 February | On this day in *Titanic* history

The following people were born on this day:
McNeill, Miss Bridget – 29 February 1880 – aged 32 when on *Titanic*; third-class passenger (ticket price: £7 15s); *Titanic* victim

Banfield, Mr Frederick James – 29 February 1884 – aged 28 when on *Titanic*; second-class passenger (ticket price: £10 10s); *Titanic* victim

(Banfield worked as a mining engineer in the USA and returned to England for three months to visit his wife, boarding *Titanic* at Southampton.)

The following person died on this day:
Coutts, Mrs Winnie 'Minnie' – 29 February 1960 – aged 36 when on *Titanic*; third-class passenger (ticket price: £15 18s); rescued in lifeboat 2

Mr Robert Hichens (quartermaster), witness at United States Senate Inquiry (Day 5)

> 'I heard by the second officer when he repeated it. He sent me with his compliments to the ship's carpenter to look out for the ship's water, that it was freezing, at 8 o'clock. Then I knew. I didn't know before, but I heard the second officer distinctly tell Mr Moody, the sixth officer to repeat through the telephone, to keep a sharp lookout for small ice until daylight, and to pass the word along for the other lookout men.'

Titanic survivor quote

Mrs Winnie 'Minnie' Coutts (née Trainer), quoted in *The Washington Post*, Saturday, 20 April 1912

> 'I rushed out on deck with the children following me. Just when I had given up hope of finding my way a seaman came along and said Hurry now; all women and children to the lifeboats. He took us to the side of the ship but I wanted a life preserver. Just then an American gentleman who had heard me asking for a life preserver stepped up to me. He raised his hat, and then slowly removed the life preserver he had strapped to himself. Take my life preserver, madam, he said. Then he reached over and put his hand on the children's heads. If I go down, please pray for me. He said there were other brave men on board the *Titanic*, for I saw them helping women into the lifeboats as our boat pulled away. After kissing those they helped into the boats the men stepped back and did everything they could to load the boats quickly. I was in the first boat that was picked up by the *Carpathia*. There were seventeen in our boat. It was frightfully cold, but neither I nor the children suffered as much as the others, because we were fully dressed.'

March

Day 61

1 March | On this day in *Titanic* history

The following people were born on this day:
Eustis, Miss Elizabeth Mussey – 1 March 1858 – aged 54 when on *Titanic*; first-class passenger (ticket price: £78 5s 4d); rescued in lifeboat 4

Moore, Mr Clarence Bloomfield – 1 March 1865 – aged 47 when on *Titanic*; first-class passenger (ticket price: £42 8s); *Titanic* victim

Simpson, Dr John Edward – 1 March 1875 – aged 37 when on *Titanic*; assistant surgeon; *Titanic* victim

O'Brien, Mr Thomas – 1 March 1886 – aged 26 when on *Titanic*; third-class passenger (ticket price: £15 10s); *Titanic* victim

Pusey, Mr William Robert Holland – 1 March 1890 – aged 22 when on *Titanic*; fireman; rescued in lifeboat 1

Björklund, Mr Ernst Herbert – 1 March 1894 – aged 18 when on *Titanic*; third-class passenger (ticket price: £7 15s); *Titanic* victim

Lane, Mr Patrick – 1 March 1896 – aged 16 when on *Titanic*; third-class passenger (ticket price: £7 15s); *Titanic* victim

The following people died on this day:
Kelly, Mrs Fanny Maria – 1 March 1920 – aged 45 when on *Titanic*; second-class passenger (ticket price: £13 10s); rescued in lifeboat 9

Harper, Mr Henry Sleeper – 1 March 1944 – aged 48 when on *Titanic*; first-class passenger (ticket price: £76 14s 7d); rescued in lifeboat 3

Rheims, Mr George Alexander Lucien – 1 March 1963 – aged 33 when on *Titanic*; first-class passenger (ticket price: £39 12s); rescued in collapsible lifeboat A

Hold, Mrs Annie Margaret – 1 March 1960 – aged 29 when on *Titanic*; second-class passenger (ticket price: £26); rescued in lifeboat 10

Lundin, Miss Olga Elida – 1 March 1973 – aged 23 when on *Titanic*; second-class passenger (ticket price: £7 17s 1d); rescued in lifeboat 10

Gilnagh, Miss Mary Katherine – 1 March 1971 – aged 17 when on *Titanic*; third-class passenger (ticket price: £17 14s 8d); rescued in lifeboat 16

Asplund, Master Edvin Rojj Felix – 1 March 1983 – aged 3 when on *Titanic*; third-class passenger (ticket price: £31 17s 9d); rescued in lifeboat 15

Titanic survivor quote

Mr George Symons (able seaman)

> 'As you stand on the C deck you look down through the gratings, and on the hatch below that is where the water was lapping around the tarpaulins and the hatch cover.'

Day 62

2 March | On this day in *Titanic* history

The following people were born on this day:

Johansson, Mr Gustaf Joel – 2 March 1879 – aged 33 when on *Titanic*; third-class passenger (ticket price: £8 13s 1d); *Titanic* victim

Blake, Mr John Douglas Stanley George – 2 March 1886 – aged 26 when on *Titanic*; mess steward; *Titanic* victim

Bryhl, Mr Kurt Arnold Gottfrid – 2 March 1887 – aged 25 when on *Titanic*; second-class passenger (ticket price: £26); *Titanic* victim

Tenglin, Mr Gunnar Isidor – 2 March 1887 – aged 25 when on *Titanic*; third-class passenger (ticket price: £7 15s 11d); rescued in lifeboat 13

Mr Hugh Woolner, witness at United States Senate Inquiry (Day 10)

> 'I went right up onto the top deck and looked around, and I took particular note of the mechanism for raising and lowering the glass windows on the A deck, and I watched the sailors winding them up with these spanners that are used for that purpose. It struck me as being rather a slow job.'

Titanic survivor quotes

Mr Harold Bride (assistant wireless telegraph operator)

> 'I swam away from the *Titanic*, saw the *Titanic* go down, and climbed on top of the collapsible boat for the second time, was picked up from there by the *Titanic*'s lifeboat, and taken on board the *Carpathia*.'

Mr Gunnar Tenglin, quoted in *Burlington Daily Gazette*, Thursday, 25 April 1912

> 'We in the steerage did not know anything about being among icebergs until the *Titanic* hit one. I lost everything I had. I had about $30 in a suitcase, concealed well, as there had been several robberies among passengers the day before the accident. When we got to New York we got $25 from some relief committee, $5 from a man who was giving away money right and left and $10 from the Salvation Army. We were also fitted out with clothing. I got a suit of clothes, overcoat and other attire and a first-class ticket to Burlington, which had been my destination, my ticket having been lost with my other effects.'

Titanic fact

The breadth of *Titanic* was 92' (28 metres).

Day 63

3 March | On this day in *Titanic* history

The following people were born on this day:
Carter, Mrs Lilian – 3 March 1867 – aged 45 when on *Titanic*; second-class passenger (ticket price: £26); *Titanic* victim

Othen, Mr Charles Alfred – 3 March 1876 – aged 36 when on *Titanic*; fireman; *Titanic* victim

Quinn, Mr John – 3 March 1876 – aged 34 when on *Titanic*; fireman; travelled to Southampton where he disembarked as the crew for the journey had already been signed up. He left with the commendation 'VG' (very good) entered against his name for 'ability' and 'good conduct'.

The following people died on this day:
Harris, Mr George – 3 March 1919 – aged 62 when on *Titanic*; second-class passenger (ticket price: £10 10s); rescued in lifeboat 15

Barrett, Mr Frederick William – 3 March 1931 – aged 29 when on *Titanic*; leading fireman; rescued in lifeboat 13

Doling, Miss Elsie – 3 March 1972 – aged 18 when on *Titanic*; second-class passenger (ticket price: £23); rescued in lifeboat unknown

Mr George Harris, mentioned in *The Daily Advocate*, 4 March 1919

> 'George Harris of 76 Garden Street died last night, in the Stamford Hospital, after a short illness. He was, on 15 April 1912, a passenger on the *Titanic*, and was asleep in his berth near the stern of the vessel. He was not awakened by the collision with the iceberg, but ten minutes after the accident was aroused by the cries of those on deck. It happened that Mr Harris was one of the fifty in boat No. 13, picked up by the *Carpathia*. In his boat there were no sailors, and Mr Harris, although then over sixty years of age, pulled an oar for hours. On 18 April he arrived at his home in this city. Since the *Titanic* experience, his health had not been good. He frequently said his nerves were never the same after his experience in the lifeboat. Of the 715 passengers that the *Carpathia* rescued, five died from exposure before the *Carpathia* made port.'

Mr Guglielmo Marconi, witness at United States Senate Inquiry (Day 1)

> '[The pay of a wireless operator] in England… is from, I should say, beginning at $4 a week to $10 or $12 a week, with board and lodging. Of course, you have not asked me this, but I might say it is fairly easy to get operators on those terms in England because it is a rate of pay which is considerably higher than what they get on the shore telegraphs; and, of course, the fact of going to sea is very attractive to a great number of young men.'

Day 64

4 March | On this day in *Titanic* history

The following people were born on this day:
Bagley, Mr Edward Henry – 4 March 1879 – aged 33 when on *Titanic*; saloon steward; *Titanic* victim

Ilmakangas, Miss Pieta Sofia – 4 March 1887 – aged 25 when on *Titanic*; third-class passenger (ticket price: £7 18s 6d); *Titanic* victim

Hämäläinen, Master Viljo Unto Johannes – 4 March 1911 – aged 1 when on *Titanic*; second-class passenger (ticket price: £14 10s); rescued in lifeboat unknown

The following people died on this day:
Hardwick, Mr Reginald – 4 March 1918 – aged 21 when on *Titanic*; kitchen porter; rescued in lifeboat 11

Watt, Miss Robertha Josephine – 4 March 1993 – aged 12 when on *Titanic*; second-class passenger (ticket price: £15 15s); rescued in lifeboat 9

Titanic survivor quote

Miss Robertha Josephine Watt

> 'We heard many pistol shots, and could see people running hopelessly up and down the decks. Some in the lifeboat were crying. One or two were hysterical. There was nothing anyone could do. We just kept on going. We didn't row much, just enough to get far enough away from the suction. Then we puttered. We had just to drift around until dawn, occasionally flicking a gentleman's cigar lighter to let the other boats see where we were… There was nothing on the lifeboat but a keg of biscuits. No water, no liquor, no light. I don't know if the first class lifeboats had all the things they needed, but if anyone was sick or collapsed in our boat, there was nothing to revive them with. It showed the disorganisation. We didn't find the rudder until we were out quite away. I had a nightie tucked into a pair of panties, and house slippers. Luckily, I had a fur lined coat. They lined them with squirrel bellies in those days, and it had a fur collar. They asked if anyone could row, and mother said she could. That's how she spent the time. Rowing or standing… We didn't get aboard *Carpathia* until about 9 am. There was a rope ladder with a belt. My mother said, "go on, you can climb that. I went up without the belt." The captain roared down, "don't let anyone come up without a belt on."'

Titanic fact

The number of injuries recorded during the construction of *Titanic* was 246.

Day 65

5 March | On this day in *Titanic* history

The following people were born on this day:
Seward, Mr Wilfred Deable – 5 March 1887 – aged 25 when on *Titanic*; chief pantry steward in second class; rescued in lifeboat 3

Mahon, Miss Bridget Delia – 5 March 1892 – aged 20 when on *Titanic*; third-class passenger (ticket price: £7 17s 7d); *Titanic* victim

Svensson, Mr Johan Cervin – 5 March 1898 – aged 14 when on *Titanic*; third-class passenger (ticket price: £9 4s 6d); rescued in lifeboat 13

Navratil, Master Edmond Roger – 5 March 1910 – aged 2 when on *Titanic*; second-class passenger (ticket price: £26); rescued in collapsible lifeboat D

The following person died on this day:
Shine, Miss Ellen Natalia – 5 March 1993 – aged 20 when on *Titanic*; third-class passenger (ticket price: £7 16s 7d); rescued in lifeboat unknown

Mr A. H. Weikman, witness at United States Senate Inquiry (Day 15)

> 'I helped to launch the boats, and there seemed to be a shortage of women. When I was on "E" deck I met the captain returning from "G" deck, who had been there with Mr Andrews, and the captain was on the bridge at that time. I did not think there was any danger. Instructions were given to get the passengers into lifebelts and get on deck from all the staterooms. I saw Mr Ismay helping to load the boats...he got in along with Mr Carter, because there were no women in the vicinity of the boat.'

Titanic survivor quote

Mr Wilfred Deable Seward

> 'I remember one of my staff coming to the cabin and telling me that the ship had bumped something. I didn't think it was anything important so I just lay on my bunk, and read on. Shortly afterwards, someone called, "You'd better get up, there's something wrong." I went on deck and was told we had hit an iceberg.
>
> 'I was ordered into a boat; there were about fifty or sixty other people with me. When we were being lowered and still a long way above the water, the boat capsized us and we were all pitched into the water.'

After Seward enlisted in August 1914, he was discharged four weeks later, his discharge certificate stating 'This man was on board the *Titanic* when she went down. He was in the water 2½ hours...Causation of the disability – immersion at sinking of the *Titanic*.'

Day 66

6 March | On this day in *Titanic* history

The following people were born on this day:
Widegren, Mr Carl / Charles Peter – 6 March 1861 – aged 51 when on *Titanic*; third-class passenger (ticket price: £7 15s); *Titanic* victim

Littlejohn, Mr Alexander James – 6 March 1872 – aged 40 when on *Titanic*; saloon steward in first class; rescued in lifeboat 13

Buckley, Miss Catherine – 6 March 1889 – aged 23 when on *Titanic*; third-class passenger (ticket price: £7 5s 8d); *Titanic* victim

Hyland, Mr James Leo – 6 March 1893 – aged 19 when on *Titanic*; steward in third class; rescued in lifeboat 11

Debreucq, Mr Maurice Emile Victor – 6 March 1894 – aged 18 when on *Titanic*; assistant waiter (restaurant staff); *Titanic* victim

The following person died on this day:
Hyman, Mr Abraham Joseph – 6 March 1956 – aged 34 when on *Titanic*; third-class passenger (ticket price: £7 17s 9d); rescued in collapsible lifeboat C

Mr James Widgery (bath steward), witness at United States Senate Inquiry (Day 7)

> '... just then some biscuits came up from the storekeeper. I helped him put one of the boxes into the bottom of the boat, and the purser took hold of my arm and said, "Get in the boat."'

Titanic survivor quotes

Mr Alexander James Littlejohn

> 'All her other lights were burning brilliantly and she looked a blaze of light from stern to stern. We watched her like this for some time, and then suddenly she gave a plunge forward and all the lights went out. Her stern went right up in the air; there were two or three explosions and it seemed to me the stern part came down again and righted itself. Immediately after there were terrible cries for help. They were awful and heart-breaking.'

Mr William Pusey (fireman)

> 'Lady Gordon said to Miss Franks, "There is your beautiful nightdress gone," and I said, "Never mind about your nightdress madam, as long as you have got your life"; and then I heard someone forward at the fore end of me say – I said we had lost our kits and that our pay was stopped from the time she [Titanic] was a wreck – "We will give you a little to start a new kit." That was all I heard.'

Day 67

7 March | On this day in *Titanic* history

The following people were born on this day:
Smith, Mr John Richard Jago – 7 March 1877 – aged 35 when on *Titanic*; postal clerk / postman; *Titanic* victim

Hogg, Mr George Alfred – 7 March 1883 – aged 29 when on *Titanic*; lookout; rescued in lifeboat 7

Hold, Mrs Annie Margaret – 7 March 1883 – aged 29 when on *Titanic*; second-class passenger (ticket price: £26); rescued in lifeboat 10

Kink, Mr Anton – 7 March 1883 – aged 29 when on *Titanic*; third-class passenger (ticket price: £22 6d); rescued in lifeboat 2

Coutin, Mr Auguste Louis – 7 March 1884 – aged 28 when on *Titanic*; entrée cook (restaurant staff); *Titanic* victim

The following person died on this day:
Johnson, Miss Eleanor Ileen – 7 March 1998 – aged 1 when on *Titanic*; third-class passenger (ticket price: £11 2s 8d); rescued in lifeboat 15

On 7 March 1990, a plaque was dedicated to the *Titanic* Musicians in Hampshire. This was to replace the one destroyed during the Second World War. Survivors present at the unveiling: Millvina Dean, Bertram Dean, Edith Haisman and Eva Hart.

Titanic survivor quotes

Mr Anton Kink, quoted in *Milwaukee Journal*, 24 April 1912

> 'A sailor took my child and handed her into one of them. My wife was also helped in by the sailors. I was touched upon the shoulder and asked to step back, whereupon my wife and child cried at the top of their voices at my being left behind. I ducked down, broke through those standing about and jumped into the boat as it was lowered.'

Mr Alfred Hogg

> 'If we had had glasses, we might've seen the berg earlier... Everything was done as far as I could see. Everybody did their best, ladies and gentlemen and sailormen...I think all the women ought to have a gold medal on their breast, God bless them. I shall always raise my hat to a woman after what I saw.'

Titanic fact

Titanic had a total of twenty-nine boilers.

Day 68

8 March | On this day in *Titanic* history

The following people were born on this day:
Angle, Mrs Florence 'Mary' Agnes – 8 March 1876 – aged 36 when on *Titanic*; second-class passenger (ticket price: £26); rescued in lifeboat 11

Cameron, Miss Clear Annie – 8 March 1877 – aged 35 when on *Titanic*; second-class passenger (ticket price: £21); rescued in lifeboat 14

Dillon, Mr Thomas Patrick – 8 March 1879 – aged 33 when on *Titanic*; trimmer; rescued in lifeboat 4

Lundström, Mr Thure Edvin – 8 March 1880 – aged 32 when on *Titanic*; second-class passenger (ticket price: £7 11s 7d); rescued in lifeboat 15

McDermott, Miss Bridget Delia – 8 March 1881 – aged 31 when on *Titanic*; third-class passenger (ticket price: £7 15s 8d); rescued in lifeboat 13

McGrady, Mr James – 8 March 1885 – aged 27 when on *Titanic*; saloon steward; *Titanic* victim

The following people died on this day:
Fortune, Mrs Mary (née McDougald) – 8 March 1929 – aged 60 when on *Titanic*; first-class passenger (ticket price: £263); rescued in lifeboat 10

Nicola-Yarred, Miss Jamila ('Amelia Garrett') – 8 March 1970 – aged 14 when on *Titanic*; third-class passenger (ticket price: £11 4s 10d); rescued in collapsible lifeboat C

Titanic survivor quote

Mr Thure Edvin Lundström, quoted in *Chicago Daily News*, 24 April 1912

> 'Mrs Sandström and her children were placed in the thirteenth boat. The other members of the party, Mrs Hulda Classen, Albin Classen, Gertrude Classen, Elida Olson and Hulda Verstrom, got into the next boat. They were never seen again and Lundström is of the opinion that the boat in which they started did not have its plugs in or else was overturned. "It was frightfully cold, but I was used to it, and I swam for many minutes. Then I was picked up by a lifeboat. There were only twenty-five people in it. I thought of all those brave young men who were drowned. Then I got extremely cold, but they let me row a while and that was good. At last we were picked up."'

Titanic fact

On 8 March 2000, an oak box made from *Titanic* driftwood sold at auction at Christie's for $18,400 (in today's money, approximately £11,975).

Day 69

9 March | On this day in *Titanic* history

The following people were born on this day:
Bowen, Miss Grace Scott – 9 March 1867 – aged 45 when on *Titanic*; first-class passenger (ticket price: £262 7s 6d); rescued in lifeboat 4

Kimball, Mr Edwin Nelson Jr. – 9 March 1870 – aged 42 when on *Titanic*; first-class passenger (ticket price: £52 11s 1d); rescued in lifeboat 5

Beauchamp, Mr George William – 9 March 1888 – aged 24 when on *Titanic*; third-class passenger (ticket price: £7 15s 8d); rescued in lifeboat 13

O'Brien, Mr Denis – 9 March 1890 – aged 22 when on *Titanic*; third-class passenger (ticket price: £7 16s 7d); rescued in lifeboat 13

Kelly, Miss Mary – 9 March 1890 – aged 22 when on *Titanic*; third-class passenger (ticket price: £7 15s); rescued in collapsible lifeboat D

The following person died on this day:
Corr, Miss Ellen (Helen) – 9 March 1980 – aged 16 when on *Titanic*; third class passenger (ticket price £7 15s); rescued in lifeboat 14 or 16

Mr Leslie Williams and Arthur Bowen, mentioned in *Chicago Examiner*, Saturday, 20 April 1912

'2 BOXERS DIE ON TITANIC

Coming From Wales for Fights

'Two professional boxers of Wales, Leslie Williams and Andy Bowen, lost their lives when the *Titanic* foundered. They were on their way to this country to enter the boxing game at the expense of Frank Torreyson of McKeesport, who was to have been their American manager. With the two pugilists was Bowen's sister, Miss G. S. Bowen, who was saved by the *Carpathia*.'

Titanic survivor quote

Mr Edwin Nelson Jr. Kimball, quoted in *Chicago Evening Post*, Tuesday, 23 April 1912

'On Sunday evening,' said Mr Kimball, in describing his experiences, 'I had just gone down from the smoking-room to my stateroom and removed my coat and was standing in the middle of the room when the ship struck the iceberg. It seemed to me like scraping and tearing, more than a shock. It was on the starboard side of the ship under our room, and the ice came in our port hole.'

Titanic fact

A first-class lunch menu from the *Titanic*, dated 14 April 1912, sold for £76,000 on 31 March 2012.

Day 70

10 March | On this day in *Titanic* history

The following people were born on this day:
Danbom, Mrs Anna Sigrid Maria – 10 March 1884 – aged 28 when on *Titanic*; third-class passenger (ticket price: £14 8s); *Titanic* victim

Panula, Master Eino Viljam – 10 March 1911 – aged 1 when on *Titanic*; third-class passenger (ticket price: £39 13s 9d); *Titanic* victim

The following people died on this day:
Caldwell, Mr Albert Francis – 10 March 1977 – aged 26 when on *Titanic*; second-class passenger (ticket price: £29); rescued in lifeboat 13

Yazbeck, Mrs Selini – 10 March 1966 – aged 15 when on *Titanic*; third-class passenger (ticket price: £14 9s 1d); rescued in collapsible lifeboat C

Mrs Anna Sigrid Maria Danbom, mentioned in *Chicago Record-Herald*, Sunday, 21 April 1912

> 'Brother Learns E. G. Danbom, Wife and Child, Perish in *Titanic* Wreck Stanton, Iowa, April 20th. Alfred Danbom of this city received from New York today a telegram confirming his fears that his brother, Ernest G. Danbom, wife and child were on the *Titanic* and that all were lost.'

Titanic survivor quotes

Mr Albert Francis Caldwell

> 'It was a carefree and happy throng that sailed with the *Titanic* on her first and last voyage… The rhythmic beat of her propellers would, as a matter of fact, not cease until the narrow Atlantic had been crossed. The weather was ideal and the sea was calm. Everyone was having a good time… The table were piled high with all the luxuries and delicacies that one would desire. All were interested in the record speed that we were making. No mention was made of the icebergs.'

Mr Hugh Woolner

> 'I stood hearing what the conjectures were. People were guessing what it might be, and one man called out, "An iceberg has passed astern," but who it was I do not know. I never have seen the man since.'

Titanic fact

Titanic had two reciprocating four-cylinder, triple-expansion, direct-acting, inverted engines: 30,000hp 77 rpm; and one low pressure Parsons turbine, 16,000hp 165rpm.

Day 71

11 March | On this day in *Titanic* history

The following people were born on this day:

Harper, Mr Henry Sleeper – 11 March 1864 – aged 48 when on *Titanic*; first-class passenger (ticket price: £76 14s 7d); rescued in lifeboat 3

McAndrew, Mr Thomas Patrick – 11 March 1874 – aged 49 when on *Titanic*; fireman; *Titanic* victim

White, Mr Arthur – 11 March 1875 – aged 37 when on *Titanic*; assistant barber in first class (victualling crew); *Titanic* victim

Del Carlo, Mr Sebastiano – 11 March 1883 – aged 29 when on *Titanic*; second-class passenger (ticket price: £27 14s 5d); *Titanic* victim

The following people died on this day:

Healy, Miss Hanora 'Nora' – 11 March 1919 – aged 33 when on *Titanic*; third-class passenger (ticket price: £7 15s); rescued in lifeboat 16

Frauenthal, Dr Henry William – 11 March 1927 – aged 49 when on *Titanic*; first-class passenger (ticket price: £133 13s);; rescued in lifeboat 15

Priest, Mr Arthur John – 11 March 1937 – aged 24 when on *Titanic*; fireman, rescued in lifeboat 5

Mr Frederick Barrett (leading fireman), witness at British Wreck Commissioner's Inquiry (Day 3)

> 'Water came pouring in two feet above the stokehold plate; the ship's side was torn from the third stokehold to the foreward end.'

Titanic survivor quotes

Dr Henry William Frauenthal, quoted in *Denver Post*, Friday, 19 April 1912

> 'From all sides came shrieks and groans and cries, and it seemed as if all the devils of hell had been let loose. Just now I am so thankful to be alive that my appreciation of the horror is dulled. I am only afraid that when I recover from the first shock it will all come back to me again, and I would rather have gone down with the boat, I think, than have to live over again, even in my imagination, the last few minutes of that fearful night. I would rather have stayed, too, than know that women went down with the *Titanic*, but I swear we thought every woman on the ship had been placed safely in the boats. It was "Women first" with all of the men, and at least it seemed as if the decks had been cleared of them, for not one was to be seen save those already lowered.'

Mr James Johnson (saloon steward)

> 'It was just dawn. We thought we saw an illumination of a boat coming; it was the Northern Lights.'

Day 72

12 March | On this day in *Titanic* history

The following people were born on this day:
Bell, Mr Joseph – 12 March 1861 – aged 51 when on *Titanic*; chief engineer; *Titanic* victim

Sinkkonen, Miss Anna – 12 March 1882 – aged 30 when on *Titanic*; second-class passenger (ticket price: £13); rescued in lifeboat 10

Alhomäki, Mr Ilmari Rudolf – 12 March 1893 – aged 19 when on *Titanic*; third-class passenger (ticket price: £7 18s 6d); *Titanic* victim

Goodwin, Miss Lillian Augusta – 12 March 1896 – aged 16 when on *Titanic*; third-class passenger (ticket price: £46 18s); *Titanic* victim

The following person died on this day:
Bailey, Mr Henry Joseph – 12 March 1943 – aged 46 when on *Titanic*; master-at-arms; rescued in lifeboat 16

Mr Joseph Bell, mentioned in *Cumberland News*, Saturday, 20 April 1912

> 'Mr Joseph Bell left the district when a youth to serve his apprenticeship as an engineer at the works of Mr Robert Stephenson, on Tyneside, which were founded by the famous engineer of that name who invented the steam engine. He has now been at sea for a number of years, and by perseverance and ability one of the very top of his profession. He had the full confidence of his employers, the White Star Line, whose latest liners, so far as their engineering departments were concerned, were built under his supervision. As the representative of the company in Belfast he saw the engines placed in both the *Laurentic* and the *Megantic*, and went out upon the first two or three voyages of each as chief engineer. Last year he went to Belfast to superintend the erection of the engines of the *Titanic*, and also went out with this vessel on her ill-fated first trip in the capacity of chief engineer.'

Titanic survivor quote

Miss Anna Sinkkonen, quoted in *New York Herald*, 20 April 1912

> 'I did notice,' she said, 'that the men were being held back and only the women were allowed to get into the boats. The officers of the ship stood guard. No, I did not hear the music…'

Titanic fact

The period of time after the *Titanic* sank before the wreck site was discovered is exactly 73 years, 4 months and 17 days.

Day 73

13 March | On this day in *Titanic* history

The following people were born on this day:

Frauenthal, Dr Henry William – 13 March 1863 – aged 49 when on *Titanic*; first-class passenger (ticket price: £133 13s); rescued in lifeboat 5

Taylor, Mr Elmer Zebley – 13 March 1864 – aged 48 when on *Titanic*; first-class passenger (ticket price: £52); rescued in lifeboat 5

Garside, Miss Ethel – 13 March 1873 – aged 39 when on *Titanic*; second-class passenger (ticket price: £13); rescued in lifeboat 12

Whilems, Mr Charles – 13 March 1881 – aged 31 when on *Titanic*; second-class passenger (ticket price: £13); rescued in lifeboat 9

Moore, Mr James – 13 March 1883 – aged 29 when on *Titanic*; fireman; rescued in lifeboat 3

Lenox-Conyngham, Master Dennis Hugh – 13 March 1902 – aged 10 when on *Titanic*; first-class passenger (ticket price: £3); cross-channel passenger, disembarked at Cherbourg

The following person died on this day:

Bonnell, Miss Caroline – 13 March 1950 – aged 30 when on *Titanic*; second-class passenger (ticket price: £164 17s 4d); rescued in lifeboat 8

Titanic survivor quote

Mr Elmer Zebley Taylor, mentioned in *Atlantic City Daily Press*, Friday, 19 April 1912

> 'Mr Taylor described the night of Sunday as one of great horror and could scarcely give many coherent details. He and Mrs Taylor were among the first who were placed in the lifeboats that were sent off from the ill-fated *Titanic*. When their boat, which contained about thirty people, started from the side of the ship they found the sea was filled with cakes of floating ice. A short time after their small boat had pulled from the side of the ship the lights of the vessel went out and then the only illumination from the sinking ship were from torches and rockets, which were sent up until a short time before the ship took her final plunge into the icy ocean. Mr Taylor stated that both he and his wife suffered much from the exposure of floating on the ocean until after day light, when they were picked up by the *Carpathia*, but that their sufferings were almost nothing as to that of many of the other passengers and sailors who were saved from the *Titanic*.'

Day 74

14 March | On this day in *Titanic* history

The following people were born on this day:
Lane, Mr Albert Edward – 14 March 1879 – aged 33 when on *Titanic*; saloon steward in first class; *Titanic* victim

Öhman, Miss Velin – 14 March 1890 – aged 22 when on *Titanic*; third-class passenger (ticket price: £7 15s 6d); rescued in collapsible lifeboat C

Kenyon, Mr Frederick Roland – 14 March 1871 – aged 41 when on *Titanic*; first-class passenger (ticket price: £51 17s 3d); *Titanic* victim

The following people died on this day:
Hosono, Mr Masabumi – 14 March 1939 – aged 41 when on *Titanic*; second-class passenger (ticket price: £13); rescued in lifeboat 10

Andersen-Jensen, Miss Carla Christine Nielsine – 14 March 1980 – aged 19 when on *Titanic*; third-class passenger (ticket price: £7 17s 1d); rescued in lifeboat 16

Titanic survivor quotes

Mr Masabumi Hosono

> 'Making my way back to the upper deck I tried to prepare myself for the last moment with no agitation, making up my mind not to leave anything disgraceful as a Japanese. But still I found myself looking for and waiting for any possible chance for survival.'

His chance came when an officer loading lifeboats shouted, "Room for two more." A man jumped in. 'I myself was deep in desolate thought that I would no more be able to see my beloved wife and children, since there was no alternative for me than to share the same destiny as the *Titanic*. But the example of the first man making a jump led me to take this last chance… After the ship sank there came back again frightful shrills and cries of those drowning in the water.'

Miss Carla Christine Nielsine Andersen-Jensen

> 'When the lifeboat I got into rowed away from *Titanic* the orchestra was still playing. One said later it was the psalm "Nearer My God to Thee". That is possible. It was not the time to listen and I did not know the psalm, since it was English.'

Titanic fact

The *Titanic* wreck lies at a depth of approximately 2.4 miles (3.75 kilometres, or 12,600 feet).

Day 75

15 March | On this day in *Titanic* history

The following people were born on this day:
Stephenson, Mrs Martha Eustis– 15 March 1860 – aged 52 when on *Titanic*; first-class passenger (ticket price: £78 5s 4d); rescued in lifeboat 4

Brandeis, Mr Emil – 15 March 1864 – aged 48 when on *Titanic*; first-class passenger (ticket price: £50 9s 11d); *Titanic* victim

Madill, Miss Georgette Alexandra – 15 March 1896 – aged 16 when on *Titanic*; first-class passenger (ticket price: £211 6s 9d); rescued in lifeboat 2

The following people died on this day:
Bishop, Mrs Helen Margaret Walton – 15 March 1916 – aged 19 when on *Titanic*; first-class passenger (ticket price: £91 1s 7d); rescued in lifeboat 7

Quick, Miss Phyllis May – 15 March 1954 – aged 2 years 8 months when on *Titanic*; second-class passenger (ticket price: £26); rescued in lifeboat 11

Titanic survivor quotes

Mrs Helen Bishop (née Walton) quoted in *Dowagiac Daily News*, Saturday, 20 April 1912

> 'We were in the first lifeboat to be lowered over the side. Someone said, "Put in the brides and grooms first." There were three newly married couples who went in that boat. Altogether, there were 28 in our boat. There might as well have been 40 or so, but the half hundred men on deck refused to leave, even though there was room for them.'
>
> 'It broke my heart to leave my little dog "Freu Freu" in my stateroom. I had purchased her in Florence, Italy, and she was the pet of the ship. The steward wouldn't let me take her to the butcher. He said she was too pretty, and she was the only one allowed to stay in the cabin. I made a little den for her in our room behind two of my suitcases, but when I started to leave her she tore my dress to bits, tugging at it. I realized, however, that there would be little sympathy for a woman carrying a dog in her arms when there were lives of women and children to be saved.'

Mr Edward J. Buley

> 'I never saw any ice until morning. We thought it was a full-rigged ship. We were right in amongst the wreckage, and we thought it was a sailing ship, until the light came on and we saw it was an iceberg.'

Day 76

16 March | On this day in *Titanic* history

The following people were born on this day:
Crabb, Mr Henry James – 16 March 1889 – aged 23 when on *Titanic*; trimmer; *Titanic* victim

Goodwin, Miss Jessie Allis Mary – 16 March 1900 – aged 12 when on *Titanic*; third-class passenger (ticket price: £46 18s); *Titanic* victim

The following person died on this day:
Hellström, Miss Hilda Maria – 16 March 1962 – aged 22 when on *Titanic*; third-class passenger (ticket price: £8 19s 3d); rescued in collapsible lifeboat C

Titanic survivor quotes

Miss Hilda Maria Hellström

> 'I was a 3rd Class passenger aboard the *Titanic*, and we were cramped for space in the lower berth. I was always curious about just what the rest of the ship was like, I was drawn to the sound of orchestral music coming from above. I made my way up through the second class level, past men in a room, smoking and playing cards. I found my way up to the main ballroom to find myself in the presence of wonderful music playing, and I remained there, unseen by anyone.'

Mr Henry Blank

> 'After we were some distance from the ship, I heard revolver shots on board, but I don't know what part of the ship they came from. I was under the impression, as were many in my boat, that everyone had escaped. When there arose a roar from the vessel herself and the screams of those passengers and crew still, I was almost overcome by the horror of the situation. Realizing that many were still aboard and left to perish has left a permanent scar. We saw the *Titanic* plunge forward and then down out of sight but not before we heard the explosions of her boilers. The sea was very calm and there was floating ice everywhere. The women in our boat began to get chilled and we men took off our coats and wrapped them about them. On the *Carpathia*, we were treated with the utmost kindness.'

Titanic fact

Titanic had 159 furnaces providing a total heating surface of 144,142 sq. feet.

Day 77

17 March | On this day in *Titanic* history

The following people were born on this day:
Stewart, Mr Albert Ankeny – 17 March 1848 – aged 64 when on *Titanic*; first-class passenger (ticket price: £27 14s 5d); *Titanic* victim

Pascoe, Mr Charles Henry – 17 March 1867 – aged 45 when on *Titanic*; able seaman; rescued in lifeboat 8

Osman, Mrs Mara – 17 March 1881 – aged 31 when on *Titanic*; third-class passenger (ticket price: £8 13s 8d); rescued in lifeboat unknown

Murphy, Miss Margaret – 17 March 1887 – aged 25 when on *Titanic*; third-class passenger (ticket price: £15 10s); rescued in lifeboat 16

Madsen, Mr Fridtjof Arne – 17 March 1888 – aged 24 when on *Titanic*; third-class passenger (ticket price: £7 2s 10d); rescued in lifeboat 13

Ings, Mr William Ernest – 17 March 1892 – aged 20 when on *Titanic*; scullion; *Titanic* victim

Sunderland, Mr Victor Francis – 17 March 1892 – aged 20 when on *Titanic*; third-class passenger (ticket price: £8 1s); rescued in collapsible lifeboat B

The following people died on this day:
Pearce, Mr John – 17 March 1933 – aged 33 when on *Titanic*; fireman; rescued in lifeboat unknown

Oliver, Mr Harry – 17 March 1944 – aged 41 when on *Titanic*; fireman; rescued in lifeboat 9

Blank, Mr Henry – 17 March 1949 – aged 39 when on *Titanic*; first-class passenger (ticket price: £31); rescued in lifeboat 7

Mr Henry Joseph Bailey, mentioned by Edwina Celia Troutt

> 'The master-at-arms in charge of our lifeboat, his name was Bailey, suddenly stood up. "Scream!" he shouted at us… There was no help we could give to those still aboard…the oars of our lifeboat trailed lifeless as our rowers slumped forward… The baby whimpered in my arms… Once again master-at-arms Bailey got to his feet. His eyes flashed in the starlight and his walrus moustache quivered as he said in the stillness, "Please. I want you to sing with me. Sing now, all of you. Please." And his deep, resonant voice rolled out with, "Pull for the shore, sailors, pull for the shore".'

Titanic fact

The average cost of a first-class parlour suite was £870/$4,350.

Day 78

18 March | On this day in *Titanic* history

The following people were born on this day:
Pascoe, Mr Charles H. – 18 March 1867 – aged 45 when on *Titanic*; able seaman; rescued in lifeboat 8

Lefebvre, Mrs Frances Marie – 18 March 1872 – aged 40 when on *Titanic*; third-class passenger (ticket price: £25 9s 4d); *Titanic* victim

Lemore, Mrs Amelia 'Milley' – 18 March 1873 – aged 39 when on *Titanic*; second-class passenger (ticket price: £10 10s); rescued in lifeboat 14

Klasén, Mrs Hulda Kristina Eugenia – 18 March 1876 – aged 36 when on *Titanic*; third-class passenger (ticket price: £12 3s 8d); *Titanic* victim

Rice, Mr Charles – 18 March 1880 aged 32 when on *Titanic*; fireman

(engineering crew); rescued in lifeboat 10

Loring, Mr Joseph Holland – 18 March 1882 – aged 30 when on *Titanic*; first-class passenger (ticket price: £45 10s); *Titanic* victim

Krins, Mr Georges Alexandre – 18 March 1889 – aged 23 when on *Titanic*; musician in *Titanic*'s orchestra; *Titanic* victim

Martin, Miss Mabel Elvina – 18 March 1892 – aged 20 when on *Titanic*; cashier (restaurant staff); *Titanic* victim

Aks, Mrs Leah – 18 March 1894 – aged 18 when on *Titanic*; third-class passenger (ticket price: £9 7s); rescued in lifeboat 13

The following people died on this day:
Omont, Mr Alfred Fernand – 18 March 1948 – aged 29 when on *Titanic*; first-class passenger (ticket price: £25 14s 10d); rescued in lifeboat 7
Sheath, Mr Frederick – 18 March 1934 – aged 20 when on *Titanic*; trimmer; rescued in lifeboat 1

Titanic survivor quote

Mr Hugh Woolner, witness at United States Senate Inquiry (Day 10)

> 'It was full up to the bow, and I said to Steffanson: "There is nobody in the bows. Let us make a jump for it. You go first." And he jumped out and tumbled in head over heels into the bow, and I jumped too, and hit the gunwale with my chest, which had on this life preserver, of course and I sort of bounced off the gunwale and caught the gunwale with my fingers, and slipped off backwards.'

Day 79

19 March | On this day in *Titanic* history

The following people were born on this day:
Oliver, Mr Harry – 19 March 1871 – aged 41 when on *Titanic*; fireman; rescued in lifeboat 9

Drew, Mrs Lulu Thorne – 19 March 1878 – aged 34 when on *Titanic*; second-class passenger (ticket price: £32 10s); rescued in lifeboat 10

Rice, Mr Charles John – 19 March 1880 – aged 32 when on *Titanic*; fireman; rescued in lifeboat 10

Pfropper, Mr Richard Paul Jozef – 19 March 1882 – aged 30 when on *Titanic*; saloon steward; rescued in lifeboat 9

Naṣr Allāh, Mrs Adāl – 19 March 1898 – aged 14 when on *Titanic*; second-class passenger (ticket price: £30 1s 5d); rescued in lifeboat unknown

Asplund, Master Edvin Rojj Felix – 19 March 1909 – aged 3 when on *Titanic*; third-class passenger (ticket price: £31 7s 9d); rescued in lifeboat 15

The following person died on this day:
Moore, Mr George Alfred – 19 March 1943 – aged 32 when on *Titanic*; able seaman; rescued in lifeboat 3

Mrs Lulu Thorne Drew, mentioned in *Brooklyn Daily Times*, Wednesday, 17 April 1912

> 'All Greenport is fearful that James V. Drew and Marshall, the 5-year-old son of William J. Drew, brother and partner of James Drew, are among the *Titanic*'s dead. These two, together with Mrs Lulu Drew, were listed among the passengers on the steamer and were on their way home from a visit to Mr Drew's mother in Constantine, Cornwall, England.
>
> James V. and William J. Drew have been in business here as monument dealers for several years. They came from England. The man who is believed to have gone down with the ship was 38 years old, a member of the Baptist Church here.'

Titanic survivor quote

Mr George Harder

> 'When I went to the porthole I saw this iceberg go by. The porthole was closed. The iceberg was, I would say, about 50 to 100 feet away. I should say it was about as high as the top deck of the boat.'

Day 80

20 March | On this day in *Titanic* history

The following people were born on this day:
Ridsdale, Miss Lucy – 20 March 1854 – aged 58 when on *Titanic*; second-class passenger (ticket price: £10 10s); rescued in lifeboat 13

Sundman, Mr Johan Julian – 20 March 1867 – aged 45 when on *Titanic*; third-class passenger (ticket price: £7 18s 6d); rescued in lifeboat 15

Taylor, Mr Percy Cornelius – 20 March 1872 – aged 40 when on *Titanic*; musician in *Titanic*'s orchestra; *Titanic* victim

Davis, Mr Stephen James – 20 March 1873 – aged 39 when on *Titanic*; able seaman; *Titanic* victim

Brown, Mr Edward – 20 March 1878 – aged 34 when on *Titanic*; saloon steward; rescued in collapsible lifeboat C

Johnson, Mr Malkolm Joackim – 20 March 1879 – aged 33 when on *Titanic*; third-class passenger (ticket price: £7 15s 6d); *Titanic* victim

Rowe, Mr George Thomas – 20 March 1881 – aged 31 when on *Titanic*; quartermaster; rescued in collapsible lifeboat C

Hakkarainen, Mrs Elin Matilda – 20 March 1888 – aged 24 when on *Titanic*; third-class passenger (ticket price: £15 17s); rescued in lifeboat 15

Aks, Mrs Leah – 20 March 1891 – aged 21 when on *Titanic*; third-class passenger (ticket price: £9 7s); rescued in lifeboat unknown

The following people died on this day:
Carter, Mr William Ernest – 20 March 1940 – aged 36 when on *Titanic*; first-class passenger (ticket price: £120); rescued in collapsible lifeboat C

Gibson, Mrs Pauline Caroline Boesen Brown – 20 March 1961 – aged 45 when on *Titanic*, first-class passenger (ticket price: £59 8s); rescued in lifeboat 7

Miss Lucy Ridsdale, quoted in *Marietta Daily Times*, 24 April 1912

> 'I saw officers and passengers manning the lifeboats and loading the women into them, and was soon helped into one myself…It was cold and with the water in the boat we suffered greatly with cold. After a long time, seemingly many hours, we were picked up by the *Carpathia*.'

Titanic fact

The amount of coal used per day was 825 tons.

Day 81

21 March | On this day in *Titanic* history

The following people were born on this day:

Gee, Mr Arthur H. – 21 March 1865 – aged 47 when on *Titanic*; first-class passenger (ticket price: £38 10s); *Titanic* victim

Peruschitz, Father Josef – 21 March 1871 – aged 41 when on *Titanic*; second-class passenger (ticket price: £13); *Titanic* victim

Widgery, Mr Isaac George – 21 March 1875 – aged 37 when on *Titanic*; bath steward; rescued in lifeboat unknown

Kink-Heilmann, Mrs Luise – 21 March 1886 – aged 26 when on *Titanic*; third-class passenger (ticket price: £22 6d); rescued in lifeboat 2

Andreasson, Mr Pål Edvin – 21 March 1892 – aged 20 when on *Titanic*; third-class passenger (ticket price: £7 17s 1d); *Titanic* victim

The following people died on this day:

Frölicher-Stehli, Mrs Margaretha Emerentia – 21 March 1955 – aged 48 when on *Titanic*; first-class passenger (ticket price: £79 4s); rescued in lifeboat 5

Kimball, Mrs Gertrude Smith Parsons – 21 March 1962 – aged 45 when on *Titanic*; first-class passenger (ticket price: £52 11s 1d); rescued in lifeboat 5

Fortune, Miss Ethel Flora – 21 March 1961 – aged 28 when on *Titanic*; first-class passenger (ticket price: £263); rescued in lifeboat 10

Mr Arthur H. Gee, mentioned in *St. Annes on the Sea Express*, Friday, 19 April 1912

> 'An interesting letter was received from Mr Gee, posted at Queenstown. The letter was: On board the R.M.S. *Titanic*, April 10th 1912. My dear – In the language of the poet, "This is a knock-out". I have never seen anything so magnificent, even in a first class hotel. I might be living in a palace. It is, indeed, an experience. We seem to be miles above the water, and there are certainly miles of promenade deck. The lobbies are so long that they appear to come to a point in the distance. Just finished dinner. They call us up to dress by bugle! It reminded me of some Russian villages where they call the cattle home from the fields by horn made from the bark of a tree. Such a dinner!!! My gracious!!!'

Titanic fact

Actress Dorothy Gibson, who survived *Titanic*, starred in a film called *Saved from the Titanic* which was released just one month after the ship sank. In the film, she was dressed in the same clothes she had worn aboard ship.

Day 82

22 March | On this day in *Titanic* history

The following people were born on this day:
Lovell, Mr John – 22 March 1875 – aged 37 when on *Titanic*; grill cook (victualling crew); *Titanic* victim

Olsen, Mr Henry Margido – 22 March 1884 – aged 28 when on *Titanic*; third-class passenger (ticket price: £22 10s 6d); *Titanic* victim

Maugé, Mr Paul Achille Maurice Germain – 22 March 1887 – aged 25 when on *Titanic*; kitchen clerk (restaurant staff); rescued in lifeboat 13

The following people died on this day:
Taylor, Mr William Henry – 22 March 1914 – aged 28 when on *Titanic*; fireman; rescued in lifeboat 15

Yūsuf, Miss Nabīah (Mary) – 22 March 1914 – aged 2 years 7 months when on *Titanic*; third-class passenger (ticket price: £22 7s 2d); rescued in collapsible lifeboat C

Mr William Henry Taylor, mentioned in *Hampshire Advertiser*, Saturday, 28 March 1914

'Mr William Henry Taylor *Titanic* Fireman rescued on boat 15.

He was charge hand employed in carrying out the work of pulling down a lairage in the docks. About 9.40 am they were just preparing to put the slugs on the "principals" to lift them, when the corner post or support, fell out and the "principal" it supported fell on deceased. The other "principals" were supported by "L" irons, but this one was not. Even if this had been supported by an "L" iron the support might have fallen out. Witness said he had been at this kind of work for some years, and had done work of a similar kind before. This particular job was carried out in the usual way and the usual precautions taken. As far as they were concerned it was quite an accident, and he blamed no one. The jury returned a verdict of "Accidental death".

Titanic survivor quote

Mr George W. Beauchamp

> 'I went aft on to the boat deck, and across to the starboard side, and stood on the deck of the ship by the boat and one foot on the boat and one foot on the lifeboat, like that, and helped the ladies and children in that were there, and the order was given by the officer then, "Lower away the boat, that will do."'

Day 83

23 March | On this day in *Titanic* history

The following people were born on this day:
Rood, Mr Hugh Roscoe – 23 March 1873 – aged 39 when on *Titanic*; first-class passenger (ticket price: £50); *Titanic* victim

Seward, Mr Frederic Kimber – 23 March 1878 – aged 34 when on *Titanic*; first-class passenger (ticket price: £26 11s); rescued in lifeboat 7

Boxhall, Mr Joseph Groves – 23 March 1884 – aged 28 when on *Titanic*; fourth officer; rescued in lifeboat 2

Sandström, Miss Marguerite Rut – 23 March 1908 – aged 4 when on *Titanic*; third-class passenger (ticket price: £16 14s); rescued in lifeboat 13

The following people died on this day:
Herman, Miss Alice – 23 March 1947 – aged 24 when on *Titanic*; second-class passenger (ticket price: £65); rescued in lifeboat 9

Hunt, Mr Albert Sylvanus – 23 March 1949 – aged 23 when on *Titanic*; trimmer; rescued in collapsible lifeboat C

Jansson, Mr Carl Olof – 23 March 1978 – aged 21 when on *Titanic*; third-class passenger (ticket price: £7 15s 11d); rescued in collapsible lifeboat A

On 23 March 1998, James Cameron's Hollywood movie *Titanic* receives eleven Oscars, and ties with the record set by *Ben Hur* in 1959.

Mr William H. Taylor, witness at United States Senate Inquiry (Day 7)

'We were supposed to go to the boat that our name was down for on the ship.'

Titanic facts

RMS *Carpathia* only heard the *Titanic*'s distress call because *Carpathia*'s wireless operator, Harold Cottam, was listening to his telegraph while getting into his pyjamas, several minutes after his shift had ended. If he had signed off when he was supposed to, there may not have been a rescue.

The British newspaper *The Daily Mail* reported '*Titanic* Sunk, No Lives Lost' in its initial 16 April 1912 story.

Day 84

24 March | On this day in *Titanic* history

The following people were born on this day:
Harris, Mr George – 24 March 1850 – aged 62 when on *Titanic*; second-class passenger (ticket price: £10 10s); rescued in lifeboat 15

Arnold-Franchi, Mr Josef – 24 March 1887 – aged 25 when on *Titanic*; third-class passenger (ticket price: £17 16s); *Titanic* victim

Bishop, Mr Dickinson H. – 24 March 1887 – aged 25 when on *Titanic*; first-class passenger (ticket price: £91 1s 7d); rescued in lifeboat 7

Turvey, Mr Charles Thomas – 24 March 1895 – aged 17 when on *Titanic*; page boy; *Titanic* victim

The following person died on this day:
Pears, Mrs Edith Wearne – 24 March 1956 – aged 22 when on *Titanic*; first-class passenger (ticket price: £66 12s); rescued in lifeboat 8

***Titanic* provisions**
14,000 gallons of drinking water used every 24 hours.

Meat, poultry and fish: 75,000lb fresh meat; 7,500lb ham and bacon; 2,500lb sausages; 25,000lb poultry; 11,000lb fresh fish; 4,000lb salted and dried fish.

Fruit and vegetables: 1,000lb grapes; 36,000 apples; 36,000 oranges; 13,000 grapefruit; 16,000 lemons; 40 tons of potatoes; 3,500lb onions; 800 asparagus bundles; 3,500lb tomatoes; 2,500lb green peas; 7,000 heads of lettuce; 10,000lb dried beans and rice.

Bakery and breakfast: 250 barrels of flour; 10,000lb cereal; 1,000 loaves of bread; 2,200lb coffee; 800lb tea; 10,000lb sugar; 1,120lb marmalade and jam.

Eggs and dairy: 40,000 fresh eggs; 1,500 gallons of fresh milk; 600 gallons of condensed milk; 1,200 quarts of ice cream; 6,000lb butter.

Titanic survivor quote

Mr Hugh Woolner, witness at United States Senate Inquiry (Day 10)

> 'I then took Mrs Candee up onto the boat deck, and there we saw preparations for lowering the boats going on. My great desire was to get her into the first boat, which I did, and we brought up a rug, which we threw in with her, and we waited to see that boat filled. It was not filled but a great many people got into it, and finally it was quietly and orderly lowered away. That was the sternmost boat No. 6 on the port side.'

Titanic fact

Titanic was also known as the 'Ship of Dreams', and White Star Line claimed she was the safest ship of all time.

Day 85

25 March | On this day in *Titanic* history

The following people were born on this day:
Roebling, Mr Washington Augustus II – 25 March 1881 – aged 31 when on *Titanic*; first-class passenger (ticket price: £50 9s 11d); *Titanic* victim

Törnquist, Mr Johan Vilhelm Henrik – 25 March 1886 – aged 26 when on *Titanic*; third-class passenger (ticket price: £7 5s); rescued in lifeboat 15

Davies, Mr Joseph – 25 March 1895 – aged 17 when on *Titanic*; third-class passenger (ticket price: £8 1s); *Titanic* victim

The following people died on this day:
Garside, Miss Ethel – 25 March 1953 – aged 39 when on *Titanic*; second-class passenger (ticket price: £13); rescued in lifeboat 12

Kreuchen, Miss Emilie – 25 March 1971 – aged 29 when on *Titanic*; first-class passenger (ticket price: £211 6s 9d); rescued in lifeboat 2

25 March 1912
All *Titanic*'s lifeboats undergo seaworthiness and passenger capability tests.

Mr Washington Augustus II Roebling, mentioned in *New York Herald*, Wednesday, 17 April 1912

> 'Trenton, N.J. Monday – Washington Roebling, II a son of Charles G. Roebling, president of the John A. Roebling Sons Company, the Roebling Construction Company and the New Jersey Wire Cloth Company, who was a passenger on board the *Titanic*, together with Stephen Blackwell, another Trenton young man, is one of the managers of the Mercer Automobile Works of this city, which manufactures the Mercer car. Mr Roebling and Mr Blackwell toured Italy, France and other countries in an automobile. Mr Roebling was born in this city and is thirty-one years old. He went to a private school here and attended the Hill Preparatory School at Pottstown, Pa. He is an ardent automobilist and golfer. Two years ago he took part in one of the important automobile races at Savannah, Ga., making second place.'

Titanic survivor quote

Mr Daniel Buckley, witness at United States Senate Inquiry (Day 13)

> 'I heard some terrible noise and I jumped out on the floor, and the first thing I knew my feet were getting wet; the water was just coming in slightly. I told the other fellows to get up, that there was something wrong and that the water was coming in. They only laughed at me. One of them says: "Get back into bed. You are not in Ireland now."'

Titanic fact

The average cost of a first-class berth was £30/$150 (or more than £3,300 today).

Day 86

26 March | On this day in *Titanic* history

The following person was born on this day:
Horswill, Mr Albert Edward James – 26 March 1879 – aged 33 when on *Titanic*; able seaman; rescued in lifeboat 1

The following person died on this day:
Davison, Mrs Mary E. – 26 March 1939 – aged 34 when on *Titanic*; third-class passenger (ticket price: £16 2s); rescued in lifeboat 16

Mr Albert Edward James Horswill, mentioned in *New York Times*, Saturday, 17 January 1914

> 'Man Who Saved 45 Persons Says He Can't Support His Wife
>
> CHICAGO, Jan. 16 – Albert Horswill, who said he was in charge of the last lifeboat to leave the *Titanic*, and saved forty-five women and children, told Judge Uhlir in court today that he was unable to support his wife. Horswill, 38 years old, married an 18-year-old girl last November and deserted her a week ago. "It would have been better if I had gone down with the *Titanic*," said Horswill. "I am working in a warehouse, but can't make enough to support one, let alone two." Judge Uhlir sent the couple away together, after telling the sailor he must support his wife.'

Titanic survivor quote

Mr Henry S. Etches, witness at United States Senate Inquiry (Day 9)

> 'Mr Andrews then told me to come down on C deck with him, and we went down the pantry staircase together. Going down he told me to be sure and make the passengers open their doors, and to tell them the lifebelts were on top of the wardrobes and on top of the racks, and to assist them in every way I could to get them on, which I endeavoured to do. We walked along C deck together. The purser was standing outside of his office, in a large group of ladies. The purser was asking them to do as he asked them, and to go back in their rooms and not to frighten themselves, but, as a preliminary caution, to put the lifebelts on, and the stewards would give them every attention. Mr Andrews said: "That is exactly what I have been trying to get them to do," and, with that, he walked down the staircase to go on lower D deck. That is the last I saw of Mr Andrews.'

Titanic fact

RMS stands for 'Royal Mail Steamer'. As well as passengers, the ship carried nearly 3,500 sacks of letters, packages and documents.

Day 87

27 March | On this day in *Titanic* history

The following people were born on this day:
Dodge, Mrs Ruth – 27 March 1878 – aged 34 when on *Titanic*; first-class passenger (ticket price: £81 17s 2d); rescued in lifeboat 5

Couper, Mr Robert Frederick William – 27 March 1882 – aged 30 when on *Titanic*; fireman; rescued in lifeboat 3

The following people died on this day:
Stap, Miss Sarah Agnes – 27 March 1937 – aged 47 when on *Titanic*; stewardess; rescued in lifeboat 11

Palmquist, Mr Oskar Leander Johansson – 27 March 1925 – aged 26 when on *Titanic*; third-class passenger (ticket price: £7 15s 6d); rescued in lifeboat 15

Phillips, Miss Kate Florence – 27 March 1964 – aged 19 when on *Titanic*; second-class passenger (ticket price: £26); rescued in lifeboat 11

Astor, Mrs Madeleine Talmage Force – 27 March 1940 – aged 18 when on *Titanic*; first-class passenger (ticket price: £247 10s 6d); rescued in lifeboat 4

Mrs Ruth Dodge, quoted in *San Francisco Bulletin*, Tuesday, 30 April 1912

> 'Reaches Home with Husband and Son after Terrible Experience at Sea.
> "The most pathetic thing was the scene on board the *Carpathia* during the rescue. As each boat drew up the survivors would peer over, straining to see the face of someone they had left behind. They were the young brides. Everybody on board, of course, had known they were brides, and they had watched them laughing and promenading with their husbands. The moans of anxiety and disappointment as each boat failed to bring up those that they were looking for were awful and finally that awful despair which fell over everyone when we knew there were no more boats to pick up..."'

Titanic survivor quote

Mr Gunnar Tenglin, mentioned in *Fort Wayne Journal Gazette*, Friday, 19 April 1912

> 'Gunnar Tenglin, a Swede, was saved in the next to the last boat which left the *Titanic*. Before getting into the boat he placed two hysterical women in another boat. Then he heard a cry, and looking up, he saw a woman standing on the upper deck. The woman, he said, dropped from her arms her baby, which Tenglin caught, and gave it to one of the women he had put in the boat. Then he got into his own boat, which was lowered, and shortly afterward came two explosions, and the plunge downward of the *Titanic*.'

Day 88

28 March | On this day in *Titanic* history

The following people were born on this day:
Osman, Mr Frank – 28 March 1885 – aged 27 when on *Titanic*; able seaman; rescued in lifeboat 2

Andrew, Mr Edgar Samuel – 28 March 1895 – aged 17 when on *Titanic*; second-class passenger (ticket price: £11 10s); *Titanic* victim

The following person died on this day:
Nysten, Miss Anna Sofia – 28 March 1977 – aged 22 when on *Titanic*; third-class passenger (ticket price: £7 15s); rescued in lifeboat 13

Mr Edgar Samuel Andrew
On 8 April 1912, Edgar wrote a moving letter to his friend Josey Cowan from Argentina:

> 'You figure Josey I had to leave on the 17th this month aboard the *Oceanic*, but due to the coal strike that steamer cannot depart, so I have to go one week earlier on board the *Titanic*. It really seems unbelievable that I have to leave a few days before your arrival, but there's no help for it, I've got to go. You figure, Josey, I am boarding the greatest steamship in the world, but I don't really feel proud of it at all, right now I wish the *Titanic* were lying at the bottom of the ocean.'

Mr Charles H. Lightoller, witness at United States Senate Inquiry (Day 5)

> 'We use the men in the crow's nest for keeping a lookout. Occasionally a man will see a light or a vessel first, particularly in daytime, when naturally we trust to them seeing. Especially all through the daytime lookout men are keeping a keen lookout, and will report a steamer long before she is in sight, apparently, by her smoke. In that instance the lookout might be very useful. In night-time, particularly in channels where there are a great many lights, we may be watching one light, and there may be another light in our course, and the man in the crow's nest will strike, say, one bell. That signifies something on the port bow, and calls our attention to it. So that the ship can approach close to us without the bridge being notified, even though the officer has not himself already seen it. The White Star Co., I may say incidentally, is the only company in the world, so far as I know, that carries six lookout men. We carry men who do nothing else, night and day, from the commencement to the finish of the voyage, except keep a lookout. They are two hours on and four off.'

Day 89

29 March | On this day in *Titanic* history

The following people were born on this day:
Baxter, Mrs Hélène de Lanaudière-Chaput – 29 March 1862 – aged 50 when on *Titanic*; first-class passenger (ticket price: £247 10s 5d); rescued in lifeboat 6

Robins, Mr Victor – 29 March 1867 – aged 45 when on *Titanic*; first-class passenger (ticket price: £247 10s 6d); *Titanic* victim

George/Joseph, Mrs Shawneene – 29 March 1874 – aged 38 when on *Titanic*; third-class passenger (ticket price: £7 4s 7d); rescued in collapsible lifeboat C

The following person died on this day:
Coutts, Master Neville Leslie – 29 March 1977 – aged 3 years 5 months when on *Titanic*; third-class passenger (ticket price: £15 18s); rescued in lifeboat 2

Titanic survivor quotes

Mrs Shawneene George/Joseph (née Whabee Abī-Saab)

> 'I was moved by the sight of the mother who had lost her son. She cried continually, hugging the smaller child to her breast. Several hours after we had been taken aboard and given clothing, I was coming from my cabin when I saw a nurse carrying a child wrapped in a blanket. She passed close to me and I recognized the child. It was Tommy! I told the nurse and she handed the little boy to me. I took him to his grief-stricken mother. The reunion was a sight I will never forget.'

Mr Olaus Abelseth, witness at United States Senate Inquiry (Day 13)

> 'I could not say, but it did not seem to be so very far. I thought I could see this mast light, the front mast light. That is what I thought I could see. A little while later there was one of the officers who came and said to be quiet, that there was a ship coming. That is all he said. He did not say what time, or anything. That is all he said. So I said to them, we had better go and get the lifebelts, as we had not brought them with us. So my cousin and I went down to get the lifebelts for all of us. When we came up again we carried the lifebelts on our arms for a while. There were a lot of steerage people there that were getting on one of these cranes that they had on deck, that they used to lift things with. They can lift about two and a half tons, I believe. These steerage passengers were crawling along on this, over the railing, and away up to the boat deck. A lot of them were doing that.'

Day 90

30 March | On this day in *Titanic* history

The following people were born on this day:
Webb, Mr Brooke Holding – 30 March 1862 – aged 50 when on *Titanic*; smoke room steward in first class; *Titanic* victim

Keane, Miss Nora Agnes – 30 March 1866 – aged 46 when on *Titanic*; second-class passenger (ticket price: £12 7s); rescued in lifeboat 10

Lightoller, Mr Charles Herbert – 30 March 1874 – aged 38 when on *Titanic*; second officer; rescued in collapsible lifeboat B

Drew, Master Marshall Brines – 30 March 1904 – aged 8 when on *Titanic*; second-class passenger (ticket price: £32 10s); rescued in lifeboat 10

The following person died on this day:
Frauenthal, Mrs Clara Heinsheimer – 30 March 1943 – aged 42 when on *Titanic*; first-class passenger (ticket price: £133 13s); rescued in lifeboat 5

Mr Guglielmo Marconi, witness at United States Senate Inquiry (Day 1)

> 'The *Carpathia* is provided with an equipment which I should call a short-distance equipment; it is an apparatus which can transmit messages, under favourable circumstances, up to about 180 or 200 miles. On the average I should say the distance is about 100 miles.'

Titanic survivor quote

Master Marshall Brines Drew

> 'When the *Titanic* struck the iceberg, I was in bed. However, for whatever reason I was awake and remember the jolt and cessation of motion. A steward knocked on the stateroom door and directed us to get dressed, put on life preservers and go to the boat deck, which we did… The steward as we passed was trying to arouse passengers who had locked themselves in for the night. Elevators were not running. We walked up to the boat deck. All was calm and orderly. An officer was in charge. "Women and children first," he said, as he directed lifeboat number 11 to be filled. There were many tearful farewells. We and Uncle Jim said goodbye… The lowering of the lifeboat 70 feet to the sea was perilous. Davits, ropes, nothing worked properly, so that first one end of the lifeboat was tilted up and then far down. I think it was the only time I was scared. Lifeboats pulled some distance away from the sinking *Titanic*, afraid of what suction might do…'

Day 91

31 March | On this day in *Titanic* history

The following people were born on this day:

Gregson, Miss Mary – 31 March 1867 – aged 45 when on *Titanic*; stewardess; rescued in lifeboat 16

Faulkner, Mr William Stephen – 31 March 1875 – aged 37 when on *Titanic*; bedroom steward in first class; rescued in lifeboat 11

Shepherd, Mr Jonathan – 31 March 1880 – aged 32 when on *Titanic*; junior assistant second engineer; *Titanic* victim

Wheelton, Mr Edneser Edward – 31 March 1883 – aged 29 when on *Titanic*; saloon steward in first class; rescued in lifeboat 11

Rentowl, Mr Philip Charles Thomas – 31 March 1886 – aged 26 when on *Titanic*; fireman; *Titanic* victim

Lehmann, Miss Bertha – 31 March 1895 – aged 17 when on *Titanic*; second-class passenger (ticket price: £12); rescued in lifeboat 12

Abbott, Mr Eugene Joseph – 31 March 1899 – aged 13 when on *Titanic*; third-class passenger (ticket price: £20 5s); *Titanic* victim

Titanic survivor quotes

Mr Edneser Edward Wheelton

> 'I shouted to Mr Murdoch "The boat is full, sir." He said "All right. Have you got your sailors in?" I said "No, sir." He told two sailors to jump in the boat.' After being lowered into the water, they rowed about 300 yards off. 'We pulled toward a light but we didn't seem to get any closer to it. A lady back of me complained of the cold and I took my coat off and gave it to her…I would like to say something about the bravery exhibited by the First Officer, Mr Murdoch. He was perfectly cool and very calm.'

Mr Thomas Jones, witness at United States Senate Inquiry (Day 7)

> 'He told me to row for the light, and land the passengers and return to the ship. I pulled for the light, and I found that I could not get near the light, and I stood by for a little while. I wanted to return to the ship, but the ladies were frightened, and I had to carry out the captain's orders and pull for that light; so I did so. I pulled for about two hours, and then it started to get daybreak, and we lost the light; and then all of a sudden we saw the *Carpathia* coming, and we turned right back and made for the *Carpathia*. That is all I know, sir.'

Titanic fact

In dry weather conditions, a riveting crew at the Harland & Wolff shipyard could complete 200 rivets per day.

April

Day 92

1 April | On this day in *Titanic* history

The following people were born on this day:
Dymond, Mr Frank – 1 April 1876 – aged 36 when on *Titanic*; fireman; rescued in lifeboat 15

Sparkman, Mr Henry William – 1 April 1876 – aged 36 when on *Titanic*; fireman; rescued in lifeboat unknown

Conway, Mr Percy Walter – 1 April 1887 – aged 23 when on *Titanic*; saloon steward in second class; *Titanic* victim

Mockler, Miss Ellen Mary – 1 April 1889 – aged 25 when on *Titanic*; third-class passenger (ticket price: £7 17s 7d); rescued in lifeboat 16

Taylor, Mr Leonard – 1 April 1893 – aged 19 when on *Titanic*; Turkish bath attendant; *Titanic* victim

Holthen, Mr Johan Martin – 1 April 1894 – aged 18 when on *Titanic*; third-class passenger (ticket price: £22 10s 6d); *Titanic* victim

Richards, Master William Rowe – 1 April 1909 – aged 3 when on *Titanic*; second-class passenger (ticket price: £18 15s); rescued in lifeboat 4

The following person died on this day:
Mockler, Miss Ellen Mary – 1 April 1984 – aged 23 when on *Titanic*; third-class passenger (ticket price: £7 17s 7d); rescued in lifeboat 16

Master William Rowe Richards, mentioned in *Cornwall Advertiser*, Thursday, 18 April 1912

'Master William Rowe Richards 3 years old, 2nd Class rescued on boat 4.

The party of seven adults and four young children from Penzance were all bound to Akron. Mr George Hocking, his mother (Mrs Guy, who is a widow), and his sister, Miss Hocking, were of St Mary-street, and with them were Mrs Emily Richards, of Newlyn, another daughter of Mrs Guy, and who with her two young children, were going to join her husband in Akron.'

Titanic survivor quote

Mr Robert Hichens

> 'There was no use stopping alongside the ship, which was gradually going by the head. We were in a dangerous place, so I told them to man the oars – ladies and all. "All of you do your best." I relieved one of the young ladies with an oar and told her to take the tiller. She immediately let the boat come athwart, and the ladies in the boat got very nervous; so I took the tiller back again and told them to manage the best way they could.'

Day 93

2 April | On this day in *Titanic* history

The following person was born on this day:
Jussila, Mr Eiriik – 2 April 1880 – aged 32 when on *Titanic*; third-class passenger (ticket price: £7 18s 6d); rescued in lifeboat 15

The following people died on this day:
Martin, Miss Mabel Elvina – 2 April 1960 – aged 20 when on *Titanic*; cashier (restaurant staff); rescued in lifeboat 6

Burke, Mr William – 2 April 1961– aged 39 when on *Titanic*; saloon steward; rescued in lifeboat unknown

2 April 1912
Titanic heads into Belfast Lough in the early morning for her 12-hour sea trials.

Mr Thomas Kilgannon, mentioned in *Western People*, Saturday, 20 April 1912

> 'The great White Star liner, *Titanic*, the largest ship in the world, which left Southampton on Friday of last week on her maiden trip to New York, collided with an iceberg off the Newfoundland coast on Tuesday last and sank in 1200 fathoms of water (over two miles in depth). Of nearly 2,000 souls on board only 670 were saved, these being mostly women and children. Few of the male passengers were saved, the lost including several big American millionaires on their way home from the continent. No such frightful disaster has ever been recorded in shipping annals. Amongst those who embarked as steerage passengers at Queenstown, according to the returns furnished by the Company's agents there were – Kate Hargadon, Ballisodare; John Mahan, Curry; Catherine McGowan, Castlebar; Anne Kate Reilly, Athenry; Martin Gallagher, Ballygar; Thos. Kilgannon, do…'

Titanic survivor quote

Mr Henry S. Etches, witness at United States Senate Inquiry (Day 9)

> 'After getting all the women that were there, they called out three times – Mr Ismay called out twice, I know, in a loud voice – "Are there any more women before this boat goes," and there was no answer. Mr Murdoch called out; and at that moment a female came up whom I did not recognize. Mr Ismay said: "Come along; jump in." She said: "I am only a stewardess." He said: "Never mind, you are a woman, take your place." That was the last woman I saw get into No. 5 boat, sir.'

Titanic fact

Stewardess Annie Robinson earned £3 and 10 shillings a month.

Day 94

3 April | On this day in *Titanic* history

The following people were born on this day:
Smith, Mr James Clinch – 3 April 1856 – aged 56 when on *Titanic*; first-class passenger (ticket price: £30 13s 11d); *Titanic* victim

Shellard, Mr Frederick William Blainey – 3 April 1857 – aged 55 when on *Titanic*; third-class passenger (ticket price: £15 2s); *Titanic* victim

Bonnell, Miss Caroline T. – 3 April 1882 – aged 30 when on *Titanic*; first-class passenger (ticket price: £164 17s 4d); rescued in lifeboat 8

Whiteley, Mr Thomas Arthur – 3 April 1894 – aged 18 when on *Titanic*; saloon steward; rescued in collapsible lifeboat B

The following people died on this day:
Foley, Mr Wilfred Cyril – 3 April 1916 – aged 21 when on *Titanic*; steward in third class; rescued in lifeboat 13

Turkula, Mrs Hedwig – 3 April 1922 – aged 63 when on *Titanic*; third-class passenger (ticket price: £9 11s 9d); rescued in lifeboat 15

Dyer-Edwardes, Mrs Clementina Georgina Lucy – 3 April 1947 – aged 53 when on *Titanic*; first-class passenger (ticket price: £3); disembarked at Cherbourg

Peñasco Y Castellana, Mrs Maria Josefa Perezde Soto y Vallejo – 3 April 1972 – aged 22 when on *Titanic*; first-class passenger (ticket price: £108 18s); rescued in lifeboat 8

***New York Times*, Wednesday, 3 April 1912**

> 'LONDON, April 2nd. The White Star liner *Titanic*, which has just been completed by Harland & Wolff, left Belfast harbor this morning and proceeded to Belfast Lough, where the compasses were adjusted and speed trials took place over a measured mile with excellent results. Later she left for Southampton, from which port she sails on 10 April on her maiden voyage to New York with a large passenger list.'

Titanic survivor quote

Mr Thomas Jones

> 'The firemen brought up their bundles, not because they thought the boat was going to sink, but because they wanted to take them out of the water, as the water was coming in.'

Day 95

4 April | On this day in *Titanic* history

The following people were born on this day:
Brookman, Mr John Cress – 4 April 1884 – aged 28 when on *Titanic*; steward in third class; *Titanic* victim

Douglas, Mrs Mary Hélène Jane Baxter 'Suzette'– 4 April 1885 – aged 27 when on *Titanic*; first-class passenger (ticket price: £247 10s 5d); rescued in lifeboat 6

The following people died on this day:
Rosenbaum, Miss Edith Louise – 4 April 1975 – aged 33 when on *Titanic*; first-class passenger (ticket price: £27 14s 5d); rescued in lifeboat 11

Cribb, Miss Laura Mae – 4 April 1974 – aged 16 when on *Titanic*; third-class passenger (ticket price: £16 2s 2d); rescued in lifeboat 12

Titanic survivor quotes

Miss Edith Louise Rosenbaum (Edith Russell)
Edith was interviewed by *Life* magazine during her stay in America and also met with historian Walter Lord who included her story in his best-selling book *A Night to Remember*, published in 1956. Russell then served as a technical adviser to producer William MacQuitty on his 1958 film adaptation of Lord's book. She was portrayed in the movie as well and attended the premiere as MacQuitty's guest of honour.

> 'Everything imaginable: swimming pool, Turkish bath, gymnasium, squash courts, cafes, tea gardens, smoking rooms, a lounge bigger than the Grand Hotel lounge; huge drawing rooms, and bedrooms larger than in the average Paris hotel. It is a monster, and I can't say I like it, as I feel as if I were in a big hotel, instead of on a cozy ship; everyone is so stiff and formal. There are hundreds of help, bell boys, stewards, stewardesses and lifts. To say that it is wonderful is unquestionable, but not the cozy ship-board feeling of former years. We are now off Queenstown. I just hate to leave Paris and will be jolly glad to get back again. Am going to take my very much needed rest on this trip, but I cannot get over my feeling of depression and premonition of trouble.'

Mr George F. Crowe, witness at United States Senate Inquiry (Day 7)

> 'I was on duty up until about 10.30 on the night of the disaster, and I turned in about 11 o'clock; it might have been a little later. About 11.40 there was a kind of shaking of the ship and a little impact, from which I thought one of the propellers had been broken off.'

Day 96

5 April | On this day in *Titanic* history

The following people were born on this day:
Dooly, Mr Patrick – 5 April 1875 – aged 37 when on *Titanic*; third-class passenger (ticket price: £7 15s); *Titanic* victim

Gustafsson, Mr Anders Vilhelm – 5 April 1875 – aged 37 when on *Titanic*; third-class passenger (ticket price: £7 18s 6d); *Titanic* victim

Slayter, Miss Hilda Mary – 5 April 1882 – aged 30 when on *Titanic*; second-class passenger (ticket price: £12 7s); rescued in lifeboat 13

The following people died on this day:
Calderhead, Mr Edward Pennington – 5 April 1961 – aged 42 when on *Titanic*; first-class passenger (ticket price: £26 5s 9d); rescued in lifeboat 5

Johansson, Mr Oscar Wilhelm – 5 April 1967 – aged 32 when on *Titanic*; third-class passenger (ticket price: £7 15s 6d); rescued in collapsible lifeboat A

Beauchamp, Mr George William – 5 April 1965 – aged 24 when on *Titanic*; fireman; rescued in lifeboat 13

5 April 1912
Titanic is 'dressed' with flags and pennants as a Good Friday tribute to the people of Southampton.

Titanic survivor quotes

Miss Hilda Mary Slayter

> 'There were many touching scenes as the boats put off. I saw Col. John Jacob Astor hand his young wife into a boat tenderly and then asked an officer whether he might also go. When permission was refused he stepped back and coolly took out his cigarette case. "Goodbye, dearie" he called gaily, as he lighted a cigarette and leaned over the rail. "I'll join you later." Another man, a Frenchman, I think, approached one of the boats about to be lowered. He had with him two little boys. An officer waved him back sternly. "Bless you" he said. "I don't want to go, but for God's sake take the boys. Their mother is waiting for them in New York." The boys were taken aboard.'

Mr Daniel Buckley, witness at United States Senate Inquiry (Day 13)

> 'Yes; I saw the lights just going out as she went down. It made a terrible noise, like thunder.'

Day 97

6 April | On this day in *Titanic* history

The following people were born on this day:
Skoog, Mr Wilhelm ('William Johansson') – 6 April 1872 – aged 40 when on *Titanic*; third-class passenger (ticket price: £27 18s); *Titanic* victim

Thompson, Mr John William – 6 April 1876 – aged 36 when on *Titanic*; fireman; rescued in collapsible lifeboat A

Cook, Mrs Selena – 6 April 1890 – aged 22 when on *Titanic*; second-class passenger (ticket price: £10 10s); rescued in lifeboat 14

The following people died on this day:
Kimball, Mr Edwin Nelson Jr. – 6 April 1927 – aged 42 when on *Titanic*; first-class passenger (ticket price: £52 11s 1d); rescued in lifeboat 5

Thomas, Mr Albert Charles – 6 April 1973 – aged 23 when on *Titanic*; saloon steward; rescued in lifeboat 15

Mrs Selena Cook (née Rogers), quoted in *Worcester Evening Gazette*, Friday, 19 April 1912

> 'New York – 19 April – Mrs Lena Rogers of Boston was saved from the *Titanic* in a boat which carried 55 women passengers. Crowded to more than its capacity, the boat was endangered of being swamped when Fourth Officer Lowe, who had it in charge, succeeded in transferring some of his passengers to one of the other boats.
>
> "As we left the *Titanic*," she said, "several men were on the point of jumping into our boat, already overcrowded. They were stopped by Officer Lowe drawing a revolver. After taking us out of the range of the *Titanic*'s suction, he transferred us to other boats that had not been completely filled and went back for more from the sinking ship. Too much praise cannot be given the officer for his work. We were in the boat for three or four hours listening to the cries and groans of those who had jumped overboard and were not rescued, and those in other boats."'

Titanic survivor quote

Mr Henry S. Etches

> 'After the *Titanic* had sunk we pulled a good distance out farther from her, after the cries were all over. We pulled away, and a light we thought was a mast headlight of a ship was across where the port bow of the *Titanic* would have been at the time. During the time the *Titanic* was there I saw no light. I was looking at the *Titanic* the whole of the time.'

Day 98

7 April | On this day in *Titanic* history

The following person was born on this day:
Peracchio, Sig. Alberto – 7 April 1892 – aged 20 when on *Titanic*; assistant waiter (restaurant staff); *Titanic* victim

The following people died on this day:
Daly, Miss Margaret Marcella – 7 April 1939 – aged 33 when on *Titanic*; third-class passenger (ticket price: £6 19s); rescued in lifeboat 15

Horswill, Mr Albert Edward James – 7 April 1962 – aged 33 when on *Titanic*; able seaman; rescued in lifeboat 1

Fortune, Miss Alice Elizabeth – 7 April 1961 – aged 24 when on *Titanic*; first-class passenger (ticket price: £263); rescued in lifeboat 10

Mr Alberto Peracchio, mentioned in *Corriere della Sera*, Friday, 19 April 1912

> 'Italians shipwrecked
>
> 'In early editions of Corriere della Sera sadly it was reported that Alberto Peracchio and his brother Sebastiano had been rescued. Many of the waiters aboard the *Titanic* have wives and children in London, we hope that they were saved. The prominent men of the colony are thinking about a subscription for their families. I have phoned to the White Star office in Southampton and learned that among the survivors picked by *Carpathia* there is the twenty-year-old waiter A. Peracchi. The fate of the others is unknown, but the fact that one is saved makes hope for the others.
>
> 'We are waiting anxiously for new lists of the survivors. Among the crew there are no other Italians. There is no Italian name on the list, apart [from] one of a fireman John Podista, that from the news is an American citizen, but native of Italy.'

Titanic survivor quote

Mr Daniel Buckley, witness at United States Senate Inquiry (Day 13)

> 'There was a girl from my place, and just when she got down into the lifeboat she thought that the boat was sinking into the water. Her name was Bridget Bradley. She climbed one of the ropes as far as she could and tried to get back into the *Titanic* again, as she thought she would be safer in it than in the lifeboat. She was just getting up when one of the sailors went out to her and pulled her down again.'

Day 99

8 April | On this day in *Titanic* history

The following people were born on this day:
Bird, Miss Ellen – 8 April 1881 – aged 31 when on *Titanic*; first-class passenger (ticket price: £221 15s 7d); rescued in lifeboat 8

Saunders, Mr William Henry – 8 April 1881 – aged 31 when on *Titanic*; fireman; *Titanic* victim

Turley, Mr Richard – 8 April 1875 – aged 37 when on *Titanic*; fireman, *Titanic* victim

Jacobsohn, Mrs Amy Frances Christy – 8 April 1888 – aged 24 when on *Titanic*; second-class passenger (ticket price: £27s); rescued in lifeboat 12

Jefferys, Mr Clifford Thomas – 8 April 1888 – aged 24 when on *Titanic*; second-class passenger (ticket price: £31 10s); *Titanic* victim

Rosblom, Mr Viktor Richard – 8 April 1894 – aged 18 when on *Titanic*; third-class passenger (ticket price: £20 4s 3d); *Titanic* victim

Kink-Heilmann, Miss Luise Gretchen – 8 April 1908 – aged 4 when on *Titanic*; third-class passenger (ticket price: £22 6d), rescued in lifeboat 2

The following people died on this day:
Flegenheim, Mrs Antoinette Wendt – 8 April 1943 – aged 48 when on *Titanic*; first-class passenger (ticket price: £31 13s 8d); rescued in lifeboat 7

Oxenham, Mr Percy Thomas – 8 April 1954 – aged 22 when on *Titanic*; second-class passenger (ticket price: £10 10s); rescued in lifeboat 13

Kink, Mr Anton – 8 April 1959 – aged 29 when on *Titanic*; third-class passenger (ticket price: £22 6d); rescued in lifeboat 2

On this day, before the maiden voyage, all loading of fresh food was completed, including 75,000lbs of meat, 40,000 eggs, 40 tons of potatoes and 7,000 heads of lettuce.

Miss Luise Gretchen Kink-Heilmann, mentioned in *Le Journal de Genève*, Sunday, 21 April 1912

> 'The Kaiser & Co Agency received a telegram from the White Star Line stating that Mr and Mrs Kink, and their child, from Zurich, who were passengers on board the *Titanic*, are listed among the survivors.'

Titanic fact

There were at least thirteen couples celebrating their honeymoons on the ship.

Day 100

9 April | On this day in *Titanic* history

The following people were born on this day:

Bailey, Mr George Frank – 9 April 1866 – aged 46 when on *Titanic*; fireman; *Titanic* victim

Futrelle, Mr Jacques Heath – 9 April 1875 – aged 37 when on *Titanic*; first-class passenger (ticket price: £53 2s); *Titanic* victim

Healy, Miss Honor – 9 April 1879 – aged 33 when on *Titanic*; third-class passenger (ticket price: £7 15s); rescued in lifeboat 16

Gavey, Mr Laurence – 9 April 1886 – aged 26 when on *Titanic*; second-class passenger (ticket price: £10 10s); *Titanic* victim

Fleming, Miss Honor – 9 April 1890 – aged 22 when on *Titanic*; third-class passenger (ticket price: £7 15s); *Titanic* victim

Gustafsson, Mr Karl Gideon – 9 April 1893 – aged 19 when on *Titanic*; third-class passenger (ticket price: £7 15s 6d); *Titanic* victim

Titanic survivor quote

Mr Jacques Heath Futrelle, mentioned in *Worcester Telegram*, Saturday, 20 April 1912

> 'Jack died like a hero,' Mrs Futrelle said, when the party became composed. 'He was in the smoking room when the crash came – the noise of the smash was terrific- and I was going to bed. I was hurled from my feet by the impact. I hardly found myself when Jack came rushing into the stateroom. The boat is going down, get dressed at once, he shouted. When we reached the deck, everything was in the wildest confusion.'
>
> 'Jacques is dead, but he died like a hero. That I know. Three or four times after the crash I rushed up to him and clasped him in my arms and begged him to get into one of the boats. "For God's sake go" he fairly screamed at me as he tried to push me away and I could see how he suffered. It's your last chance, go. Then one of the ship's officers forced me into a lifeboat and I gave up all hope that he could be saved.
>
> "Hurry up May, you're keeping the others waiting" were his last words as he lifted me into a lifeboat and kissed me goodbye. I was in one of the last lifeboats to leave the ship. We had not put on many minutes when the *Titanic* disappeared. I almost saw as I saw her sink that I could see Jack standing where I had left him and waving at me.'

Day 101

10 April | On this day in *Titanic* history

The following people were born on this day:

Cardeza, Mrs Charlotte Wardle Drake – 10 April 1854 – aged 58 when on *Titanic*; first-class passenger (ticket price: £512 6s); rescued in lifeboat 3

Banewer, Mr Solomon – 10 April 1870 – aged 42 when on *Titanic*; second-class passenger (ticket price: £13); *Titanic* victim

Mason, Mr Frank Archibald Robert – 10 April 1880 – aged 32 when on *Titanic*; fireman (engineering crew); rescued in lifeboat 3 or 9

Ḥannā, Mr Mubārik Sulaymān Abī Āsī – 10 April 1885 – aged 27 when on *Titanic*; third-class passenger (ticket price: £7 4s 7d); rescued in lifeboat 15

Tu'mah, Mrs Hinnah – 10 April 1885 – aged 27 when on *Titanic*; third-class passenger (ticket price: £15 4s 11d); rescued in collapsible lifeboat C

Naughton, Miss Hannah – 10 April 1891 – aged 21 when on *Titanic*; third-class passenger (ticket price: £7 15s); *Titanic* victim

Crease, Mr Ernest James – 10 April 1892 – aged 20 when on *Titanic*; third-class passenger (ticket price: £8 3s 2d); *Titanic* victim

The following people died on this day:

Johnson, Master Harold Theodor – 10 April 1968 – aged 4 when on *Titanic*; third-class passenger (ticket price: £11 2s 8d); rescued in lifeboat 15

Harper, Miss Annie Jessie – 10 April 1986 – aged 6 when on *Titanic*; second-class passenger (ticket price: £33); rescued in lifeboat 11

The Western People, 1912.

> 'The Largest Vessel in the World.
>
> THIRD CLASS ACCOMMODATION.
>
> 'The dining saloon smoke room, general room, and promenade all leave nothing to be desired. The vessel will accommodate in all about 3,500 passengers and crew. The *Titanic* as an instrument of commerce, like her sister vessel, the *Olympic* represents the highest skill and perfection in naval architecture and marine engineering. The success of the *Olympic* has already been so gratifying that the popularity of the *Titanic* is assured. The *Titanic* will be in charge of Commander Edward J. Smith, R.D., R.N.R., who is being transferred from the *Olympic* and his successor to the latter steamer is Commander Herbert J. Haddock.'

Titanic fact

Lookout George Alfred Hogg earned £5 and 5 shillings a month.

Day 102

11 April | On this day in *Titanic* history

The following people were born on this day:
Hocking, Mrs Eliza – 11 April 1858 – aged 54 when on *Titanic*; second-class passenger (ticket price: £23); rescued in lifeboat 4

Haggan, Mr John – 11 April 1877 – aged 35 when on *Titanic*; fireman; rescued in lifeboat 3

Kvillner, Mr Johan Henrik Johannesson – 11 April 1881 – aged 31 when on *Titanic*; second-class passenger (ticket price: £10 10s); *Titanic* victim

Phillips, Mr John George – 11 April 1887 – aged 25 when on *Titanic*; telegraphist (victualling crew); *Titanic* victim

The following people died on this day:
Beckwith, Mr Richard Leonard – 11 April 1933 – aged 37 when on *Titanic*; first-class passenger (ticket price: £52 11s 1d); rescued in lifeboat 5

McKay, Mr Charles Donald – 11 April 1956 – aged 30 when on *Titanic*; saloon steward; rescued in lifeboat 11

Doling, Mrs Ada Julia Elizabeth – 11 April 1962 – aged 34 when on *Titanic*; second-class passenger (ticket price: £23); rescued in lifeboat unknown

***Manchester Guardian*, 11 April 1912**

> 'The White Star Liner *Titanic*, which left Southampton for New York yesterday on her first Atlantic voyage, surpasses in size and luxury, but especially in luxury, anything else afloat.'

Mr Joseph Bruce Ismay, concerning the plans, drawings, and specifications

> 'Messrs. Harland & Wolff prepare the plans. They are then submitted to us, to the directors of the White Star Line or to the manager of the White Star Line. They are carefully gone through with the representatives from the shipbuilders. They try to make suggestions to improve those plans. They are taken back and thoroughly thrashed out again, and they are submitted, I should be afraid to say how often. You see, when you build a ship you have to start building her probably five or six years before you want her.'

Titanic survivor quote

Mr John Hart (steward in third class)

> 'I heard two or three say they preferred to remain on the ship than be tossed about on the water like a cockle shell.'

Day 103

12 April | On this day in *Titanic* history

The following people were born on this day:
Funk, Miss Annie Clemmer – 12 April 1874 – aged 38 when on *Titanic*; second-class passenger (ticket price: £13); *Titanic* victim

Ketchley, Mr Henry – 12 April 1876 – aged 36 when on *Titanic*; saloon steward; *Titanic* victim

The following people died on this day:
Hemming, Mr Samuel Ernest – 12 April 1928 – aged 43 when on *Titanic*; lamp trimmer; rescued in lifeboat 4

Slayter, Miss Hilda Mary – 12 April 1965 – aged 30 when on *Titanic*; second-class passenger (ticket price: £12 7s,); rescued in lifeboat 13

Karlsson, Mr Einar Gervasius – 12 April 1958 – aged 21 when on *Titanic*; third-class passenger (ticket price: £7 15s 11d); rescued in lifeboat 13

Dorking, Mr Edward Arthur – 12 April 1954 – aged 18 when on *Titanic*; third-class passenger (ticket price: £8 1s); rescued in collapsible lifeboat B

Hirvonen, Miss Hildur Elisabeth – 12 April 1956 – aged 2 when on *Titanic*; third-class passenger (ticket price: £12 5s 9d); rescued in lifeboat 15

Mr Edward Arthur Dorking, mentioned in *Buda Plain Dealer*, Friday, 17 May 1912

> 'Edward Dorking, the young man from Oglesby who survived the terrible experiences of the *Titanic* disaster, will speak in Opera Hall in Buda, Saturday evening, 18 May… the young man recalls vividly the details of the great calamity which startled the world. The tale at first-hand is thrilling and, with the personal touch the speaker gives it, is sure to hold close attention from start to finish. Though the young man has no particular liking for speaking from the stage, he realises that this is the best way at present to earn some money to recompense him for the loss of all of his personal effects which went down with the ill-fated liner.'

Titanic survivor quote

Mr Edward J. Buley

> 'I think that is what the majority thought, that the ship would float. They thought she would go down a certain distance and stop there.'

Titanic fact

Titanic had a heated swimming pool (the first ever built into a vessel).

Day 104

13 April | On this day in *Titanic* history

The following people were born on this day:

Scott, Mr Archibald – 13 April 1870 – aged 42 when on *Titanic*; fireman; *Titanic* victim

Fox, Mr Stanley Hubert – 13 April 1874 – aged 38 when on *Titanic*; second-class passenger (ticket price: £13); *Titanic* victim

Wheat, Mr Joseph Thomas – 13 April 1882 – aged 30 when on *Titanic*; assistant second steward; rescued in lifeboat 11

Baggott, Mr Allen Marden – 13 April 1884 – aged 28 when on *Titanic*; saloon steward (victualling crew); rescued in lifeboat 9

Slade, Mr Alfred Edward – 13 April 1887 – aged 25 when on *Titanic*; fireman; disembarked at Southampton (he was listed as having deserted the ship)

Saturday, 13 April, 6.00 pm. *Titanic* received her second ice warning from the *Rappahannock*. The *Rappahannock* was travelling from Halifax to London; it had damaged her bow and rudder in heavy ice only 10 miles to the west of *Titanic*'s position.

Titanic survivor quote

Mr John Podesta speaking about **Mr Alfred Edward Slade**

> 'I got up on the morning of April 10th and made off down to the ship for eight o'clock muster, as is the case on all sailing days, which takes about an hour. As the ship is about to sail at about twelve o'clock noon most of us firemen and trimmers go ashore again until sailing time. So off we went [with] several others I knew on my watch, which was 4 to 8. My watch-mate, whose name was William Nutbean and I went off to our local public-house for a drink in the Newcastle Hotel. We left about eleven fifteen making our way toward the docks. Having plenty of time we dropped into another pub called the Grapes, meeting several more ship-mates inside. So having another drink about six of us left about ten minutes to twelve and got well into the docks and toward the vessel. With me and my mate were three brothers named Slade: Bertram, Tom and Alfred.
>
> 'We were at the top of the main road and a passenger train was approaching us from another part of the docks. I heard the Slades say, "Oh, let the train go by." But me and Nutbean crossed over and managed to board the liner. Being a long train, by the time it passed, the Slades were too late, and the gangway was down leaving them behind. So it seemed they did not have to go.'

Day 105

14 April | On this day in *Titanic* history

The following people were born on this day:

Howard, Mrs Ellen Truelove – 14 April 1851 – aged 61 when on *Titanic*; second-class passenger (ticket price: £26); *Titanic* victim

Osborne, Mr William Edward – 14 April 1879 – aged 33 when on *Titanic*; saloon steward; *Titanic* victim

Turner, Mr Leopold Olerenshaw – 14 April 1884 – aged 28 when on *Titanic*; saloon steward in first class; *Titanic* victim

Fredericks, Mr Walter Francis – 14 April 1891 – aged 21 when on *Titanic*; trimmer (engineering crew); rescued in lifeboat 15

Hood, Mr Ambrose – 14 April 1891 – aged 21 when on *Titanic*; second-class passenger (ticket price: £73 10s); *Titanic* victim

Sjöblom, Miss Anna Sofia – 14 April 1894 – aged 18 when on *Titanic*; third-class passenger (ticket price: £6 9s 11d); rescued in lifeboat 16

Rush, Mr Alfred George John – 14 April 1896 – aged 17 when on *Titanic*; porter for third-class passengers; *Titanic* victim

Pålsson, Master Paul Folke – 14 April 1906 – aged 6 when on *Titanic*; third-class passenger (ticket price: £21 1s 6d); *Titanic* victim

Skoog, Miss Margit Elizabeth – 14 April 1910 – aged 2 when on *Titanic*; third-class passenger (ticket price: £27 18s); *Titanic* victim

The following people died on this day:

Pallàs I Castelló, Sr. Emili – 14 April 1940 – aged 29 when on *Titanic*; second-class passenger (ticket price: £13 17s 2d); rescued in lifeboat 9

Thayer, Mrs Marian Longstreth Morris – 14 April 1944 – aged 39 when on *Titanic*; first-class passenger (ticket price: £110 17s 8d); rescued in lifeboat 4

Dean, Master Bertram Vere – 14 April 1992 – aged 1 when on *Titanic*; third-class passenger (ticket price: £20 11s 6d); rescued in lifeboat 10

Titanic survivor quote

Mr Edward J. Buley, witness at United States Senate Inquiry (Day 7)

> 'It must have been a boat, sir. It was too low down in the sea for a star. Then we were quite convinced afterwards, because we saw it go right by us when we were in the lifeboats. We thought she was coming toward us to pick us up.'

Day 106

15 April | On this day in *Titanic* history

The following people were born on this day:
Woody, Mr Oscar Scott – 15 April 1868 – aged 41 when on *Titanic*; postal clerk / postman (victualling crew); *Titanic* victim

Colley, Mr Edward Pomeroy – 15 April 1875 – aged 37 when on *Titanic*; first-class passenger (ticket price: £25 11s 9d); *Titanic* victim

Vartanian, Mr David – 15 April 1890 – aged 22 when on *Titanic*; third-class passenger (ticket price: £7 4s 6d); rescued in lifeboat 15

Burke, Mr Jeremiah – 15 April 1893 – aged 19 when on *Titanic*; third-class passenger (ticket price: £11 4s 10d); *Titanic* victim

Nīqūla Yārid, Miss Jamīlah – 15 April 1898 – aged 14 when on *Titanic*; third-class passenger (ticket price: £11 4s 10d); rescued in collapsible lifeboat C

The following person died on this day:
Cassebeer, Eleanor Genevieve Fosdick – 15 April 1969 – aged 36 when on *Titanic*; first-class passenger (ticket price: £27 14s 5d); rescued in lifeboat 5

Mr Joseph Bruce Ismay

> 'The first time I remember Capt. Smith being commander of one of our ships was when he was in command of one of our cargo boats called the *Cufic*, a great many years ago. He was in command of the *Olympic*, he was in the *Adriatic*, the *Baltic*, and the old *Britannic*. I can not remember them all, sir. We have a record in the office of every ship he has commanded…I was there to inspect the ship and see if there were any defects in her, with the idea of not repeating them in the other ship which we are now building at Belfast.'

Titanic fact

These are the numbers of passengers and crew, lost and saved:

	Lost	Saved	Total
First-Class Passengers	123	201	324
Second-Class Passengers	166	118	284
Third-Class Passengers	528	181	709
Crew	679	212	891
	1496	712	2208

Day 107

16 April | On this day in *Titanic* history

The following people were born on this day:
Benville, Mr Edward – 16 April 1864 – aged 47 when on *Titanic*; fireman; *Titanic* victim

Brito, Mr José Joaquim de – 16 April 1871 – aged 40 when on *Titanic*; second-class passenger (ticket price: £13); *Titanic* victim

Sweet, Mr George Frederick – 16 April 1897 – aged 14 when on *Titanic*; second-class passenger (ticket price: £65); *Titanic* victim

Andersson, Miss Sigrid Elisabeth – 16 April 1900 – aged 11 when on *Titanic*; third-class passenger (ticket price: £31 5s 6d); *Titanic* victim

Nicola-Yarred, Master Elias ('Louis Garrett') – 16 April 1900 – aged 11 when on *Titanic*; third-class passenger (ticket price: £11 4s 10d); rescued in collapsible lifeboat C

Andersson, Miss Ingeborg Constanzia – 16 April 1902 – aged 9 when on *Titanic*; third-class passenger (ticket price: £31 5s 6d); *Titanic* victim

The following person died on this day:
Stewart, Mr John – 16 April 1946 – aged 28 when on *Titanic*; veranda steward; rescued in lifeboat 15

Mr George Frederick Sweet, mentioned in *Newark Evening News*, Saturday, 20 April 1912

> 'Special Service of the NEWS
> 'Mrs Herman said that a boy named Sweet, fifteen years old, who accompanied Mr Herman and whose first name they did not know, was also lost in the wreck.'

Titanic survivor quote

Mr George F. Crowe, witness at United States Senate Inquiry (Day 7)

> 'She sank around half-past 2, from statements made by a man that was supposed to have jumped from the poop of the ship – that is, the quarter deck – into the water. He had a watch on, and as his watch stopped at 20 minutes past 2, he said she was in a sinking condition then and her stern on end – a man named Burnett, a storekeeper aboard ship.'

Titanic fact

Titanic had a 50-phone switchboard complete with operator for intra-ship calls.

Day 108

17 April | On this day in *Titanic* history

The following people were born on this day:
Morgan, Mr John Pierpont – 17 April 1837 – financier/banker, owner of International Mercantile Marine Company and of the White Star Line

Sincock, Miss Maude – 17 April 1891 – aged 20 when on *Titanic*; second-class passenger (ticket price: £36 15s); rescued in lifeboat 11

The following people died on this day:
Wolff, Mr Gustav Wilhelm – 17 April 1913 aged 78. – co-founder of Harland & Wolff, shipbuilder

Jewell, Mr Archie – 17 April 1917 – aged 23 when on *Titanic*; lookout; rescued in lifeboat 7

Titanic survivor quotes

Miss Maude Sincock

> 'I am saved but have lost everything… I have not a penny and no clothes. I was thrown on board a little boat in my nightdress and boots. I had no stockings on. We were in this little boat in the middle of the ocean for six hours. And I was nearly frozen when we were picked up. I shall be a pretty sight when I land. We were rescued by a passing ship, the *Carpathia*. The *Titanic* struck just before midnight and was underwater about 2 o'clock. There were over 1000 persons on board when she foundered... We could hear the screams from the men as the *Titanic* was sinking. I think there are hundreds drowned… I don't know what I shall do when I get to New York... I am frightened to death nearly, and I am afraid I shall catch my death of cold by the time I get to Hancock. I will write again as soon as possible and tell you more news. I don't know where they are going to put us when we get to New York.
>
> Your loving daughter Maud'

Mr Samuel Hemming (lamp trimmer)

> 'I pulled my head above the gunwale, and I said, "Give us a hand in, Jack." Foley was in the boat. I saw him standing up in the boat. He said, "Is that you, Sam?" I said, "Yes;" and him and the women and children pulled me in the boat.'

Titanic fact

Titanic had a veranda cafe with live palm trees.

Day 109

18 April | On this day in *Titanic* history

The following people were born on this day:
Peuchen, Major Arthur Godfrey – 18 April 1859 – aged 52 when on *Titanic*; first-class passenger (ticket price: £30 10s); rescued in lifeboat 6

McIntyre, Mr William – 18 April 1890 – aged 21 when on *Titanic*; trimmer; rescued in collapsible lifeboat A or B

The following people died on this day:
Schabert, Mrs Emma Mock – 18 April 1961 – aged 35 when on *Titanic*; first-class passenger (ticket price: £57 15s); rescued in lifeboat 11

Māmā, Mr Hannā Mikā'īl – 18 April 1952 – aged 18 when on *Titanic*; third-class passenger (ticket price: £7 4s 7d); rescued in lifeboat 15

Nilsson, Miss Helmina Josefina – 18 April 1971 – aged 26 when on *Titanic*; third-class passenger (ticket price: £7 17s 1d); rescued in lifeboat 13

On 18 April at 9.00 pm, the *Carpathia* arrives in New York.

Titanic survivor quotes

Major Arthur Godfrey Peuchen, quoted in *Washington Herald*, Sunday, 21 April 1912

> 'I cannot repeat too strongly that J. Bruce Ismay knew of the presence of icebergs, but deliberately "took a chance", for reasons which he can best explain,' said Major Peuchen before leaving for Canada. 'It will be easy to prove also that while the *Titanic* was in peril in fact, within two hours of the actual collision this man Ismay and Capt. Smith were members of a dinner party in the first cabin saloon.' The exact words of Mrs Ryerson as reported by Major Peuchen, and later repeated privately to the Senatorial Investigating committee by Mrs Douglas, were: 'Late Sunday afternoon (just a few hours before the collision), while the *Titanic* was proceeding at a very great speed, I went to Mr Ismay and said: "Oh, Mr Ismay, I have heard that the wireless has reported a large number of icebergs in the path of our ship. Are you not going to order her to slow down?" He replied to me: "On the contrary, Mrs Ryerson, we are going to go along faster than we have been going." And we did go faster.'

Titanic fact

An additional safety benefit that came from a higher fare was that rooms were closer to the lifeboats.

Day 110

19 April | On this day in *Titanic* history

The following people were born on this day:
Gold, Mrs Jane Kate Coulson – 19 April 1866 – aged 45 when on *Titanic*; stewardess; rescued in lifeboat 11

Prichard, Mrs Alice Maud – 19 April 1875 – aged 36 when on *Titanic*; stewardess; rescued in lifeboat 11

Cairns, Mr Alexander Milne – 19 April 1883 – aged 28 when on *Titanic*; first-class passenger (ticket price: £31); *Titanic* victim

Collander, Mr Erik Gustaf – 19 April 1884 – aged 27 when on *Titanic*; second-class passenger (ticket price: £13); *Titanic* victim

The following people died on this day:
Stengel, Mr Charles Emil Henry – 19 April 1914 – aged 54 when on *Titanic*; first-class passenger (ticket price: £55 8s 10d); rescued in lifeboat 1

De Mulder, Mr Theodoor – 19 April 1954 – aged 30 when on *Titanic*; third-class passenger (ticket price: £9 10s); rescued in lifeboat 11

19 April 1912: the United States Senate Inquiry into *Titanic*'s sinking begins.

Mrs Jane Kate Coulson Gold, quoted in *The Bathurst Times*, Saturday, 6 September 1913
Last women to leave the *Titanic*.

> 'I was not a bit afraid, somehow. I was quite cool. There was really nothing in it. After I entered the boat Mr Ismay got in, making the 78th person in the boat, which was then lowered 80ft. into sea. It was dangerously overcrowded. We pulled away a bit, and saw the *Titanic* go down at 20 minutes to 2 am on the 15th. Some survivors say not, but to me just before she sank she seemed to break up. We heard terrible cries, but could not see where they came from too well. They must have been from those in the freezing water. The cries of terror and distress blended, and sounded like shouting at a football match sounds from a distance. I heard shouting like that over at Mosman a few days ago. I went quite – oh, to think of it! To think of it!'

Mr Daniel Buckley, witness at United States Senate Inquiry (Day 13)

> 'We sighted the lights of the big steamer, the *Carpathia*. All the women got into a terrible commotion and jumped around. They were hallooing and the sailors were trying to keep them sitting down, and they would not do it. They were standing up all the time.'

Day 111

20 April | On this day in *Titanic* history

The following people were born on this day:
Crafton, Mr John Bertram – 20 April 1853 – aged 58 when on *Titanic*; first-class passenger (ticket price: £26 11s); *Titanic* victim

Minahan, Dr William Edward – 20 April 1867 – aged 44 when on *Titanic*; first-class passenger (ticket price: £90); *Titanic* victim

Strandén, Mr Juho Niilosson – 20 April 1881 – aged 30 when on *Titanic*; third-class passenger (ticket price: £7 18s 6d); rescued in lifeboat 9

The following people died on this day:
Leader, Dr Alice Farnham – 20 April 1944 – aged 49 when on *Titanic*; first-class passenger (ticket price: £25 18s 7d); rescued in lifeboat 8

Duff-Gordon, Sir Cosmo Edmund – 20 April 1931 – aged 49 when on *Titanic*; first-class passenger (ticket price: £39 12s); rescued in lifeboat 1

Duff-Gordon, Lady Lucy Christiana – 20 April 1935 – aged 48 when on *Titanic*; first-class passenger (ticket price: £56 18s 7d); rescued in lifeboat 1

West, Mrs Ada Mary – 20 April 1953 – aged 33 when on *Titanic*; second-class passenger (ticket price: £27 15s); rescued in lifeboat 10

20 April 1910 – 15-year-old Samuel Joseph Scott becomes the ship's first fatality, when he falls 23ft from one of *Titanic*'s timber construction platforms.

20 April 1912 – Harland & Wolff shipyard closed for a day, out of respect for Thomas Andrews and the yardmen who perished in the sinking.

Titanic survivor quote

Mr Harold S. Bride (assistant wireless telegraph operator)

> 'On Mr Phillips's request I started to gather up his spare money and put on another coat, and made general preparations for leaving the ship.'

Titanic fact

In 1912, the Football Association Charity Shield was played at White Hart Lane, home of Tottenham Hotspur. The two teams were the Football League champions, Blackburn Rovers, and the winners of the Southern League championship, Queens Park Rangers. The game ended in a 2–1 win for Blackburn Rovers, and all the proceeds of the match were donated to the *Titanic* Relief Fund.

Day 112

21 April | On this day in *Titanic* history

The following people were born on this day:

Douglas, Mr Walter Donald – 21 April 1865 – aged 46 when on *Titanic*; first-class passenger (ticket price: £106 8s 6d); *Titanic* victim

Thayer, Mr John Borland – 21 April 1862 – aged 49 when on *Titanic*; first-class passenger (ticket price: £110 17s 8d); *Titanic* victim

Francatelli, Miss Laura Mabel – 21 April 1880 – aged 31 when on *Titanic*; first-class passenger (ticket price: £56 18s 7d); rescued in lifeboat 1

Johnson, Mrs Aliina Vilhelmina – 21 April 1884 – aged 27 when on *Titanic*; third-class passenger (ticket price: £11 2s 8d); rescued in lifeboat 15

Lucas, Mr William Arthur – 21 April 1886 – aged 25 when on *Titanic*; able seaman; rescued in collapsible lifeboat D

Edbrooke, Mr Francis Samuel Jacob – 21 April 1888 – aged 23 when on *Titanic*; steward in third class; *Titanic* victim

The following people died on this day:

Douglas, Mrs Mahala Dutton – 21 April 1945 – aged 48 when on *Titanic*; first-class passenger (ticket price: £106 8s 6d); rescued in lifeboat 2

Abī Sa'b, Mrs Sha'nīnah – 21 April 1947 – aged 38 when on *Titanic*; third-class passenger (ticket price: £7 4s 7d,); rescued in collapsible lifeboat C

Earnshaw, Mrs Olive Potter – 21 April 1958 – aged 23 when on *Titanic*; first-class passenger (ticket price: £83 3s 2d); rescued in lifeboat 7

The New York Times Sunday, 21 April 1912

> 'All of the crew of the *Titanic*, except the saved officers and about twenty of the crew who are to be witnesses, sailed for their homes yesterday morning on the Red Star liner *Lapland*. So many of them were saved, Second Officer Lightoller has testified, because of their powers of endurance, for of the persons picked up in the water by the lifeboats five out of every six were either firemen or stewards. Eliminating the stewardesses, who were saved with the passengers, and thirty men who were picked up clinging to an overturned boat, 158 were rescued.'

Mr Harold Thomas Cottam (Marconi operator, RMS *Carpathia*)

Message from *Titanic*: "Come at once. Come at once. It is a distress message; C. Q. D."

Day 113

22 April | On this day in *Titanic* history

The following people were born on this day:
Rosblom, Mrs Helena Wilhelmina – 22 April 1870 – aged 41 when on *Titanic*; third-class passenger (ticket price: £20 4s 3d); *Titanic* victim

Richards, Mrs Emily – 22 April 1887 – aged 24 when on *Titanic*; second-class passenger (ticket price: £18 15s); rescued in lifeboat 4

Yvois, Miss Henriette Virginie – 22 April 1889 – aged 22 when on *Titanic*; second-class passenger (ticket price: £13); *Titanic* victim

Mr Joseph G. Boxhall, witness at United States Senate Inquiry (Day 3)

> 'I said, "The captain wants you to sound the ship." He said, "The ship is making water," and he went on the bridge to the captain, and I thought I would go down forward again and investigate; and then I met a mail clerk, a man named Smith, and he asked where the captain was. I said, "He is on the bridge." He said, "The mail hold is full" or "filling rapidly." I said, "Well, you go and report it to the captain and I will go down and see," and I proceeded right down into the mail room.'

Titanic survivor quote

Mrs Emily Richards (née Hocking), quoted in *Akron Beacon Journal*, Saturday, 20 April 1912

> 'Once the boat started away some of the women stood up and the seamen with their hands full with the oars simply put their feet on them and forced them back into the sitting position. We had not gone far away by the time the ship went down and after that there were men floating in the water around and seven of them were picked up by us in the hours that followed between that and daybreak.
>
> 'Some of those seven were already mad with exposure and babbled gibberish and kept trying to get up and overturn the boat. The other men had to sit upon them to hold them down. Two of the men picked up were so overcome with the cold of the water that they died before we reached the *Carpathia* and their dead bodies were taken aboard.'

Titanic fact

Captain E. J. Smith earned £105 per month (about £11,000 today) whereas a stewardess earned £3 10 shillings per month (around £375 today).

Day 114

23 April | On this day in *Titanic* history

The following people were born on this day:
Lahtinen, Rev William – 23 April 1876 – aged 22 when on *Titanic*; second-class passenger (ticket price: £26); *Titanic* victim

Riihivuori, Miss Susanna Juhantytär – 23 April 1889 – aged 22 when on *Titanic*; third-class passenger (ticket price: £39 13s 9d); *Titanic* victim

The following people died on this day:
Taylor, Mrs Juliet Cummings – 23 April 1927 – aged 49 when on *Titanic*; first-class passenger (ticket price: £52); rescued in lifeboat 5

Stanley, Miss Amy Zillah Elsie – 23 April 1955 – aged 24 when on *Titanic*; third-class passenger (ticket price: £7 11s); rescued in collapsible lifeboat C

Titanic survivor quotes

Mr Arthur G. Peuchen

> 'When I came on deck first, on this upper deck, there were, it seems to me, about 100 stokers came up with their dunnage bags, and they seemed to crowd this whole deck in front of the boats. One of the officers – I do not know which one, but a very powerful one – came along and drove these men right off that deck. It was a splendid act. He drove them, every man, like a lot of sheep, right off the deck.'

Mr Robert Hichens (quartermaster)

> 'In every ship that I have been in we always have had every quartermaster, or whatever he may be, seaman, fireman, steward, always have their boat stations, and they would have a proper muster every Saturday or whenever it may be – it lies at the Captain's discretion whenever he liked to give us a drill, and everybody is mustered in front of their boats, but I never saw it like that on the *Titanic*.'

Titanic fact

Swedish first-class passenger and businessman Mauritz Håkan Björnström-Steffansson owned one of the most highly valued single objects on board: a masterpiece of French neoclassical painting entitled 'La Circassienne au Bain', for which he later claimed $100,000 in compensation.

Day 115

24 April | On this day in *Titanic* history

The following people were born on this day:
Wennerström, Mr August – 24 April 1884 – aged 27 when on *Titanic*; third-class passenger (ticket price: £7 15s 11d); rescued in collapsible lifeboat A

Hensford, Mr Herbert George Ernest – 24 April 1885 – aged 26 when on *Titanic*; assistant butcher; *Titanic* victim

Denbuoy, Mr Albert Joseph – 24 April 1886 – aged 25 when on *Titanic*; second-class passenger (ticket price: £31 10s); *Titanic* victim

Prideaux, Mr John Arthur (Jack) – 24 April 1888 – aged 23 when on *Titanic*; steward in third class; *Titanic* victim

Titanic survivor quotes

Mr August Wennerström

> 'One of our friends, a man by the name of Johan Lundahl who had been home to the old country on a visit and was going back to the United States said to us, "Goodbye friends; I'm too old to fight the Atlantic." He went to the smoking room and there on a chair was awaiting his last call. So did an English lady. She sat down by the piano and, with her child on her knee, she played the piano until the Atlantic grave called them both.'

Charles Joughin (chief baker)

> 'They were getting the boat ready for getting the passengers in, and Mr Wilde shouted out for the stewards to keep the people back, to keep the men back, but there was no necessity for it. The men kept back themselves, and we made a line and passed the ladies and children through.
>
> 'I myself and three or four other chaps went on the next deck and forcibly brought up women and children, there are only about ten stairs to go up. We threw them in. The boat was standing off about a yard and a half from the ship's side, with a slight list. We could not put them in; we could either hand them in or just drop them in.'

Titanic fact

Among its many amenities, the *Titanic* had its own daily newspaper, the *Atlantic Daily Bulletin*. The paper was printed on board the ship every day and included news, advertisements, horse-racing results, stock prices, society gossip and the menu for the day.

Day 116

25 April | On this day in *Titanic* history

The following people were born on this day:
Marconi, Guglielmo – 25 April 1874 – formed The Wireless Telegraph & Signal Company Limited in 1900; it was renamed Marconi's Wireless Telegraph Company Limited

Andersen, Mr Albert Karvin – 25 April 1878 – aged 33 when on *Titanic*; third-class passenger (ticket price: £22 10s 6d); *Titanic* victim

Scarrott, Mr Joseph George – 25 April 1878 – aged 33 when on *Titanic*; able seaman; rescued in lifeboat 14

Turpin, Mrs Dorothy Ann – 25 April 1885 – aged 26 when on *Titanic*; second-class passenger (ticket price: £21); *Titanic* victim

Ford, Miss Robina Maggie – 25 April 1904 – aged 7 when on *Titanic*; third-class passenger (ticket price: £34 7s 6d); *Titanic* victim

Panula, Master Urho Abraham – 25 April 1909 – aged 2 when on *Titanic*; third-class passenger (ticket price: £39 13s 9d); *Titanic* victim

The following people died on this day:
Newell, Miss Madeleine – 25 April 1969 – aged 31 when on *Titanic*; first-class passenger (ticket price: £113 5s 6d); rescued in lifeboat 6

Boxhall, Mr Joseph Groves – 25 April 1967 – aged 28 when on *Titanic*; fourth officer; rescued in lifeboat 2

Willard, Miss Constance – 25 April 1964 – aged 21 when on *Titanic*; first-class passenger (ticket price: £26 11s); rescued in lifeboat 8

Mr Guglielmo Marconi, mentioned in *New York Times*, Thursday, 18 April 1912

> 'Guglielmo Marconi First LECTURE IN AMERICA: letter from Thomas Alva Edison.
>
> "I regret my inability to be present at your lecture, but hasten to congratulate you upon the success of your beautiful invention the wireless telegraph and on the splendid work your system has done in saving human life in disasters on the sea."

Mr Joseph G. Boxhall, witness at United States Senate Inquiry (Day 3)

> 'Seven on either side. I never counted them, but I think there were 7. There were 14 lifeboats and 2 sea boats. They were equally divided.'

Titanic fact

The top speed of the *Titanic* was 23 knots (more than 26 miles per hour).

Day 117

26 April | On this day in *Titanic* history

The following people were born on this day:
Bristow, Mr Harry – 26 April 1873 – 25 April 1885 – aged 38 when on *Titanic*; saloon steward (victualling crew); *Titanic* victim

Fitzpatrick, Mr Cecil William – 26 April 1890 – aged 21 when on *Titanic*; mess steward; rescued in lifeboat 8

Spinner, Mr Henry John – 26 April 1880 – aged 31 when on *Titanic*; third-class passenger (ticket price: £8 1s); *Titanic* victim

Morrow, Mr Thomas Rowan – 26 April 1881– aged 30 when on *Titanic*; third-class passenger (ticket price: £7 15s); *Titanic* victim

Bradley, Mr Thomas Henry – 26 April 1883– aged 28 when on *Titanic*; able seaman; *Titanic* victim

The following people died on this day:
Phillimore, Mr Harold Charles William – 26 April 1967– aged 23 when on *Titanic*; saloon steward in second class; rescued in lifeboat 14

Graham, Miss Margaret Edith – 26 April 1976 – aged 19 when on *Titanic*; first-class passenger (ticket price: £153 9s 3d); rescued in lifeboat 3

Mr Harry Bristow (part of a letter sent home on 9 April 1912)

> 'I have earned my first days pay on the *Titanic* 4/- & been paid & I may say spent it do you know dearie...my uniform will cost £1 17s 6d, coat plus waistcoat and cap and Star regulation collars and paper front (don't laugh dearie it's quite true) two white jackets etc. so it won't leave me very much to take up, my pay is £3 15s plus tips. I'm in the first class saloon so I may pick up a bit. I've been scrubbing the floor today in Saloon, about a dozen of us. I lost myself a time or two, she is such an enormous size I expect it will take me a couple of trips before I begin to know my way about here… your ever loving Harry'

Mr Joseph Bruce Ismay, chairman and managing director of White Star Line

> 'Naturally, in a ship of that size, there were a great many minor defects on board the ship, which he was rectifying. I think there were probably three or four apprentices on board from Messrs. Harland & Wolff's shipbuilding yard, who were there to right any small detail which was wrong.'

Titanic fact

The first-class smoking lounge was for men only.

Day 118

27 April | On this day in *Titanic* history

The following people were born on this day:
Cribb, Mr John Hatfield – 27 April 1868 – aged 43 when on *Titanic*; third-class passenger (ticket price: £16 2s); *Titanic* victim

Taylor, Mr Charles William Frederick – 27 April 1871 – aged 40 when on *Titanic*; able seaman; *Titanic* victim

Renouf, Mr Peter Henry – 27 April 1878 – aged 33 when on *Titanic*; second-class passenger (ticket price: £21); *Titanic* victim

Chambers, Mr Norman Campbell – 27 April 1884 – aged 27 when on *Titanic*; first-class passenger (ticket price: £53 2s); rescued in lifeboat 5

Lacey, Mr Bertie William – 27 April 1892 – aged 19 when on *Titanic*; assistant saloon steward in second class; *Titanic* victim

The following people died on this day:
Wilkes, Mrs Ellen – 27 April 1955 – aged 47 when on *Titanic*; third-class passenger (ticket price: £7); rescued in lifeboat 16

Smyth, Miss Julia – 27 April 1977 – aged 18 when on *Titanic*; third-class passenger (ticket price: £7 14s 8d); rescued in lifeboat 13

Mr Herbert John Pitman (third officer), witness at United States Senate Inquiry (Day 4)

> 'My duties comprised working out celestial observations, finding the deviation of the compass, general supervision around the decks, and looking after the quartermasters; also relieving the bridge if necessary.'

Mr Peter Henry Renouf, mentioned in *Elizabeth Daily Journal*, Friday, 26 April 1912

> 'Mrs Peter Reniff is Left Destitute
> With the vessel sank the entire savings of the family as well as all of Mrs Reniff's clothing with the exception of that which she wore when leaving the ship. Her jewellery and many valuable gifts, which she was bringing to this country from her native land, England, were also lost. According to the woman there were no lights of another vessel to be seen when the *Titanic* was sinking, as claimed by some of the ship's officers. The first lights I saw, she said, were those of the *Carpathia*, as the boat steamed toward us at daybreak. The enormous loss of life, she continued, is due only to the fact that the number of lifeboats was inadequate …'

Day 119

28 April | On this day in *Titanic* history

The following person was born on this day:
Nichols, Mrs Emily – 28 April 1863 – aged 48 when on *Titanic*; first-class passenger (ticket price: £4); disembarked at Queenstown

Feltham, Mr George William – 28 April 1870 – aged 41 when on *Titanic*; Vienna baker; *Titanic* victim

Shannon, Mr Andrew John – 28 April 1876 – aged 35 when on *Titanic*; third-class passenger (ticket price: £7 5s); *Titanic* victim

Kerr, Mr Thomas Russell – 28 April 1885 – aged 26 when on *Titanic*; fireman; *Titanic* victim

The following person died on this day:
al-Ba'qlīnī, Miss Hilānah – 28 April 1939 – aged 9 months when on *Titanic*; third-class passenger (ticket price: £19 5s 2d); rescued in collapsible lifeboat C

Titanic survivor quotes

Mr Harold Bride, mentioned in *The New York Times*, Sunday, 28 April 1912

> 'Two Marconi Men Worked to the Last. [Harold] Bride said he fixed a lifebelt onto [Jack] Phillips while the latter continued to send out wireless messages about the *Titanic*'s distress.'

Mr Walter Hurst

> 'Well, I saw the for'ard part of the boat deck dip under water so I jumped overboard swimming away from the ship, turned around to look at her but down came the funnel and smashed into the water right in front of my face. I got a gush of wind and dirt through that nearly blinded me and I felt the cap go from my head, and one slipper off of my foot, but it didn't take my attention off of this boat... the collapsible boat that washed straight over off the deck within a few yards of me but I managed to get onto it and I was followed quickly by the second officer and a few others anyhow she soon got filled up. There was an old man stood next to me, he was complaining all the night long about his head was so cold... mine was pretty cold too but a friend of mine, a shipmate named Lindsay, gave me a drink from the bottle. I thought I was on a good thing [thinking] it was whisky or brandy instead it was essence of peppermint, it nearly choked me.'

Day 120

29 April | On this day in *Titanic* history

The following person was born:
Bradshaw, Mr John Albert Perkin – 29 April 1869 – aged 42 when on *Titanic*; plate steward in first class; *Titanic* victim

The following people died on this day:
Swift, Mrs Margaret Welles Barron – 29 April 1948 – aged 46 when on *Titanic*; first-class passenger (ticket price: £25 18s 7d); rescued in lifeboat 8

Louch, Mrs Alice Adelaide – 29 April 1949 – aged 43 when on *Titanic*; second-class passenger (ticket price: £26); rescued in lifeboat unknown

Bride, Mr Harold Sydney – 29 April 1956 – aged 22 when on *Titanic*; assistant telegraphist; rescued in collapsible lifeboat B

Mr Joseph G. Boxhall, witness at United States Senate Inquiry (Day 3)

> 'He was working down the port side most of the time right from the first. I never saw Mr Lightoller on the starboard side. Whenever I did see him it was on the port side.'

Titanic survivor quote

Mr Harold Sydney Bride

> 'I went to the place I had seen the collapsible boat on the boat deck, and to my surprise I saw the boat and the men still trying to push it off. I guess there wasn't a sailor in the crowd. They couldn't do it. I went up to them and was just lending a hand when a large wave came awash of the deck. The big wave carried the boat off. I had hold of an oarlock and I went off with it. The next I knew I was in the boat.
>
> 'But that was not all. I was in the boat and the boat was upside down and I was under it. And I remember I realised I was wet through, and that whatever happened I must not breathe, for I was underwater. I knew I had to fight for it and I did. How I got out from under the boat I do not know, but I felt a breath of air at last… I felt I simply had to get away from the ship. She was a beautiful sight then. Smoke and sparks were rushing out of her funnel. There must have been an explosion, but we heard none. We only saw the big stream of sparks. The ship was gradually turning on her nose, just like a duck does that goes down for a dive.'

Day 121

30 April | On this day in *Titanic* history

The following person was born on this day:
Scavino, Sig. Candido – 30 April 1869 – aged 42 when on *Titanic*; waiter (restaurant staff); *Titanic* victim

Shutes, Miss Elizabeth Weed – 30 April 1871 – aged 40 when on *Titanic*; first-class passenger (ticket price: £123 8s 3d); rescued in lifeboat 3

Compton, Miss Sara Rebecca – 30 April 1872 – aged 39 when on *Titanic*; first-class passenger (ticket price: £83 3s 2d); rescued in lifeboat 14

Lawrence, Mr Arthur – 30 April 1875 – aged 36 when on *Titanic*; saloon steward in first class; *Titanic* victim

The following people died on this day:
De Messemaeker, Mrs Anna – 30 April 1918 – aged 36 when on *Titanic*; third-class passenger (ticket price: £17 8s); rescued in lifeboat 13

Minahan, Miss Ida E. 'Daisy' – 30 April 1919 – aged 46 when on *Titanic*; first-class passenger (ticket price: £90); rescued in lifeboat 14

Titanic survivor quotes

Miss Sara Rebecca Compton, mentioned in *Newark Evening News*, Friday, 19 April 1912

> 'When we waved goodbye to my son, said Mrs Compton, we did not realise the great danger, but thought we were only being sent out in the boats as a precautionary measure. When Captain Smith handed us life preservers, he said cheerily: They will keep you warm if you do not have to use them. Then the crew began clearing the boats and putting the women into them. My daughter and I were lifted in the boat commanded by the fifth officer. There was a moan of agony and anguish from those in our boat when the *Titanic* sank, and we insisted that the officer head back for the place where the *Titanic* had disappeared. We found one man with a life preserver on him struggling in the cold water, and for a moment I thought that he was my son.'

Mr Samuel Hemming, witness at United States Senate Inquiry (Day 7)

> 'We were moving around, constantly, sir. Sometimes the stern of the boat would be toward the *Titanic*, and sometimes the bow of the boat would be toward the *Titanic*. One moment we would be facing one way, and a few moments later we would be facing another way; first the bow, and then the stern toward the ship.'

May

Day 122

1 May | On this day in *Titanic* history

The following people were born on this day:
Čačić, Miss Marija – 1 May 1875 – aged 36 when on *Titanic*; third-class passenger (ticket price: £3 11s 3d); *Titanic* victim

Charman, Mr John James – 1 May 1886 – aged 25 when on *Titanic*; saloon steward in second class (victualling crew); *Titanic* victim

Phillimore, Mr Harold Charles William – 1 May 1888 – aged 23 when on *Titanic*; saloon steward in second class; rescued in lifeboat 14

Rotta, Sig. Angelo Mario – 1 May 1888 – aged 23 when on *Titanic*; waiter (restaurant staff); *Titanic* victim

Borebank, Mr John James – 1 May 1870 – aged 41 when on *Titanic*; first-class passenger (ticket price: £26 11s); *Titanic* victim

The following people died on this day:
Weller, Mr William Clifford – 1 May 1954 – aged 30 when on *Titanic*; able seaman; rescued in lifeboat 7

Wheelton, Mr Edneser Edward – 1 May 1949 – aged 28 when on *Titanic*; saloon steward in first class; rescued in lifeboat 11

Sloper, Mr William Thompson – 1 May 1955 – aged 28 when on *Titanic*; first-class passenger (ticket price: £35 10s); rescued in lifeboat 7

Mr William Thompson Sloper, mentioned in *Chicago Examiner*, Friday, 19 April 1912

> 'Resolved to die after having done his utmost to aid in placing the women and children of the *Titanic* aboard the lifeboats, Councilman William T. Sloper, clad in a white night robe, was himself taken for a woman and thrust into one of the last lifeboats lowered away. His father, Senator Andrew J. Sloper of New Britain, received a message today telling of his son's rescue and departed at once for New York. With him went Mrs Albert S. Cook of this city to meet her brother, Richard L. Beckwith, who with his wife and stepdaughter, was also saved.'

Mr John Pitman, witness at United States Senate Inquiry (Day 4)

Senator Smith: 'If I recollect what you have said, you saw the icebergs, or evidence of ice, when you were on the watch from 6 to 8 o'clock Sunday evening?'
Mr John Pitman: 'I saw none whatever, sir, until I was in the boat, and that was about half-past 3 Monday morning. That was the first ice I saw.'

Day 123

2 May | On this day in *Titanic* history

The following person was born on this day:
Knowles, Mr Thomas – 2 May 1867 – aged 44 when on *Titanic*; firemen's messman; rescued in collapsible lifeboat C

Stengel, Mrs Annie May – 2 May 1868 – aged 43 when on *Titanic*; first-class passenger (ticket price: £55 8s 10d); rescued in lifeboat 5

Howard, Miss May Elizabeth – 2 May 1885 – aged 26 when on *Titanic*; third-class passenger (ticket price: £8 1s); rescued in collapsible lifeboat C

Wendel, Mr Olof Edvin – 2 May 1891 – aged 20 when on *Titanic*; third-class passenger (ticket price: £7 17s 1d); *Titanic* victim

The following people died on this day:
Silvey, Mrs Alice – 2 May 1958 – aged 39 when on *Titanic*; first-class passenger (ticket price: £55 18s); rescued in lifeboat 11

Holverson, Mrs Mary Aline – 2 May 1918 – aged 35 when on *Titanic*; first-class passenger (ticket price: £52); rescued in lifeboat 8

Barber, Miss Ellen Mary – 2 May 1963 – aged 27 when on *Titanic*; first-class passenger (ticket price: £78 17s); rescued in lifeboat 6

Titanic survivor quotes

Miss May Elizabeth Howard, quoted in *Rochester Democrat and Chronicle*, 24 April 1912

> 'One of the officers grabbed Mrs Goldsmith and myself and pushed us to the edge of the ship where the lifeboat was being filled with women and children. An officer there shouted "All men back, women come first." There was no attempt of any of the men to get into the boat and when we left the *Titanic* our boat would easily have carried fifteen more had they cared to leave the ship, but they did not realise that the ship was in danger and refused to enter the small boats…The lifeboat in which I left the *Titanic* was the last but one of those picked up by the *Carpathia*."

Mr Thomas Knowles (part of a letter by Mr Knowles' daughter)

> '"If you want a chance, stay here, or we shall all cling together as drowning rats." Only a few stayed. He could not swim, strange to say, but as the *Titanic* gradually sank, he stood on the rails, water over his feet, and Captain Smith pacing the bridge…'

Day 124

3 May | On this day in *Titanic* history

The following person was born on this day:
Morley, Mr Henry Samuel – 3 May 1873 – aged 38 when on *Titanic*; second-class passenger (ticket price: £26); *Titanic* victim

The following people died on this day:
Bowen, Miss Grace Scott – 3 May 1945 – aged 45 when on *Titanic*; first-class passenger (ticket price: £262 6s); rescued in lifeboat 4

Davidson, Mrs Orian – 3 May 1979 – aged 27 when on *Titanic*; first-class passenger (ticket price: £52); rescued in lifeboat 3

Smith, Mrs Mary Eloise – 3 May 1940 – aged 18 when on *Titanic*; first-class passenger (ticket price: £60); rescued in lifeboat 6

O'Dwyer, Miss Ellen – 3 May 1917 – aged 24 when on *Titanic*; third-class passenger (ticket price: £7 17s 7d); rescued in lifeboat 10

Mrs Mary Eloise Smith (née Hughes) quoted in *Washington Times*, Friday, 19 April 1912

> 'Lucian is getting so anxious to get home and drive the car and fool around on the farm...We leave here Sunday....By boat to Brindisi [Italy], by rail to Nice and Monte Carlo, then to Paris and via Cherbourg either on the Lusitania or the new *Titanic*...I will love so much to tell my Sunday School class when I get home...' NEW YORK, 19 April – Not only did J. Bruce Ismay, managing director of the White Star Line, get into one of the first lifeboats to be launched, but he was escorted and assisted by several seamen, while the women had to tumble in and take care of themselves, according to Mrs Lucien P. Smith, daughter of Congressman Hughes of West Virginia, and widowed bride of Lucien P. Smith.

Mr Joseph Bruce Ismay, witness at United States Senate Inquiry (Day 11)

> 'Yes, sir. I sent the message on Monday morning, very shortly after I got on board the *Carpathia*. The captain came down to me and said, "Don't you think, sir, you had better send a message to New York, telling them about this accident?" I said, "Yes." I wrote it out on a slip of paper, and I turned to the commander of the *Carpathia* and I said, "Captain, do you think that is all I can tell them?" He said, "Yes." Then he took it away from the room.'

Titanic fact

On board, there were two barber's shops, two libraries and one gymnasium.

Day 125

4 May | On this day in *Titanic* history

The following person was born on this day:
Drew, Mr James Vivian – 4 May 1869 – aged 42 when on *Titanic*; second-class passenger (ticket price: £32 10s); *Titanic* victim

McLaren, Mrs Hypatia – 4 May 1869 – aged 42 when on *Titanic*; stewardess; rescued in lifeboat 5

Barlow, Mr George – 4 May 1872 – aged 39 when on *Titanic*; bedroom steward (victualling crew); *Titanic* victim

Lefebvre, Miss Mathilde – 4 May 1899 – aged 12 when on *Titanic*; third-class passenger (ticket price: £25 9s 4d); *Titanic* victim

The following person died on this day:
Cherry, Miss Gladys – 4 May 1965 – aged 30 when on *Titanic*; first-class passenger (ticket price: £86 10s); rescued in lifeboat 8

Titanic survivor quote

Miss Gladys Cherry, quoted in *The Henley Free Press*, 7 June 1912

> 'Thomas Jones, a native of Anglesey, who was an able seaman on the *Titanic*, has received the following letter, dated from the Great Northern Hotel, New York:
>
> "I feel I must write and tell you how splendidly you took charge of our boat on the fatal night. There were only four English people in it – my cousin Lady Rothes, her maid, you and myself – and I think you were wonderful.
>
> "The dreadful regret I shall always have, and I know you share with me, is that we ought to have gone back to see whom we could pick up; but if you remember, there was only an American lady, my cousin, self and you who wanted to return. I could not hear the discussion very clearly, as I was at the tiller; but everyone forward and the three men refused; but I shall always remember your words: 'ladies, if any of us are saved, remember, I wanted to go back. I would rather drown with them than leave them.' You did all you could, and being my own countryman, I wanted to tell you this.
>
> "Yours very truly, Gladys Cherry."

Titanic fact

The *Titanic*'s launch was delayed by six weeks because her sister ship *Olympic* needed repairs in the same dry dock. That delay put seasonal icebergs right in the *Titanic*'s path.

Day 126

5 May | On this day in *Titanic* history

The following people were born on this day:
Hogeboom, Mrs Anna Louisa – 5 May 1860 – aged 51 when on *Titanic*; first-class passenger (ticket price: £77 19s 2d); rescued in lifeboat 10

Haines, Mr Albert – 5 May 1880 – aged 31 when on *Titanic*; apprentice electrician; rescued in lifeboat 9

Watson, Mr Ennis Hastings – 5 May 1893 – aged 18 when on *Titanic*; boatswain mate; *Titanic* victim

The following person died on this day:
Jessop, Miss Violet Constance – 5 May 1971 – aged 24 when on *Titanic*; stewardess (victualling crew); rescued in lifeboat unknown

***Titanic* survivor quotes**

Miss Violet Constance Jessop

> 'I was ordered up on deck. Calmly, passengers strolled about. I stood at the bulkhead with the other stewardesses, watching the women cling to their husbands before being put into the boats with their children. Some time after, a ship's officer ordered us into the boat 16 first to show some women it was safe. As the boat was being lowered the officer called: "Here, Miss Jessop. Look after this baby." And a bundle was dropped on to my lap.
>
> 'I was still clutching the baby against my hard cork lifebelt I was wearing when a woman leaped at me and grabbed the baby, and rushed off with it. It appeared that she put it down on the deck of the *Titanic* while she went off to fetch something, and when she came back the baby had gone. I was too frozen and numb to think it strange that this woman had not stopped to say "thank you".
>
> 'I leapt into the water but was sucked under the ship's keel which struck my head. I escaped, but years later when I went to my doctor because of a lot of headaches, he discovered I had once sustained a fracture of the skull!'

Mr Joseph G. Boxhall, witness at United States Senate Inquiry (Day 3)

> 'Well, the *Titanic*'s lights seem to have disappeared some considerable time before I boarded the *Carpathia*, because I saw the *Carpathia*'s lights for some considerable time.'

Day 127

6 May | On this day in *Titanic* history

The following person was born on this day:
Fortune, Mrs Mary – 6 May 1851 – aged 60 when on *Titanic*; first-class passenger (ticket price: £263); rescued in lifeboat 10

Davies, Mr Gordon Raleigh – 6 May 1879 – aged 32 when on *Titanic*; bedroom steward in first class; *Titanic* victim

Jupe, Mr Boykett Herbert – 6 May 1881 – aged 30 when on *Titanic*; electrician; *Titanic* victim

Kilgannon, Mr Thomas – 6 May 1890 – aged 21 when on *Titanic*; third-class passenger (ticket price: £7 14s 9d); *Titanic* victim

The following people died on this day:
McGough, Mr George Francis 'Paddy' – 6 May 1940 – aged 36 when on *Titanic*; able seaman; rescued in lifeboat unknown

Asplund, Miss Lillian Gertrud – 6 May 2006 – aged 5 when on *Titanic*; third-class passenger (ticket price: £31 7s 9d); rescued in lifeboat 15

Titanic survivor quote

Mrs Mary Fortune, mentioned in *New York Times*, Saturday, 20 April 1912

> 'Mrs Mary Fortune and her three daughters were placed in the tenth boat that was lowered away. They said it was "terribly overcrowded." Four members of the crew who had been assigned to their boat, were transferred to another as soon as the craft struck the water.
>
> 'The women saw changes being made in the loading of several other boats, and finally saw that the collapsible craft were proving unseaworthy. Some began to leak as soon as they stood away from the side of the liner, Mrs Fortune asserted, and their occupants had to be taken into other boats. Mrs Fortune was of the opinion that no discrimination was made between passengers of the first, second, or third class in making allotments for seats in the boats. That rule applied only to the women, however. When the ship struck several men in the steerage tried to rush the officers in charge of the lifeboats, Mrs Fortune remembered. At first the officers were able to keep them off by slugging them, but the passengers grew more terrified. Then the officers made use of their revolvers, first to fire in the air, and then to aim at the bodies of the men.'

Titanic fact

Carpathia had to travel 56 miles to circumnavigate the ice-field to rescue survivors.

Day 128

7 May | On this day in *Titanic* history

The following person was born on this day:
Asplund, Mr Carl Oscar Vilhelm Gustafsson – 7 May 1871 – aged 40 when on *Titanic*; third-class passenger (ticket price: £31 7s 9d); *Titanic* victim

Gale, Mr Harry – 7 May 1874 – aged 37 when on *Titanic*; second-class passenger (ticket price: £21); *Titanic* victim

Pears, Mr Thomas Clinton – 7 May 1882 – aged 29 when on *Titanic*; first-class passenger (ticket price: £66 12s); *Titanic* victim

Cooper, Mr Harry – 7 May 1886 – aged 25 when on *Titanic*; fireman; *Titanic* victim

Pachera, Sig. Jean Baptiste Stanislas – 7 May 1892 – aged 19 when on *Titanic*; assistant larder cook (restaurant staff); *Titanic* victim

Allison, Master Hudson Trevor – 7 May 1911 – aged 11 months when on *Titanic*; first-class passenger (ticket price: £151 16s); rescued in lifeboat 11

The following people died on this day:
Lucas, Mr William Watson – 7 May 1944 – aged 31 when on *Titanic*; saloon steward in first class; rescued in collapsible lifeboat A

Nutbean, Mr William – 7 May 1947 – aged 30 when on *Titanic*; fireman; rescued in lifeboat (thought to be 3)

Mr Thomas Clinton Pears, mentioned in *Chicago Daily Journal*, Tuesday, 16 April 1912

> 'Sadly the news was incorrect and Thomas actually died in the *Titanic* disaster.
>
> A wireless dispatch received today by the firm of Pears, soap makers, and timed 1:20 yesterday, said merely "All well." It was unsigned but was believed to be from Thomas Pears, who with his wife was among the *Titanic*'s passengers. It came via the liner Potsdam, but there was no indication as to where it was originally sent from.
>
> It may have been dispatched before the *Titanic* sank, but, nevertheless, it gave welcome relief to the family, who believe that Mr Pears is aboard one of the steamers which reached the scene of the disaster yesterday.'

Titanic fact

The *New York Tribune* ran a headline on Sunday, 27 November 1910, asking this question, 'How can we dock this marine monster when she reaches the port of New York?'

Day 129

8 May | On this day in *Titanic* history

The following person was born on this day:
Finch, Mr Henry Herman – 8 May 1893– aged 18 when on *Titanic*; steward in third class; *Titanic* victim

Mr Joseph Bruce Ismay, chairman and managing director of White Star Line

> 'Capt. Smith was a man who had a very, very clear record. I should think very few commanders crossing the Atlantic have as good a record as Capt. Smith had, until he had the unfortunate collision with the *Hawke*.'

Arthur Rostron, captain of the *Carpathia*

> 'Devoutly thankful I was that the long race was over…but with that feeling was the veritable ache which the now certain knowledge of the liner's loss brought. No sign of her.'

Titanic survivor quote

Mrs Elizabeth Shutes (Liz)

> 'Suddenly a strange quivering ran under me, apparently the whole length of the ship. Startled by the very strangeness of the shivering motion, I sprang up. Margaret was pretending to eat a sandwich, her hands shook so much that the bread kept parting company with the chicken. I saw she was frightened and for the first time, I was too.'
>
> 'Then, and not until then, did I realise the horror of an accident at sea. We left from the sun deck, seventy-five feet above the water. Mr Case and Mr Roebling, brave American men, saw us to the lifeboat, made no effort to save themselves, but stepped back on deck. Later they went to an honoured grave … Rough seamen all giving different orders. No officer aboard. As only one side of the ropes worked, the lifeboat at one time was in such a position that it seemed we must capsize in mid-air.'

'The first touch of our lifeboat on that black sea came to me as a last goodbye to life, and so we put off – a tiny boat on a great sea – rowed away from what had been a safe home for five days. But surely the outline of that great, good ship was growing less. The bow of the boat was getting black. Light after light was disappearing.'

'The life preservers helped to keep us warm, but the night was bitter cold, and it grew colder and colder, and just before dawn, the coldest, darkest hour of all, no help seemed possible.'

'Across the water swept that awful wail, the cry of those drowning people. The stars slowly disappeared, and in their place came the faint pink glow of another day. Then I heard, "A light, a ship."'

Day 130

9 May | On this day in *Titanic* history

The following person was born on this day:
Julian, Mr Henry Forbes – 9 May 1861 – aged 50 when on *Titanic*; first-class passenger (ticket price: £26); *Titanic* victim

Bliss, Mrs Emma – 9 May 1866 – aged 45 when on *Titanic*; stewardess (victualling crew); rescued in lifeboat 15

King, Mr Thomas Walter – 9 May 1869 – aged 42 when on *Titanic*; master-at-arms; *Titanic* victim

Franklin, Mr Thomas Parnham – 9 May 1874 – aged 37 when on *Titanic*; first-class passenger (ticket price: £26 11s); *Titanic* victim

Ball, Mrs Ada E. – 9 May 1875 – aged 36 when on *Titanic*; second-class passenger (ticket price: £13); rescued in lifeboat 10

Foley, Mr Wilfred Cyril – 9 May 1890 – aged 21 when on *Titanic*; steward in third class; rescued in lifeboat 13

The following person died on this day:
Hewlett, Mrs Mary Dunbar – 9 May 1917 – aged 56 when on *Titanic*; second-class passenger (ticket price: £16); rescued in lifeboat 13

Titanic survivor quote

Mrs Mary Dunbar Hewlett

> 'I got on my under clothes & my fur lined coat & a hat & went on deck & there an officer was ordering people to climb up a ladder on to the upper deck so I went with the crowd & when I reached the top there were eight or ten stewards there who said I must get into a boat that was on the davits – however they insisted & I was put into boat No. 13 – with about 50 people – mostly men of the unemployed class & stokers, stewards & cooks – not one real seaman amongst them there were only about 12 women in the boat & no compass, lanterns or water.'

Mr Daniel Buckley

> 'One of the firemen that was working on the *Titanic* told me, when I got on board the *Carpathia* and he was speaking to me, that he did not think it was any iceberg; that it was only that they wanted to make a record, and they ran too much steam and the boilers bursted. That is what he said.'

Day 131

10 May | On this day in *Titanic* history

The following person was born on this day:

Howard, Mr Benjamin – 10 May 1848 – aged 63 when on *Titanic*; second-class passenger (ticket price: £26); *Titanic* victim

Warren, Mr Frank Manley – 10 May 1848 – aged 63 when on *Titanic*; first-class passenger (ticket price: £75 5s); *Titanic* victim

Leader, Dr Alice – 10 May 1862 – aged 49 when on *Titanic*; first-class passenger (ticket price: £25 18s 7d); rescued in lifeboat 8

Bidois, Miss Rosalie – 10 May 1865 – aged 46 when on *Titanic*; personal maid to Mrs Madeleine Talmage Astor who was a first-class passenger (ticket price: £247 10s 6d); rescued in lifeboat 4

Cardeza, Mr Thomas Drake Martinez – 10 May 1875 – aged 36 when on *Titanic*; first-class passenger (ticket price: £512 6s 7d); rescued in lifeboat 3

Nichols, Mr Walter Henry – 10 May 1876 – aged 35 when on *Titanic*; assistant saloon steward in second class; rescued in lifeboat 15

Shaughnessy, Mr Patrick – 10 May 1886 – aged 25 when on *Titanic*; third-class passenger (ticket price: £7 15s); *Titanic* victim

Fortune, Miss Alice Elizabeth – 10 May 1887 – aged 24 when on *Titanic*; first-class passenger (ticket price: £263); rescued in lifeboat 10

Peracchio, Sig. Sebastiano – 10 May 1894 – aged 17 when on *Titanic*; assistant waiter (restaurant staff); *Titanic* victim

The following person died on this day:

Baclini, Mrs Latifa – 10 May 1962 – aged 23 when on *Titanic*; third-class passenger (ticket price: £19 5s 2d); rescued in collapsible lifeboat C

Titanic survivor quote

Mr Thomas Cardeza, quoted in *New York Sun*, Saturday, 27 April 1912

> 'The officer called for other women, but there were none thereabout. Then he called for men passengers. There were only about six just there, of whom I was one, and we got in. The boat was still not filled, so the officer put in some of the crew. There were in the boat, as near as I could judge, between forty-three and forty-five persons.'

Titanic fact

Most third-class cabins contained four to six bunks.

Day 132

11 May | On this day in *Titanic* history

The following people were born on this day:
Flegenheim, Mrs Antoinette – 11 May 1863 – aged 48 when on *Titanic*; first-class passenger (ticket price: £31 13s 8d); rescued in lifeboat 7

Abbing, Mr Anthony – 11 May 1870 – aged 41 when on *Titanic*; third-class passenger (ticket price: £7 11s); *Titanic* victim

Giles, Mr John Robert – 11 May 1879 – aged 32 when on *Titanic*; second baker; *Titanic* victim

Biddlecombe, Mr Reginald Charles – 11 May 1880 – aged 31 when on *Titanic*; fireman; *Titanic* victim

Murphy, Miss Nora – 11 May 1880 – aged 31 when on *Titanic*; third-class passenger (ticket price: £15 10s); rescued in lifeboat 16

Svensson, Mr Olof – 11 May 1887 – aged 24 when on *Titanic*; third-class passenger (ticket price: £7 15s 11d); *Titanic* victim

Greenfield, Mr William Bertram – 11 May 1888 – aged 23 when on *Titanic*; first-class passenger (ticket price: £63 7s 2d); rescued in lifeboat 7

Yūsuf, Master Michael (Shafīq Peter) – 11 May 1907 – aged 4 when on *Titanic*; third-class passenger (ticket price: £22 7s 2d); rescued in collapsible lifeboat D

Dumfries and Galloway Standard and Advertiser, Saturday, 11 May 1912

> 'Dear Mrs Murdoch, – I am writing on behalf of the surviving officers to express our deep sympathy in this, your awful loss. Words cannot convey our feelings – much less a letter… You will better understand when I say that I was working the port side of the ship, and Mr Murdoch was principally engaged on the starboard side of the ship, filling and launching the boats. Having got my boat down off the top of the house, and there being no time to open it, I left it and ran across to the starboard side, still on top of the quarters. … Mr Murdoch died like a man, doing his duty. Call on us without hesitation for anything we can do for you. Yours very sincerely, Signed C. H. Lightoller, 2nd Officer, G Groves, Boxhall, 4th Officer, H. J. Pitman, 3rd Officer, H. G. Lowe, 5th Officer.'

Titanic fact

Second Officer Charles Lightoller and passenger Jack Thayer both wrote books describing their survival experiences aboard the *Titanic*.

Day 133

12 May | On this day in *Titanic* history

The following people were born on this day:
Jonsson, Mr Karl – 12 May 1886 – aged 25 when on *Titanic*; third-class passenger (ticket price: £7 17s 1d); rescued in lifeboat unknown

Krekorian, Mr Neshan – 12 May 1886 – aged 25 when on *Titanic*; third-class passenger (ticket price: £7 4s 7d); rescued in lifeboat 10

Gustafsson, Mr Alfred Ossian – 12 May 1892 – aged 19 when on *Titanic*; third-class passenger (ticket price: £9 16s 11d); *Titanic* victim

The following people died on this day:
Stone, Mrs Martha Evelyn – 12 May 1924 – aged 61 when on *Titanic*; first-class passenger (ticket price: £80); rescued in lifeboat 6

Lowe, Mr Harold Godfrey – 12 May 1944 – aged 23 when on *Titanic*; fifth officer; rescued in lifeboat 14

Titanic survivor quote

Mr Harold Godfrey Lowe
When asked by Senator Smith what an iceberg is composed of, the reply from Fifth Officer Lowe was, 'Ice, I suppose, sir.'

Miss Caroline Bonnell

> 'The *Titanic* was fading in the distance, but her lights were quite visible. About twenty minutes after we were put in the boat we noticed that the giant ship was sinking low in the water. Then we realised for the first time that it was in danger, and our lark turned into a frightened party of women. Lower and lower sank the *Titanic*. The faint strains of a band came to us. Then all of a sudden the lower lights seemed to go out. Only the lights on the upper deck were visible. And then we saw the ship sink, this great unsinkable liner. It didn't plunge, as far as we could see, but seemed to settle lower and lower into the water and went down gently, grandly, to its grave. Then the full horror of the thing came over us. We were frightened. But the men in the boat tried to reassure us. They told us that those left behind on the boat would surely leave it, that they would be picked up in a short time.'

Titanic fact

When the *Titanic* first hit the iceberg, large chunks of ice landed on the forward deck and, being carefree, passengers tossed the ice around in a spontaneous game of ice football. They were oblivious to the impending disaster.

Day 134

13 May | On this day in *Titanic* history

The following people were born on this day:

Van Der Hoef, Mr Wyckoff – 13 May 1850 – aged 61 when on *Titanic*; first-class passenger (ticket price: £33 10s); *Titanic* victim

Noel, Major Gerard Thomas – 13 May 1856 – aged 55 when on *Titanic*; first-class passenger (ticket price: £3); disembarked at Cherbourg

Hart, Mrs Esther Ada – 13 May 1863 – aged 48 when on *Titanic*; second-class passenger (ticket price: £26 5s); rescued in lifeboat 14

Porter, Mr Walter Chamberlain – 13 May 1865 – aged 46 when on *Titanic*; first-class passenger (ticket price: £52); *Titanic* victim

Olsson, Mr Nils Johan Göransson – 13 May 1883 – aged 28 when on *Titanic*; third-class passenger (ticket price: £7 17s 1d); *Titanic* victim

Clark, Mr Walter Miller – 13 May 1884 – aged 27 when on *Titanic*; first-class passenger (ticket price: £136 15s 7d); *Titanic* victim

The following person died on this day:

Podesta, Mr Alfred John Alexander – 13 May 1968 – aged 24 when on *Titanic*; fireman; rescued in lifeboat 3

Mr Joseph Bruce Ismay, witness at United States Senate Inquiry (Day 11)

> 'We have given instructions that no ship belonging to the I.M.M. Co. is to leave any port unless she has sufficient boats on board for the accommodation of all the passengers and the whole of the crew.'

Titanic survivor quote

Mrs Ben Hart, quoted in *Ilford Graphic*, Friday, 10 May 1912

> 'And now about 8 o'clock the *Carpathia* came into sight and we were all aboard by 8.30. I cannot say much of my life on board this vessel. It was no matter for a ship to take on another 700 people, many of them but lightly clad, most of them ill, and all suffering severely from shock; all was done for us that could be done: but I could neither rest nor sleep. My little Eva was still suffering from her vomiting attack and I found my hands full in nursing her; but when at night she was asleep, I could do nothing but walk the corridor, up and down, up and down, and thinking, thinking all the time. So much did I walk about at night that the kind hearted sailors christened me The Lady of the Watch.'

Day 135

14 May | On this day in *Titanic* history

The following people were born on this day:

Weir, Colonel John – 14 May 1852 – aged 61 when on *Titanic*; first-class passenger (ticket price: £26 11s); *Titanic* victim

Rostron, Captain Arthur Henry – 14 May 1869 – Captain of RMS *Carpathia* (rescue ship)

Akerman, Mr Joseph Francis – 14 May 1874 – aged 37 when on *Titanic*; assistant pantryman steward; *Titanic* victim

Johnston, Mr Andrew Emslie – 14 May 1876 – aged 35 when on *Titanic*; third-class passenger (ticket price: £23 9s); *Titanic* victim

Waelens, Mr Achille – 14 May 1889 – aged 22 when on *Titanic*; third-class passenger (ticket price: £9); *Titanic* victim

The following person died on this day:

Jonsson, Mr Karl – 14 May 1956 – aged 25 when on *Titanic*; third-class passenger (ticket price: £7 17s 1d); rescued in lifeboat unknown

Mr Colonel John Weir, mentioned in *San Francisco Bulletin*, Saturday, 20 April 1912

> 'Colonel John Weir, a well-known mining man of the Pacific Coast, was one of the victims of the *Titanic* horror. It has been learned by his friends that he took passage on the ill-fated steamer on an unexpected trip to California, probably on urgent business connected with mining interests in this State. The *Titanic* passenger list shows the name of J. Weir, and he was not among the rescued.
>
> Colonel Weir was very well known in San Francisco, where he lived many years. He was president of the Utah and Nevada Mining and Smelting Company at one time.'

Mr Karl Jonsson, mentioned in *Chicago Daily Tribune*, Friday, 26 April 1912

> 'Two men and a girl from Sweden, Oscar Hedmann, Karl Johnson, and Anna Sjöblom, spent several hours in Chicago yesterday on their way west. All found themselves in New York without a cent. They were taken to an immigration home and clothed and fed. Finally Hedmann wired to a former employer in St. Paul, who sent some money, which Hedmann divided so that the three could get west.'

Day 136

15 May | On this day in *Titanic* history

The following person was born on this day:

Tannūs, Mr Bashīr – 15 May 1881 – aged 30 when on *Titanic*; third-class passenger (ticket price: £6 8s 9d); *Titanic* victim

Ridout, Mr Walter George – 15 May 1882 – aged 29 when on *Titanic*; saloon steward in second class; *Titanic* victim

Niklasson, Mr Samuel – 15 May 1883 – aged 28 when on *Titanic*; third-class passenger (ticket price: £8 1s); *Titanic* victim

Pedersen, Mr Olaf – 15 May 1883 – aged 28 when on *Titanic*; third-class passenger (ticket price: £7 15s 6d); *Titanic* victim

Evans, Mr Frank Oliver – 15 May 1884 – aged 27 when on *Titanic*; able seaman; rescued in lifeboat 10

The following people died on this day:

Lemore, Mrs Amelia – 15 May 1950 – aged 39 when on *Titanic*; second-class passenger (ticket price: £10 10s); rescued in lifeboat 14

Mannion, Miss Margaret – 15 May 1970 – aged 28 when on *Titanic*; third-class passenger (ticket price: £7 14s 9d); rescued in lifeboat 16

Østby, Miss Helen Ragnhild – 15 May 1978 – aged 22 when on *Titanic*; first-class passenger (ticket price: £61 19s 7d); rescued in lifeboat 5

Titanic survivor quote

Miss Helen Ragnhild Østby

> 'There had been no panic. But at the very end, we could see and hear that the people on board were realizing there was no place to go. As the ship began to stand on end, we heard a big rumbling, rattling noise as if everything was being torn from its moorings inside the ship. All of a sudden that stopped, and she stood on end very quietly for a minute, then went down like an arrow.'

Mr Joseph G. Boxhall (fourth officer), witness at British Wreck Commissioner's Inquiry (Day 13)

> 'I had not enough people; my boat was rather deep. I had only one man who seemed to understand boat orders. I was pulling the stroke oar and trying to steer the boat at the same time myself.'

Titanic fact

Titanic survivor Lawrence Beesley, known for his book *The Loss of the Titanic*, had a son, Alec, who married Dodie Smith, author of *The Hundred and One Dalmatians.*

Day 137

16 May | On this day in *Titanic* history

The following person was born on this day:
Hays, Mr Charles Melville – 16 May 1856 – aged 55 when on *Titanic*; first-class passenger (ticket price: £93 10s); *Titanic* victim

Simmons, Mr Frederick Charles – 16 May 1887 – aged 24 when on *Titanic*; saloon steward; *Titanic* victim

Mr Charles Melville Hays, mentioned in *Chicago Tribune*, Tuesday, 7 May 1912

> 'Only one of the seventeen persons whose bodies were recovered by the cable ship *Minia* in the vicinity of the *Titanic* tragedy died from drowning, in the opinion of the cable ship's physician. The other sixteen perished from exposure. Among the bodies preserved was that of Charles M. Hays, president of the Grand Trunk railway. It was shipped to his home at Montreal today on a special train.'

Titanic survivor quote

Mr Olaus Abelseth, witness at United States Senate Inquiry (Day 13)

> 'My brother-in-law said to me, "We had better jump off or the suction will take us down." I said, "No. We won't jump yet. We ain't got much show anyhow, so we might as well stay as long as we can." So he stated again, "We must jump off." But I said, "No; not yet." So, then, it was only about 5 feet down to the water when we jumped off. It was not much of a jump. Before that we could see the people were jumping over. There was water coming onto the deck, and they were jumping over, then, out in the water.'
>
> 'My brother-in-law took my hand just as we jumped off; and my cousin jumped at the same time. When we came into the water, I think it was from the suction – or anyway we went under, and I swallowed some water. I got a rope tangled around me, and I let loose of my brother-in-law's hand to get away from the rope. I thought then, "I am a goner." That is what I thought when I got tangled up in this rope.'
>
> 'But I came on top again, and I was trying to swim, and there was a man – lots of them were floating around – and he got me on the neck like that (illustrating) and pressed me under, trying to get on top of me. I said to him, "Let go." Of course, he did not pay any attention to that, but I got away from him. Then there was another man, and he hung on to me for a while, but he let go. Then I swam; I could not say, but it must have been about 15 or 20 minutes. It could not have been over that. Then I saw something dark ahead of me. I did not know what it was, but I swam toward that, and it was one of those collapsible boats.'

Day 138

17 May | On this day in *Titanic* history

The following people were born on this day:

Svensson, Mr Johan – 17 May 1837 – aged 74 when on *Titanic*; third-class passenger (ticket price: £7 15s 6d); *Titanic* victim

Davidson, Mr Thornton – 17 May 1880 – aged 31 when on *Titanic*; first-class passenger (ticket price: £52); *Titanic* victim

Ide, Mr Harry John – 17 May 1880 – aged 31 when on *Titanic*; bedroom steward in first class; *Titanic* victim

Gibson, Miss Dorothy Winifred – 17 May 1889 – aged 22 when on *Titanic*; first-class passenger (ticket price: £59 8s); rescued in lifeboat 7

Jansson, Mr Carl Olof – 17 May 1890 – aged 21 when on *Titanic*; third-class passenger (ticket price: £7 15s 11d); rescued in collapsible lifeboat A

The following people died on this day:

Eustis, Miss Elizabeth Mussey – 17 May 1936 – aged 54 when on *Titanic*; first-class passenger (ticket price: £78 5s 4d); rescued in lifeboat 4

Hirvonen, Mrs Helga Elisabeth Lindqvist – 17 May 1961 – aged 22 when on *Titanic*; third-class passenger (ticket price: £12 5s 9d); rescued in lifeboat 15

Silverthorne, Mr Spencer Victor – 17 May 1964 – aged 35 when on *Titanic*; first-class passenger (ticket price: £26 5s 9d); rescued in lifeboat 5

Titanic survivor quotes

Miss Dorothy Winifred Gibson, quoted in *New York Dramatic Mirror*, Wednesday, 1 May 1912

> 'A pause of silence held everything and everybody spellbound, until the stern shot back into sight and immediately sank again. Then, there burst out the most ghastly cries, shrieks, yells and moans that a mortal could ever imagine. No one can describe the frightful sounds, that gradually died away to nothing.'

Mr Arthur G. Peuchen, witness at United States Senate Inquiry (Day 4)

> 'The quartermaster (Hichens) who was in charge of our boat told us to row as hard as we could to get away from this suction, and just as we got a short distance away this stowaway made his appearance. He was an Italian by birth, I should think, who had a broken wrist or arm, and he was of no use to us to row. He got an oar out, but he could not do much, so we got him to take the oar in.'

Mr Samuel Hemming, witness at United States Senate Inquiry (Day 7)

> 'The captain was there, and he sung out: "Everyone over to the starboard side, to keep the ship up as long as possible."'

Day 139

18 May | On this day in *Titanic* history

The following people were born on this day:
Crisp, Mr Albert Hector – 18 May 1872 – aged 39 when on *Titanic*; saloon steward; *Titanic* victim

Dean, Mrs Eva Georgetta – 18 May 1879 – aged 32 when on *Titanic*; third-class passenger (ticket price: £20 11s 6d); rescued in lifeboat 10

Butt, Mr William John – 18 May 1881 – aged 30 when on *Titanic*; fireman; *Titanic* victim

Davis, Miss Mary – 18 May 1883 – aged 28 when on *Titanic*; second-class passenger (ticket price: £13); rescued in lifeboat unknown

al-Ba'qlīnī, Mrs Laṭīfah – 18 May 1888 – aged 23 when on *Titanic*; third-class passenger (ticket price: £19 5s 2d); rescued in collapsible lifeboat C

O'Sullivan, Miss Bridget Mary – 18 May 1890 – aged 21 when on *Titanic*; third-class passenger (ticket price: £7 12s 7d); *Titanic* victim

Vine, Mr Herbert Thomas Gordon – 18 May 1894 – aged 17 when on *Titanic*; assistant restaurant controller; *Titanic* victim

Panula, Mr Ernesti Arvid – 18 May 1895 – aged 16 when on *Titanic*; third-class passenger (ticket price: £39 13s 9d); *Titanic* victim

The following people died on this day:
O'Leary, Miss Hanora 'Nora' – 18 May 1975 – aged 16 when on *Titanic*; third-class passenger (ticket price: £7 16s 7d); rescued in lifeboat 13

Yūsuf, Master Michael (Shafīq Peter) – 18 May 1991 – aged 4 when on *Titanic*; third-class passenger (ticket price: £22 7s 2d); rescued in collapsible lifeboat D

Titanic survivor quotes

Miss Mary Davis, mentioned in *Evening Star* (Washington DC) Monday, 22 April 1912

> 'She said that after she was placed in a lifeboat she was shoved overboard, but was quickly rescued by one of the sailors who manned the boat. She said she was unconscious when taken onboard the *Carpathia* and remained in that condition for many hours.'

Mr Arthur G. Peuchen, witness at United States Senate Inquiry (Day 4)

> 'While the lights were burning, I saw her bow pointing down and the stern up; not in a perpendicular position, but considerable.'

Day 140

19 May | On this day in *Titanic* history

The following people were born on this day:
Lee, Mr Reginald Robinson – 19 May 1870 – aged 41 when on *Titanic*; lookout; rescued in lifeboat 13

Bedford, Mr William Barnett – 19 May 1880 – aged 31 when on *Titanic*; assistant roast cook (victualling crew); *Titanic* victim

Bishop, Mrs Helen – 19 May 1892 – aged 19 when on *Titanic*; first-class passenger (ticket price: £91 1s 7d); rescued in lifeboat 7

Wirz, Mr Albert – 19 May 1894 – aged 17 when on *Titanic*; third-class passenger (ticket price: £8 13s 3d); *Titanic* victim

The following people died on this day:
Evans, Mr Frank Oliver – 19 May 1952 – aged 27 when on *Titanic*; able seaman; rescued in lifeboat 10

Prentice, Mr Frank Winnold – 19 May 1982 – aged 23 when on *Titanic*; storekeeper; rescued in lifeboat 4

Mr Frank Winnold Prentice, mentioned in *Chicago Daily News*, Friday, 19 April 1912

> 'A passenger who was picked up in a drowning condition caused grim amusement on the *Carpathia* by demanding a bath as soon as the doctors were through with him. Storekeeper Prentice, the last man off the *Titanic* to reach the *Carpathia*, swam about in the icy water for hours, but was restored. He said he had leaped from the *Titanic*'s poop-deck.'

Titanic survivor quotes

Mrs J. Stuart White, witness at United States Senate Inquiry (Day 12)

> 'We simply rowed away. We had the order, on leaving the ship, to do that. The officer who put us in the boat – I do not know who he was – gave strict orders to the seamen, or the men, to make for the light opposite and land the passengers and get back just as soon as possible. That was the light that everybody saw in the distance.'

Mr John Collins (assistant cook in first-class galley), witness at United States Senate Inquiry (Day 7)

> 'Her bow was in the water. She exploded in the water. She exploded once in the water, and her stern end was up out of the water; and with the explosion out of the water it blew her stern up.'

Day 141

20 May | On this day in *Titanic* history

The following people were born on this day:
Moor, Mrs Beila – 20 May 1882 – aged 29 when on *Titanic*; third-class passenger (ticket price: £12 9s 6d); rescued in lifeboat 14

Bristow, Mr Robert Charles – 20 May 1873 – aged 38 when on *Titanic*; steward in third class; *Titanic* victim

Aubart, Mme Léontine Pauline – 20 May 1887 – aged 24 when on *Titanic*; first-class passenger (ticket price: £69 6s); rescued in lifeboat 9

The following person died on this day:
Taylor, Mr Elmer Zebley – 20 May 1949 – aged 29 when on *Titanic*; third-class passenger (ticket price: £12 9s 6d); rescued in lifeboat 14

Titanic survivor quotes

Mr Elmer Zebley Taylor, quoted in *Atlantic City Daily Press*, Saturday, 20 April 1912

> 'We were about a mile away when the *Titanic* went down – at least that was what the officer who was in charge of our boat said, but it didn't seem that far. We could see her plainly. She went down very slowly and there was no explosion at all. There was a horrible noise, but I couldn't quite make out what it was. She went down bow first, and at the last moment broke or buckled… As I say, the scenes on leaving the ship were not very exciting, because no one expected her to sink. I didn't hear any shots or anything of that sort. We were in the boat about four hours before the *Carpathia* sighted us.'

Mr Charles H. Lightoller, witness at United States Senate Inquiry (Day 5)

> 'On the port side on deck, I can say, as far as my own observations went, from my own endeavour and that of others to obtain women, there were none. I can give you the name of a man who will give testimony, who was working with me, one of our best men, a man I picked out especially to man the falls for lowering away. He went from the port side to the starboard side of the deck, as I did, and after that, when she went under water forward, instead of taking to the water he walked aft the whole length of the boat deck previous to sliding down the aft fall on the port side, and in the whole length of the deck and in crossing the bridge he saw two women. They were standing amidships on the bridge perfectly still. They did not seem to be endeavouring to get to one side or the other to see if there were any boats or not. The whole length of the boat deck, so far as he went, he did not see any women.'

Day 142

21 May | On this day in *Titanic* history

The following people were born on this day:
Porteus, Mr Thomas Henry – 21 May 1878 – aged 33 when on *Titanic*; assistant butcher (victualling crew); *Titanic* victim

Mäkinen, Mr Kalle Edvard – 21 May 1882 – aged 29 when on *Titanic*; third-class passenger (ticket price: £7 18s 6d); *Titanic* victim

Dean, Master Bertram Vere – 21 May 1910 – aged 1 when on *Titanic*; third-class passenger (ticket price: £20 11s 6d); rescued in lifeboat 10

The following people died on this day:
Savage, Mr Charles Joseph – 21 May 1957 – aged 23 when on *Titanic*; steward; rescued in lifeboat 11

Salomon, Mr Abraham Lincoln – 21 May 1959 – aged 29 when on *Titanic*; first-class passenger (ticket price: £26); rescued in lifeboat 1

Björnström-Steffansson, Mr Mauritz Hokan – 21 May 1962 – aged 28 when on *Titanic*; first-class passenger (ticket price: £26 11s); rescued in collapsible lifeboat D

Krekorian, Mr Neshan – 21 May 1978 – aged 25 when on *Titanic*; third-class passenger (ticket price: £7 4s 7d); rescued in lifeboat 10

Sincock, Miss Maude – 21 May 1984 – aged 20 when on *Titanic*; second-class passenger (ticket price: £36 15s); rescued in lifeboat 11

Titanic survivor quotes

Miss Maude Sincock, quoted in *New York Herald*, Saturday, 20 April 1912

> 'I was put into the last boat that left the *Titanic*,' she said. 'I should say there were fifty women in the boat and two men. One of the men was a sailor, the other a man whose sister was in the boat. I don't know why they let him in. Oh yes, I know too – he could row – rowed well. He was a young man, smooth shaven...Yes, I am sure it was the last boat to go...'

Mr Joseph Bruce Ismay, chairman and managing director of White Star Line

> 'Mr Andrews dined with me one night. We had no conversation, really, in regard to the ship. Indeed, the only plan which Mr Andrews submitted to me was a plan where he said he thought the writing room and reading room was unnecessarily large, and he said he saw a way of putting a stateroom in the forward end of it. That was a matter which would have been taken up and thoroughly discussed after we got back to England.'

Day 143

22 May | On this day in *Titanic* history

The following people were born on this day:
Goodwin, Mr Charles Edward – 22 May 1897 – aged 14 when on *Titanic*; third-class passenger (ticket price: £46 18s); *Titanic* victim

Hansen, Mr Henry Damsgaard – 22 May 1891 – aged 20 when on *Titanic*; third-class passenger (ticket price: £7 17s 1d); *Titanic* victim

Mr Charles Edward Goodwin, mentioned in *Bath Journal*, Saturday, 20 April 1912

> 'Frederick Goodwin and his wife and six children were on board. They had been living in Melksham for some years. Goodwin was a machine hand at the foundry, and about 40 years of age. The eldest of the children, a girl, was 16 or 17 years of age, and the youngest was a baby of a few months. The Goodwins have relatives in the States, some of whom had recently been on a visit to Melksham, and returned not long ago. Presumably they induced Goodwin to leave the old country; he was going to a situation near the Falls, already found for him, and his relatives had provided a house and furniture for the family. The supplementary list of survivors issued yesterday contained no Goodwins, so that the whole family would seem to have perished.'

Mr Harold Cottam (Marconi operator of the *Carpathia*)

> 'About 1 o'clock in the morning, after I've taken the press or listened to the press I took a batch of messages for her and with the intention of redirecting them on to her I called her up and the only reply I got that she'd struck ice, I said was it serious? and she said "yes this is a CQD old man, here's the position, report it and get here as soon as you can." So I took the position and on the scrap of paper and rushed up to the bridge with it. When I got on the bridge I contacted the officer of the watch and the information didn't seem as though it had sunk in as fast as I thought it ought to so I rushed down off the ladder – and knocked on the captain's cabin and as I saw a light I rushed in and he said "Who the hell" or words to that effect and so I said "Well, the *Titanic* struck ice sir, and she's in distress, I've got the position here." So he said "Well, give it me" and he put a dressing gown on and went so he said "Will you confirm this? Go aft and confirm it if you can."'

Titanic fact

Of the twelve dogs on board the *Titanic*, only three survived its sinking: a Pekingese and two Pomeranians.

Day 144

23 May | On this day in *Titanic* history

The following people were born on this day:
Bourke, Mr John – 23 May 1869 – aged 42 when on *Titanic*; third-class passenger (ticket price: £15 10s); *Titanic* victim

Otter, Mr Richard – 23 May 1873 – aged 38 when on *Titanic*; second-class passenger (ticket price: £13); *Titanic* victim

Schabert, Mrs Emma – 23 May 1876 – aged 35 when on *Titanic*; first-class passenger (ticket price: £57 15s); rescued in lifeboat 11

The following person died on this day:
Bowker, Miss Ruth Harwood – 23 May 1956 – aged 31 when on *Titanic*; cashier; rescued in lifeboat 6

Titanic survivor quotes

Mr Frederick Clench (able seaman), witness at United States Senate Inquiry (Day 7)

> 'No. 4 was my boat. We were sent there to uncover the boat, and an officer came along and drafted me on the other side, the port side. I went to No. 16 on the port side – the after boat, and started getting out the boat falls to let them down; I got out the two falls and coiled them down on the deck. When I was putting the plug in the boat in readiness to be lowered they were swinging the boat out.'

Mrs Emma Schabert, mentioned in *Worcester Telegram*, Saturday, 20 April 1912

> 'New York, 19 April – One version of the deaths of John Jacob Astor and William T. Stead was told by Philip Mock, who, with his sister, Mrs Paul Schabert, were among the survivors.
>
> "Many men were hanging on to rafts in the sea," said Mr Mock. "William T. Stead, the author, and Col. John Jacob Astor clung to a raft. Their feet became frozen and they were compelled to release their hold. Both were drowned."
>
> 'According to a surgeon of the New York hospital who went aboard the *Carpathia* after she docked, four bodies were buried at sea from that steamer. The chief steward of the *Carpathia* explained the large number of the crew saved by saying that the majority of them jumped from the *Titanic* and were picked up by the boats.'

Titanic fact

Third-class passengers could hear the loud roar of the ship's engines in their cabins at all times.

Day 145

24 May | On this day in *Titanic* history

The following people were born on this day:
Weisz, Mrs Mathilde Françoise – 24 May 1874 – aged 37 when on *Titanic*; second-class passenger (ticket price: £26); rescued in lifeboat 10

Bassani, Mrs Albina – 24 May 1880 – aged 31 when on *Titanic*; first-class passenger (ticket price: £76 5s 10d); rescued in lifeboat 8

Lewis, Mr Arthur Ernest Read – 24 May 1884 – aged 27 when on *Titanic*; steward; rescued in lifeboat 15

O'Connell, Mr Patrick Denis – 24 May 1893 – aged 18 when on *Titanic*; third-class passenger (ticket price: £7 14s 8d); *Titanic* victim

West, Miss Barbara Joyce – 24 May 1911 – aged 10 months when on *Titanic*; second-class passenger (ticket price: £27 15s); rescued in lifeboat 11

The following people died on this day:
Shelley, Mrs Imanita Parrish – 24 May 1954 – aged 24 when on *Titanic*; second-class passenger (ticket price: £26); rescued in lifeboat 12

Sägesser, Mlle Emma – 24 May 1964 – aged 24 when on *Titanic*; first-class passenger (ticket price: £69 6s); rescued in lifeboat 9

Titanic survivor quotes

Mr Alfred Olliver (quartermaster), witness at United States Senate Inquiry (Day 7)

> 'When I was doing this bit of duty I heard three bells rung up in the crow's nest, which I knew that it was something ahead; so I looked, but I did not see anything. I happened to be looking at the lights in the standing compass at the time. That was my duty, to look at the lights in the standing compass, and I was trimming them so that they would burn properly. When I heard the report, I looked, but could not see anything, and I left that and came and was just entering on the bridge just as the shock came. I knew we had touched something.'

Mrs Imanita Shelley, witness at United States Senate Inquiry, Wednesday, 15 May 1912

> 'After coming loose from the ship the orders were to pull out toward the other boats and get as far away from the probable suction which would ensue if the steamer should sink. Orders were also given to keep in sight of the green light of the ship's boat which had been sent out ahead to look for help. That on reaching a distance of about 100 yards from the *Titanic* a loud explosion or noise was heard, followed closely by another, and the sinking of the big vessel began.'

Day 146

25 May | On this day in *Titanic* history

The following person was born on this day:
Mishellany, Mr Abraham Mansoor – 25 May 1858 – aged 53 when on *Titanic*; printer steward; *Titanic* victim

The following people died on this day:
Picard, Mr Benoît – 25 May 1941 – aged 33 when on *Titanic*; third-class passenger (ticket price: £8 1s); rescued in lifeboat 9

Daniels, Mr Sidney Edward – 25 May 1983 – aged 18 when on *Titanic*; steward in third class; rescued in lifeboat unknown

Mr Sidney Edward Daniels, mentioned in *Daily Sketch*, Saturday, 4 September 1915

> 'The only man who had not been to the front – and he was very angry that he had not – was an Army Service Corps man, who had driven up a Lewisham contingent in a motor-omnibus. He, to make up for having no fighting, was a survivor from the *Titanic* (Sid Daniels). He described the scene wonderfully well and ended his story with the graphic touch that when he got home again he "couldn't help but laugh to think of all he had been through."

Titanic survivor quote

Master Marshall Brines Drew

> 'As row by row of the porthole lights of the *Titanic* sank into the sea this was about all one could see. When the *Titanic* upended to sink, all was blacked out until the tons of machinery crashed to the bow... As this happened hundreds and hundreds of people were thrown into the sea. It isn't likely I shall ever forget the screams of these people as they perished in water said to be 28 degrees...At this point in my life I was being brought up as a typical British kid. You were not allowed to cry. You were a "little man". So as a cool kid I lay down in the bottom of the lifeboat and went to sleep. When I awoke it was broad daylight as we approached the *Carpathia*. Looking around over the gunwale it seemed to me like the Arctic. Icebergs of huge size ringed the horizon for 360 degrees.'

Titanic fact

On the night of the disaster, the temperature was 31F (-6°C).

Day 147

26 May | On this day in *Titanic* history

The following people were born on this day:

Futrelle, Mrs Lily May – 26 May 1876 – aged 35 when on *Titanic*; first-class passenger (ticket price: £53 2s); rescued in collapsible lifeboat D

Meyer, Mr August – 26 May 1880 – aged 31 when on *Titanic*; second-class passenger (ticket price: £13); *Titanic* victim

Wright, Miss Marion – 26 May 1885 – aged 26 when on *Titanic*; second-class passenger (ticket price: £13 10s); rescued in lifeboat 9

Padron Manent, Mr Julian – 26 May 1885 – aged 26 when on *Titanic*; second-class passenger (ticket price: £13 17s 3d); rescued in lifeboat 9

Laroche, Mr Joseph Philippe Lemercier – 26 May 1886 – aged 25 when on *Titanic*; second-class passenger (ticket price: £41 11s 7d); *Titanic* victim

The following people died on this day:

Simonius-Blumer, Mr Colonel (Oberst Alfons) – 26 May 1920 – aged 56 when on *Titanic*; first-class passenger (ticket price: £35 10s); rescued in lifeboat 3

Harder, Mr George Achilles – 26 May 1959 – aged 25 when on *Titanic*; first-class passenger (ticket price: £55 8s 10d); rescued in lifeboat 5

Titanic survivor quotes

Mrs Lily May Futrelle, quoted in *Atlanta Constitution*, Friday, 26 April 1912

> 'After I left the ship two collapsible lifeboats were launched. In the last one was Mrs Henry B. Harris, wife of the theatrical manager; Mrs Thorn and Mrs Hoyt, that I remember...the ship at that time had sunk to within 15 feet of the first deck...A minute afterward Mr Hoyt leaped into the water, and was found by the boat after eight minutes, almost frozen to death...the *Carpathia* picked us up about 6:30 o'clock...'

Mr Arthur G. Peuchen

> 'Every woman on the port side was given an opportunity. In fact, we had not enough women to put into the boats. We were looking for them. I can not understand why we did not take some men. The boats would have held more.'

Titanic fact

The *Titanic* musicians continued playing for two hours from 12.20 am; sadly, they all went down with the ship.

Day 148

27 May | On this day in *Titanic* history

The following people were born on this day:
Stagg, Mr John Henry – 27 May 1873 – aged 38 when on *Titanic*; saloon steward; *Titanic* victim

Nye, Mrs Elizabeth – 27 May 1882 – aged 29 when on *Titanic*; second-class passenger (ticket price: £10 10s); rescued in lifeboat 11

Vicat, Sig. Alphonse Jean Eugene – 27 May 1890 – aged 21 when on *Titanic*; cook; *Titanic* victim

The following person died on this day:
Mellinger, Miss Madeleine Violet – 27 May 1976 – aged 13 when on *Titanic*; second-class passenger (ticket price: £19 10s); rescued in lifeboat 14

Mr George A. Hogg (lookout)

> 'I dressed myself, and we relieved the lookout at 12 o'clock, me and my mate Evans. We stopped about 20 minutes, and lifted up the back cover of the nest, the weather cover, and I saw people running about with lifebelts on. I went to the telephone then, to try to ring up on the bridge and ask whether I was wanted in the nest, when I saw this. I could get no answer on the telephone. I went straight to the boat deck. I assisted in starting to uncover the boats. Then I was sent for a Jacob's ladder. I was told to drop it. As I got past the No. 7 boat on the starboard side, Mr Murdoch, chief officer, said: "See that those plugs are in that boat." I put the plugs in, and I said: "The plugs are all correct," and I jumped out again.'

Mr Edward J. Buley (able seaman), witness at British Wreck Commissioner's Inquiry (Day 16)

> 'I believe, after the collision you found some ice on the deck?' – 'Yes, on the well deck.' 'Was there much?' – 'A couple of tons.'

Titanic survivor quote

Mrs Elizabeth Nye, quoted in *Folkestone Herald*, Saturday, 4 May 1912

> 'I have lost everything I had on board. The only thing I saved was my watch Dad gave me eleven years ago. But all my treasures and clothes and some money have gone. I have only the scanty clothes that I stand up in, including my big coat, which has been a blessing...We expect to land on Wednesday night, or next morning. I shall be so thankful, for I feel so ill on this boat... But still, I thank God I am alive...I could tell you much more of the horrors of Sunday night, but ... I can't bear to think of it all now. ...tell my friends I am safe. You must have all been anxious.'

Day 149

28 May | On this day in *Titanic* history

The following people were born on this day:
Humblen, Mr Adolf Mathias Nicolai Olsen – 28 May 1869 – aged 42 when on *Titanic*; third-class passenger (ticket price: £7 13s); *Titanic* victim

Cherrett, Mr William Victor – 28 May 1887 – aged 24 when on *Titanic*; fireman; *Titanic* victim

O'Connor, Mr Patrick – 28 May 1888 – aged 23 when on *Titanic*; third-class passenger (ticket price: £7 15s); *Titanic* victim

Jerwan, Mrs Marie Marthe – 28 May 1888 – aged 23 when on *Titanic*; second-class passenger (ticket price: £13 15s 10d); rescued in lifeboat 11

The following people died on this day:
Pusey, Mr William Robert Holland – 28 May 1918 – aged 22 when on *Titanic*; fireman; rescued in lifeboat 1

Oliva Y Ocana, Doña Fermina – 28 May 1969 – aged 39 when on *Titanic*; first-class passenger (ticket price: £108 18s); rescued in lifeboat 8

Caton, Miss Annie – 28 May 1947 – aged 33 when on *Titanic*; Turkish bath stewardess; rescued in lifeboat 11

Wells, Mrs 'Addie' Dart – 28 May 1954 – aged 29 when on *Titanic*; second-class passenger (ticket price: £23); rescued in lifeboat 14

Mr Charles H. Lightoller, witness at United States Senate Inquiry (Day 5)

> 'I should judge the forepeak tank was pierced. My reason for saying that is that the lamp trimmer's and storekeeper's room was on the starboard side there, and they were both turned in. They felt the shock. They turned out and had a look around under the forecastle head, and there seemed to be nothing doing, and they went back to bed. They were, I suppose, closer to the point of impact than anyone else in the ship.'

Titanic survivor quote

Miss Annie Caton, quoted in *Hastings and St. Leonards Observer*, Saturday, 4 May 1912

> 'Well, we then drifted on and on, tossing up and down and I cannot say how thankful we were when daylight came. After about eight hours we sighted the '*Carpathia*,' and made for it, and when they picked us up we were in a half-frozen condition. The doctors attended to us all, and rubbed a little life into our limbs, and wrapped us up in blankets, and also gave us brandy.'

Day 150

29 May | On this day in *Titanic* history

The following people were born on this day:
Harper, Reverend John – 29 May 1872 – aged 39 when on *Titanic*; second-class passenger (ticket price: £33); *Titanic* victim
Hunt, Mr Tom – 29 May 1884 – aged 27 when on *Titanic*; fireman; *Titanic* victim

The following person died on this day:
Burns, Miss Elizabeth Margaret – 29 May 1921 – aged 41 when on *Titanic*; nursemaid to Master Robert Douglas Spedden, first-class passenger (ticket price: £134 10s); rescued in lifeboat 3

Miss Elizabeth Margaret Burns, mentioned in *Newark Evening News*, Friday, 19 April 1912

> 'Miss Elizabeth M. Burns, trained nurse, who resided in this city, is among the survivors. There was some doubt as to her safety at first, because her name was wirelessed to land on the rescued list as "Mrs G. M. Burns." With Mr and Mrs F. O. Spedden and their son, Robert Douglas, of Tuxedo Park, N. Y., Miss Burns was one of the first to leave the *Carpathia*. She has been in the employ of the Speddens for six years and has travelled with them extensively.'

Who were the oldest and youngest passengers?
Millvina Dean aged 2 months, 13 days – the youngest female aboard *Titanic*; third-class passenger, who survived in lifeboat 10.

Master Gilbert Sigvard Emanuel Danbom aged 4 months, 29 days – the youngest male aboard the ship; third-class passenger, who sadly died in the sinking. His body was never recovered.

Mrs Mary Eliza Compton aged 64 years, 8 months, 8 days – the oldest female on *Titanic*; first-class passenger, who survived in lifeboat 14.

Mr Johan Svensson aged 74 years, 10 months, 29 days – the oldest male on board; third-class passenger, who perished in the sinking, and whose body was not recovered.

Joseph G. Boxhall (fourth officer), witness at British Wreck Commissioner's Inquiry (Day 13)

> 'The first officer said, "An iceberg, Sir. I hard-a-starboarded and reversed the engines, and I was going to hard-a-port round it but she was too close. I could not do any more. I have closed the watertight doors." The Commander asked him if he had rung the warning bell, and he said "Yes."'

Day 151

30 May | On this day in *Titanic* history

The following people were born on this day:
Burke, Reverend Henry P. – 30 May 1869 – aged 42, minister on board *Carpathia*

Behr, Mr Karl Howell – 30 May 1885 – aged 26 when on *Titanic*; first-class passenger (ticket price: £30); rescued in lifeboat 5

Clark, Mrs Virginia Estelle – 30 May 1885 – aged 26 when on *Titanic*; first-class passenger (ticket price: £136 15s 7d); rescued in lifeboat 4

The following person died on this day:
Cottam, Harold Thomas – 30 May 1984 – aged 21 when a telegraphist on *Carpathia*

Mr Karl Howell Behr, quoted in *Newark Evening News*, Saturday, 20 April 1912

> 'Karl H. Behr, the tennis player, who went to Australia in 1910 with the American team and was one of the *Titanic*'s survivors, tells that he was with a party of four, whom he hurried to the *Titanic*'s top deck at the first alarm.
>
> "I knew exactly where the lifeboats were," explained Mr Behr today, "so we went to the top deck. All was perfectly calm. We waited while the first boat was being filled and lowered. We went in the second boat. At that time we supposed there were plenty of lifeboats for all the passengers. One of the ladies asked Mr Ismay whether the men could go with her. I heard Mr Ismay reply quietly, 'Why, certainly, madam.'"
>
> "We all got go into the boat. Even then it was not filled, and Mr Ismay ordered an officer and two or three more of the crew to join us. We were apparently the last passengers on the top deck."'

Titanic survivor quote

Mr Alfred Crawford
'After we struck I went out and saw the iceberg passing along the starboard side. Then I went back and went around to all the staterooms to see that all the passengers were up and called all those; and as I was going around Mr and Mrs Bishop [Dickinson and Helen] came out and asked me what was the matter. I said we had run into a piece of ice. I told them to go back to their rooms and dress; to put on as much of their clothes as they could; that I did not think there was any immediate danger. Afterwards a gentleman – a Mr Stewart – came down and asked me to help dress him, and to tie his shoes, and I did so. He went on deck and came back again and told me that it was serious; that they had told passengers to put on lifebelts.'

Day 152

31 May | On this day in *Titanic* history

The following people were born on this day:
Pirrie, Lord William James – 31 May 1847 – aged 64 at the time *Titanic* sailed; shipbuilder

Thomas, Mr Benjamin James – 31 May 1881 – aged 30 when on *Titanic*; saloon steward; rescued in lifeboat 15

Allaria, Sig. Battista Antonio – 31 May 1889 – aged 22 when on *Titanic*; assistant waiter; *Titanic* victim

The following people died on this day:
Nīqūla Yārid, Master Ilyās – 31 May 1981 – aged 11 when on *Titanic*; third-class passenger (ticket price: £11 4s 10d); rescued in collapsible lifeboat C

Dean, Miss Elizabeth Gladys 'Millvina' – 31 May 2009 – aged 2 months when on *Titanic*; third-class passenger (ticket price: £20 11s 6d); rescued in lifeboat 10

The Times, 1 June 1911

'Launch of the *Titanic*: Vessel successfully takes the water

Belfast, May 31. In brilliant weather and in the presence of thousands of spectators, the new White Star liner *Titanic* was successfully launched from the yard of Messrs. Harland & Wolff (Limited) a few minutes after noon this morning. Her sister ship, the *Olympic*, left for the Mersey in the afternoon, so that there was an opportunity of seeing almost side by side the two largest vessels afloat, which have been between them, when loaded to their full draft of 34ft 6in, an aggregate displacement of about 120,000 tons.'

Mr Stanley Lord (captain of the *Californian*)

> 'I saw several empty boats, some floating planks, a few deck chairs, and cushions; but considering the size of the disaster, there was very little wreckage. It seemed more like an old fishing boat had sunk.'

Titanic survivor quote

Mrs Millvina Dean

> 'I think it is fantastic and very, very interesting. Children still ask me all the time about the *Titanic*. I think it has become a piece of history, particularly because people said it was unsinkable and it was so luxurious. It is still a romantic enigma after all these years, but I still love the sea.'

June

Day 153

1 June | On this day in *Titanic* history

The following people were born on this day:
Woodford, Mr Frederick Ernest – 1 June 1871 – aged 40 when on *Titanic*; greaser; *Titanic* victim

Thresher, Mr George Terrill – 1 June 1886 – aged 25 when on *Titanic*; fireman; rescued in lifeboat 9

Bricoux, Mr Roger Marie – 1 June 1891 – aged 20 when on *Titanic*; musician in *Titanic*'s orchestra; *Titanic* victim

Preston, Mr Thomas Charles Alfred – 1 June 1891 – aged 20 when on *Titanic*; trimmer; *Titanic* victim

Mr Roger Marie Bricoux, mentioned in *The New York Times*, Sunday, 21 April 1912

> 'Last night the members of the orchestra of the *Celtic* went ashore, and their first question was about the musicians of the *Titanic*, whom many of them knew intimately.
>
> "Not one of them saved?" said John S. Carr, cellist on the *Celtic*. "We knew most of them well. They were Englishmen, you know. Every one of them, I think. Nearly all the steamship companies hire their musicians abroad, and the men go from one ship to another often. The musicians on the *Titanic* were picked from a number of other White Star ships. Most of them had bunked with some of the *Celtic* players."
>
> There were five other musicians on the *Titanic*. Herbert Taylor was the pianist, a skilled musician, about forty years of age. His wife is an actress. The cellist was George Woodward, unmarried, of Leeds, England. The other three men were Brailey, Krins and Breicoux, who formed a trio which played in the second cabin and when the other men were off duty.'

Titanic survivor quote

Mr Walter T. Brice

> 'I went aft and began shipping the rudder, and during the time I was shipping the rudder the boat was being filled. They lowered the boat, sir. We had a bit of difficulty in keeping the boat clear of an outlet, a big body of water coming from the ship's side. The after block got jammed, but I think that must have been on account of the trip not being pushed right down to disconnect the block from the boat. We managed to keep the boat clear from this body of water coming from the ship's side.'

Day 154

2 June | On this day in *Titanic* history

The following people were born on this day:
Harder, Mr William – 2 June 1872 – aged 39 when on *Titanic*; window cleaner; rescued in lifeboat 14

Hartley, Mr Wallace Henry – 2 June 1878 – aged 33 when on *Titanic*; musician in *Titanic*'s orchestra; *Titanic* victim

Wilson, Mr Bertie – 2 June 1884 – aged 27 when on *Titanic*; senior assistant second engineer; *Titanic* victim

Olliver, Mr Alfred John – 2 June 1884 – aged 27 when on *Titanic*; quartermaster; rescued in lifeboat 5

Rogers, Mr Michael Joseph – 2 June 1885 – aged 26 when on *Titanic*; saloon steward; *Titanic* victim

Fischer, Mr Eberhard Thelander – 2 June 1893 – aged 18 when on *Titanic*; third-class passenger (ticket price: £7 15s 11d); *Titanic* victim

The following people died on this day:
Francatelli, Miss Laura Mabel – 2 June 1967 – aged 31 when on *Titanic*; first-class passenger (ticket price: £56 18s 7d); rescued in lifeboat 1

Drew, Mrs Lulu Thorne – 2 June 1970 – aged 34 when on *Titanic*; second-class passenger (ticket price: £32 10s); rescued in lifeboat 10

Dick, Mr Albert Adrian – 2 June 1970 – aged 31 when on *Titanic*; first-class passenger (ticket price: £57); rescued in lifeboat 3

Williams, Mr Richard Norris II – 2 June 1968 – aged 21 when on *Titanic*; first-class passenger (ticket price: £61 7s 7d); rescued in collapsible lifeboat A

Mr Joseph Bruce Ismay, chairman and managing director of White Star Line

> 'I think the result of this horrible accident is that the whole question of life-saving appliances on board vessels and ships will be very carefully gone through and receive the most full and careful consideration to see what is the best thing to be done.'

Titanic survivor quote

Mrs Laura Mabel Francatelli

> 'Oh, at daybreak, when we saw the lights of that ship, about 4 miles away, we rowed like mad, and passed icebergs like mountains, at last about 6.30 the dear *Carpathia* picked us up…'

Day 155

3 June | On this day in *Titanic* history

The following people were born on this day:
Risien, Mr Samuel Beard – 3 June 1842 – aged 69 when on *Titanic*; third-class passenger (ticket price: £14 10s); *Titanic* victim

Dodge, Dr Washington – 3 June 1859 – aged 52 when on *Titanic*; first-class passenger (ticket price: £81 17s 2d); rescued in lifeboat 13

Connolly, Miss Catherine – 3 June 1889 – aged 23 when on *Titanic*; third-class passenger (ticket price: £7 15s); rescued in lifeboat 13

The following people died on this day:
Brown, Mr Edward – 3 June 1926 – aged 34 when on *Titanic*; saloon steward; rescued in lifeboat unknown

Byström, Mrs Karolina – 3 June 1964 – aged 39 when on *Titanic*; second-class passenger (ticket price: £13); rescued in lifeboat unknown

Mr Charles H. Lightoller, witness at United States Senate Inquiry (Day 5)

> 'It is that Chief Officer Wilde was at the starboard collapsible boat in which Mr Ismay went away, and that he told Mr Ismay, "There are no more women on board the ship." Wilde was a pretty big, powerful chap, and he was a man that would not argue very long. Mr Ismay was right there. Naturally he was there close to the boat, because he was working at the boats and he had been working at the collapsible boat, and that is why he was there, and Mr Wilde, who was near him, simply bundled him into the boat.'

Titanic survivor quote

Dr Washington Dodge, quoted in *San Francisco Bulletin*, Saturday, 20 April 1912

> 'When the steerage passengers came up many of them had knives, revolvers and clubs and sought to fight their way to the two unlaunched, collapsible boats. Many of these were shot by the officers.
>
> 'Only one of the rafts floated, and even that did not float above the water's edge. From 40 to 50 persons who had jumped overboard clambered aboard it and stood upon it, locked arm and arm together until it was submerged to a depth of at least 18 inches. They all tried to hold together, but when the *Carpathia*'s boat reached them there were only 16 left.'

Titanic fact

There was a distance of 58 miles to the rescue vessel *Carpathia*, at the time of the distress call.

Day 156

4 June | On this day in *Titanic* history

The following people were born on this day:
Barkworth, Mr Algernon Henry – 4 June 1864 – aged 47 when on *Titanic*; first-class passenger (ticket price: £30); rescued in collapsible lifeboat B

Calderhead, Mr Edward Pennington – 4 June 1869 – aged 42 when on *Titanic*; first-class passenger (ticket price: £26 5s 9d); rescued in lifeboat 5

O'Brien, Mrs Hannah – 4 June 1884 – aged 27 when on *Titanic*; third-class passenger (ticket price: £15 10s); rescued in lifeboat unknown

Devaney, Miss Margaret Bedelia – 4 June 1891 – aged 20 when on *Titanic*; third-class passenger (ticket price: £7 17s 7d); rescued in collapsible lifeboat C

Titanic survivor quotes

Mr Algernon Henry Barkworth, quoted in *The Times*, Saturday, 20 April 1912

> 'The following further statements have been made by survivors – Mr A. H. Barkworth, of Tranby House, East Yorkshire, said he was sitting in the smoking room when the boat struck the iceberg. He saw Mr W. T. Stead on deck. He described how the forecastle was full of powdered ice. He noted the foremast was listing heavily to starboard. As Captain Smith was telling the women to put on their lifebelts he went down to his cabin and changed his clothes. All the boats had left his deck. He put on his lifebelt and fur coat and jumped overboard. While he was swimming hard to get away he was struck by wreckage and a huge wave passed over his head. Swimming about, he found a boat which was rather crowded. He clutched at it and was helped on board. After that they helped another man in. Two men died after being helped into the small boat.'

Mr Edward Wheelton, witness at United States Senate Inquiry (Day 7)

> 'We lowered away. Everything went very nice, very smoothly, until we touched the water. We pushed away from the ship's side and had just a slight difficulty in hoisting the afterblock. We were not there a minute. The sailor got at the block and loosened the tackle. We pulled away from the ship. We pulled away until we were about 300 yards away from the ship. I looked around into the boat. I saw the boat was pretty well crowded, because I had some little difficulty in rowing, because the passengers were so close together, and I kept hitting my hands against the passengers. We bent to our oars until the ship eventually went down.'

Day 157

5 June | On this day in *Titanic* history

The following people were born on this day:
Turkula, Mrs Hedwig – 5 June 1882 – aged 64 when on *Titanic*; third-class passenger (ticket price: £9 11s 9d); rescued in lifeboat 15

Charters, Mr David – 5 June 1891 – aged 20 when on *Titanic*; third-class passenger (ticket price: £7 14s 8d); *Titanic* victim

Allison, Miss Helen Loraine – 5 June 1909 – aged 2 when on *Titanic*; first-class passenger (ticket price: £151 16s); *Titanic* victim

The following people died on this day:
Saalfeld, Mr Adolphe – 5 June 1926 – aged 47 when on *Titanic*; first-class passenger (ticket price: £30 10s); rescued in lifeboat 3

De Messemaeker, Mr Guillaume Joseph – 5 June 1955 – aged 36 when on *Titanic*; third-class passenger (ticket price: £17 8s); rescued in lifeboat 15

Noss, Mr Henry – 5 June 1961 – aged 30 when on *Titanic*; fireman; rescued in lifeboat unknown

Mr Charles H. Lightoller, witness at United States Senate Inquiry (Day 5)

> 'As a general rule there would be enough men in a boat without the quartermaster having to take an oar. If an officer goes in a boat the quartermaster takes an oar, and if an officer is not in a boat the quartermaster takes the tiller.'

Titanic survivor quotes

Mr Harold Godfrey Lowe (fifth officer)

> 'I have not the remotest idea of time from the time she went down until we boarded the *Carpathia*. All I know is that when we boarded the *Carpathia* in the morning it was six o'clock, and that is the only time I know of.'

Mr George Symons (able seaman)

> 'My mate turned round from time to time and said, "It is very cold here." I said, "Yes; by the smell of it there is ice about." He asked me why, and I said, "As a Rule you can smell the ice before you get to it."'

Titanic facts

Chief baker Charles Joughin helped load bread into the lifeboats.

Titanic's plunge pool was 6 feet (2 m) deep.

Day 158

6 June | On this day in *Titanic* history

The following people were born on this day:
Marsh, Mr Frederick Charles – 6 June 1872 – aged 39 when on *Titanic*; fireman; *Titanic* victim

Saunders, Mr William Ernest – 6 June 1879 – aged 32 when on *Titanic*; saloon steward; *Titanic* victim

Shulver, Mr Christopher Arthur – 6 June 1883 – aged 28 when on *Titanic*; fireman; rescued in lifeboat unknown

Willard, Miss Constance – 6 June 1890 – aged 21 when on *Titanic*; first-class passenger (ticket price: £26 11s); rescued in lifeboat 8

The following people died on this day:
Pirrie, Lord William James – 6 June 1924 – aged 64 when *Titanic* set sail; shipbuilder

Cardeza, Mr Thomas Drake Martinez – 6 June 1952 – aged 36 when on *Titanic*; first-class passenger (ticket price: £512 6s 7d); rescued in lifeboat 3

Haines, Mr Albert – 6 June 1933 – aged 31 when on *Titanic*; boatswain mate; rescued in lifeboat 9

Drew, Master Marshall Brines – 6 June 1986 – aged 8 when on *Titanic*; second-class passenger (ticket price: £32 10s); rescued in lifeboat 10

Titanic survivor quotes

Miss Constance Willard, quoted in *Chicago Tribune*, Sunday, 21 April 1912

> 'In our boat there were seven men, about twenty women, and several children. The night was dark. Twenty minutes after leaving the *Titanic* we heard an explosion and the vessel appeared to split in two and sank. Then a foreign woman in our boat began singing a hymn, and we all joined, although few knew the words. All around us we heard crying and sobbing for perhaps three minutes.'

Mr Alfred Olliver (quartermaster)

> 'The iceberg was about the height of the boat deck; if anything, just a little higher. It was almost alongside of the boat, sir. The top did not touch the side of the boat, but it was almost alongside of the boat. The sound was like she touched something; a long grinding sound, like.'

Day 159

7 June | On this day in *Titanic* history

The following people were born on this day:
Hendrickson, Mr Charles Osker – 7 June 1883 – aged 28 when on *Titanic*; leading fireman; rescued in lifeboat 1

Adams, Mr Robert John – 7 June 1885 – aged 26 when on *Titanic*; fireman; *Titanic* victim

Heininen, Miss Wendla Maria – 7 June 1888 – aged 23 when on *Titanic*; third-class passenger (ticket price: £7 18s 6d); *Titanic* victim

McGovern, Miss Mary – 7 June 1891 – aged 20 when on *Titanic*; third-class passenger (ticket price: £7 17s 7d); rescued in lifeboat 13

Aks, Master Frank Philip – 7 June 1911 – aged 10 months when on *Titanic*; third-class passenger (ticket price: £9 7s); rescued in lifeboat unknown

Master Philip Aks, mentioned in *Washington Post*, Thursday, 25 April 1912.

> 'Mrs Aks and Child, *Titanic* Survivors, Reach Norfolk
>
> Mrs Leah Aks and her infant, survivors from the *Titanic* disaster, has arrived here, but the woman was too ill to talk and had to go to bed. Mrs Aks came to America to join her husband, S. Aks, a tailor, who had never before seen their child.
>
> 'It was the Aks infant that was recovered by the mother on the *Carpathia* after she had given it up as lost. The infant having been taken by a frantic man fell into the lap of a woman survivor in a lifeboat as it was tossed over the side of the *Titanic*. The mother, who fainted, was placed in another lifeboat.'

Titanic survivor quote

Mr George Hogg (lookout)

> 'I saw the lights of the *Carpathia*. I said: "It is all right, now, ladies. Do not grieve. We are picked up. Now, gentlemen, see what you can do in pulling these oars for this light." It was practically daylight then. Then the passengers could see for themselves that there was a ship there. I pulled up and went alongside, and I assisted in putting a bowline around all the ladies, to haul them up aboard. After I saw all aboard the boat, me and my friend went aboard, and I put some blankets around myself and went to sleep.'

Titanic fact

Nearly 300 feet of the *Titanic*'s hull was damaged in the collision, allowing water to flood six of the sixteen major watertight compartments.

Day 160

8 June | On this day in *Titanic* history

The following person was born on this day:
Troutt, Miss Edwina Celia – 8 June 1884 – aged 27 when on *Titanic*; second-class passenger (ticket price: £10 10s); rescued in lifeboat 16

The following person died on this day:
Osman, Mr Frank – 8 June 1938 – aged 27 when on *Titanic*; able seaman; rescued in lifeboat 2

On 9 June 1911 in Southampton, the new White Star Liner RMS *Olympic* was being prepared for her maiden voyage, and vast quantities of food, coal, linen and china were loaded aboard.

Fourteen passengers on *Titanic* were from Addergoole Parish in County Mayo, Ireland. Eleven died and three survived, and collectively they became known locally as The Addergoole Fourteen. On 15 April each year a bell ringing service takes place at Saint Patrick's Church, Lahardane, commencing at 2.20 am and lasting for one hour. This tradition started in 2002.

Titanic survivor quote

Miss Edwina Celia Troutt, quoted in *Newark Evening News*, Saturday, 20 April 1912

> '"I can never forget the experience," said Miss Trout today. "As we pulled away from the ship the men in our boat sang 'Pull for the Shore'. At the same time we could hear the strains of the band on the doomed vessel, playing 'Nearer, My God, to Thee.' Oh, it was awful!" Miss Trout said that she had not realized the seriousness of the situation on the fated liner at first and was reluctant to enter a boat. A man with several children appealed for some one to care for one of the tots and she volunteered to do so. She was then hustled into a boat. Upon reaching the *Carpathia*, she said, she had the satisfaction of restoring the baby to its parents. They had escaped with their other children in another boat. It was one of the last boats to leave the ship into which she was put, Miss Trout declared. The *Titanic* sank within a half hour. She did not suffer with cold, she explained, for she and the baby in her arms were well wrapped in a blanket. Miss Trout speaks in praise of the treatment given her by those on the *Carpathia*. The parents of the baby she carried were steerage passengers from southern Europe. She did not know their names, she concluded.'

Day 161

9 June | On this day in *Titanic* history

The following people were born on this day:
Salussolia, Sig. Giovanni – 9 June 1886 – aged 25 when on *Titanic*; glass man (restaurant staff); *Titanic* victim

Golder, Mr Martin William – 9 June 1876 – aged 35 when on *Titanic*; fireman; *Titanic* victim

Pugh, Mr Alfred – 9 June 1891 – aged 20 when on *Titanic*; steward in third class; rescued in lifeboat 14

The following person died on this day:
Kennedy, Mr John – 9 June 1918 – aged 24 when on *Titanic*; third-class passenger (ticket price: £7 15s); rescued in lifeboat unknown

Titanic survivor quotes

Mr Albert Haines, witness at United States Senate Inquiry (Day 7)

> 'It being Sunday night, the men did not work Sunday night, and the men were in the mess room, and I was outside, sir. If it had been any other night, we would have been washing the decks.'

Mr Henry S. Etches

> 'There were two firemen in the bow of the boat, Mr Pitman was standing in the centre on the deck waiting, and one steward was standing with myself at the forward fall. Olliver, the sailor, the quartermaster who had been assisting, stepped back with myself into the crowd of gentlemen. Mr Murdoch said to me: "Are you the steward appointed to this boat?" I said, "Yes, sir; No. 5 boat is mine." He said: "Then jump in and assist those men with the forward fall." I took my place. At the same moment Mr Pitman called out: "Is there a sailor in the boat?" I looked around and I said: "No, sir." He said to this man Olliver, who was standing on the deck: "Are you a sailor?" He said: "Yes, sir." He said: "Take your place in this boat;" and he jumped in. Mr Murdoch then stepped up and said, "Are you the officer going in this boat?" He said: "Yes, sir." Then he said: "Take your place," held out his hand and shook hands and said, "Goodbye and good luck." and he took his place and the order was given to lower the boat.'

Day 162

10 June | On this day in *Titanic* history

The following people were born on this day:
Fryer, Mr Albert Ernest – 10 June 1882 – aged 29 when on *Titanic*; trimmer; rescued in lifeboat unknown

Holland, Mr Thomas – 10 June 1883 – aged 28 when on *Titanic*; reception steward; *Titanic* victim

Abelseth, Mr Olaus Jørgensen – 10 June 1886 – aged 25 when on *Titanic*; third-class passenger (ticket price: £7 13s); rescued in collapsible lifeboat A

O'Leary, Miss Hanora 'Nora' – 10 June 1895 – aged 16 when on *Titanic*; third-class passenger (ticket price: £7 16s 7d); rescued in lifeboat 13

Caldwell, Master Alden Gates – 10 June 1911 – aged 10 months when on *Titanic*; second-class passenger (ticket price: £29); rescued in lifeboat 13

Master Alden Gates Caldwell, mentioned in *Monmouth Daily Atlas*, Wednesday, 24 April 1912

> 'Both he and Mrs Caldwell were fully dressed, but the baby had but his nightdress on beneath a steamer robe which had been wrapped about him. Mrs Caldwell said the women stood while some of the men appropriated the seats. One man in particular, a big fellow, occupied the choice seat in the middle of the boat and utilized Mrs Caldwell as a back rest the entire time they were afloat. "He was a coward," said Mrs Caldwell.'

Titanic survivor quote

Miss Ida Daisy Minahan

> 'When the lifeboat was filled there were no seamen to man it. The officer in command of No.14 called for volunteers in the crowd who could row. Six men offered to go. At times when we were being lowered we were at an angle of 45° and expected to be thrown into the sea. As we reached the level of each deck men jumped into the boat until the officer threatened to shoot the next man who jumped. We landed in the sea and rowed to a safe distance from the sinking ship. The officer counted our number and found us to be 48. The officer commanded everyone to feel in the bottom of the boat for a light. We found none. Nor was there bread or water in the boat. The officer whose name I learned afterwards to be Lowe, was continually making remarks such as, "A good song to sing would be 'Throw Out the Life Line'," and "I think the best thing for you women to do is to take a nap."'

Day 163

11 June | On this day in *Titanic* history

The following people were born on this day:

Smith, Mr Richard William – 11 June 1854 – aged 57 when on *Titanic*; first-class passenger (ticket price: £26); *Titanic* victim

Bourke, Miss Mary – 11 June 1871 – aged 40 when on *Titanic*; third-class passenger (ticket price: £7 15s); *Titanic* victim

Graham, Mr George Edward – 11 June 1873 – aged 38 when on *Titanic*; first-class passenger (ticket price: £30); *Titanic* victim

Johanson, Mr Jakob Alfred – 11 June 1877 – aged 34 when on *Titanic*; third-class passenger (ticket price: £6 9s 11d); *Titanic* victim

Fellowes, Mr Alfred James – 11 June 1883 – aged 28 when on *Titanic*; assistant boots steward in first class; *Titanic* victim

Eagle, Mr Alfred James Jacob – 11 June 1884 – aged 27 when on *Titanic*; trimmer; *Titanic* victim

Shiers, Mr Alfred Charles – 11 June 1886 – aged 25 when on *Titanic*; fireman; rescued in lifeboat 5

Silvén, Miss Lyyli Karoliina – 11 June 1894 – aged 17 when on *Titanic*; second-class passenger (ticket price: £13); rescued in lifeboat 12

Mallet, Master André Clement – 11 June 1910 – aged 1 when on *Titanic*; second-class passenger (ticket price: £37 1d); rescued in lifeboat 10

The following person died on this day:

Newell, Miss Marjorie Anne – 11 June 1992 – aged 23 when on *Titanic*; first-class passenger (ticket price: £113 5s 6d); rescued in lifeboat unknown

Titanic survivor quote

Mrs Mahala Douglas

> 'All the women told of insufficient seamen to man the boats; all women rowed; some had to bail water from their boats. Mrs Smith was told to watch a cork in her boat, and if it came out to put her finger in place of it.'

Titanic fact

The broad corridor that was used to get from one end of the ship to the other was called Scotland Road.

Day 164

12 June | On this day in *Titanic* history

The following people were born on this day:

Rosenbaum, Miss Edith Louise (Edith Russell) – 12 June 1879 – aged 32 when on *Titanic*; first-class passenger (ticket price: £27 14s 5d); rescued in lifeboat 11

Payne, Mr Vivian Ponsonby – 12 June 1888 – aged 23 when on *Titanic*; clerk; *Titanic* victim

Dick, Mrs Vera – 12 June 1894 – aged 17 when on *Titanic*; first-class passenger (ticket price: £57); rescued in lifeboat 3

Navratil, Master Michel Marcel – 12 June 1908 – aged 3 when on *Titanic*; second-class passenger (ticket price: £26); rescued in collapsible lifeboat D

The following people died on this day:

Coffey, Mr John – 12 June 1957 – aged 23 when on *Titanic*; fireman; left *Titanic* at Queenstown, Ireland

Devaney, Miss Margaret Delia – 12 June 1974 – aged 20 when on *Titanic*; third-class passenger (ticket price: £7 17s 7d); rescued in collapsible lifeboat C

Thomas/Tannous, Master Assad Alexander – 12 June 1931 – aged 5 months when on *Titanic*; third-class passenger (ticket price: £8 10s 4d); rescued in lifeboat 16

Titanic survivor quotes

Miss Edith Louise Rosenbaum, quoted in *New York Times*, Tuesday, 23 April 1912

> 'Miss Rosenbaum told a reporter for THE TIMES yesterday that she was among the last who, preferring her chances on the sinking vessel to the risk of the lifeboats in an open sea, departed from the doomed liner. She was forced to take to the lifeboat only by the physical force of Mr Ismay, who, she declares, caught her by the arm on the boat deck and thrust her down a narrow stairway to A deck, where she was heaved over safely into the life craft.'

Mr Joseph Bruce Ismay, chairman and managing director of White Star Line

> 'The ship had quite a list to port. Consequently this canvas boat, this collapsible boat, was getting hung up on the outside of the ship, and she had to rub right along her, and we had to try to shove her out, and we had to get the women to help to shove to get her clear of the ship. The ship had listed over that way.'

Day 165

13 June | On this day in *Titanic* history

The following people were born on this day:
Duff-Gordon, Lady Lucy Christiana, – 13 June 1863 – aged 48 when on *Titanic*; first-class passenger (ticket price: £56 18s 7d); rescued in lifeboat 1

Wilkes, Mrs Ellen – 13 June 1864 – aged 47 when on *Titanic*; third-class passenger (ticket price: £7); rescued in lifeboat 16

Simmons, Mr Andrew George James – 13 June 1880 – aged 31 when on *Titanic*; scullion; rescued in lifeboat unknown

Ford, Miss Dollina Margaret – 13 June 1891 – aged 20 when on *Titanic*; third-class passenger (ticket price: £34 7s 6d); *Titanic* victim

Titanic survivor quote

Lady Lucy Christiana Duff-Gordon, quoted in *New York Herald*, Friday, 19 April 1912

> 'I recall that I was pushed along toward one of the boats and helped in. The boat was lowered part way down on the davits. Just as we were about to clear the ship a man made a rush to get aboard and was shot. He was apparently killed instantly and his body fell into the boat at our feet. No one made an effort to move the body, and it remained beneath our feet until we were picked up by the *Carpathia*.'

Mrs J. Stuart White

> 'Before we cut loose from the ship two of the seamen with us – the men, I should say; I do not call them seamen; I think they were dining-room stewards – before we were cut loose from the ship they took out cigarettes and lighted them on an occasion like that! That is one thing that we saw. All of those men escaped under the pretence of being oarsmen. The man who rowed me took his oar and rowed all over the boat, in every direction. I said to him, "Why don't you put the oar in the oarlock?" He said, "Do you put it in that hole?" I said "Certainly." He said, "I never had an oar in my hand before." I spoke to the other man and he said; "I have never had an oar in my hand before, but I think I can row." Those were the men that we were put to sea with at night – with all these magnificent fellows left on board, who would have been such a protection to us. Those were the kind of men with whom we were put out to sea that night.'

Day 166

14 June | On this day in *Titanic* history

The following people were born on this day:

Endres, Miss Caroline Louise – 14 June 1872 – aged 39 when on *Titanic*; first-class passenger (ticket price: £247 10s 6d); rescued in lifeboat 4

Urbini, Sig. Roberto – 14 June 1891 – aged 20 when on *Titanic*; waiter; *Titanic* victim

The following people died on this day:

Lenox-Conyngham, Miss Harriet Alice Katherine – 14 June 1956 – aged 34 when on *Titanic*; first-class passenger (ticket price: £1 10s); left *Titanic* at Cherbourg

Hyland, Mr James Leo – 14 June 1974 – aged 19 when on *Titanic*; steward in third class; rescued in lifeboat 11

14 June 1911 was the maiden voyage of the RMS *Olympic*, sister ship of *Titanic*.

Mr Leo James Hyland, quoted in *Southern Evening Echo*, Tuesday, 24 November 1970

> '*Titanic* survivor 77-year-old Mr Lee James Hyland, of Sarnia Court, Boniface Crescent, Lordshill, Southampton who recalls Fred Allen, a boy crew member, mentioned in reference the playing of the *Titanic* Hymn. Allen, who lost his life in the disaster was last seen by Mr Hyland jumping off one of the deck housings onto the boat deck. "We had been ordered to boat Stations at the time, I saw him go forward but did not see him again. Allen was either a lift or bell boy and I knew him well." Mr Hyland was in charge of passengers on the lower decks [and] got away in one of the boats, [but] had to get into the boat to free the tackle which had jammed, he said. Mr Hyland went back to sea after the disaster, working in all types of ships from tramps to liners before retiring a few years ago.'

Titanic survivor quote

Edward J. Buley, witness at United States Senate Inquiry (Day 7)

Senator Fletcher: 'Were any ladies on the deck when you left?'

Mr Edward J. Buley: 'No, sir. Ours was the last boat up there, and they went around and called to see if there were any, and they threw them in the boat at the finish, because they didn't like the idea of coming in.'

Senator Fletcher: 'Pushed them in, you mean?'

Mr Edward J. Buley: 'Threw them in. One young lady slipped, and they caught her by the foot on the deck below, and she came up then and jumped in.'

Day 167

15 June | On this day in *Titanic* history

The following people were born on this day:
Sutton, Mr Frederick – 15 June 1850 – aged 61 when on *Titanic*; first-class passenger (ticket price: £32 6s 5d); *Titanic* victim

Lines, Mrs Elizabeth Lindsey – 15 June 1861 – aged 50 when on *Titanic*; first-class passenger (ticket price: £39 8s); rescued in lifeboat 9

Cotton, Mr Alfred – 15 June 1876 – aged 35 when on *Titanic*; trimmer; *Titanic* victim

Bessant, Mr Edward – 15 June 1880 – aged 31 when on *Titanic*; baggage steward in first class; *Titanic* victim

Ådahl, Mr Mauritz Nils Martin – 15 June 1881 – aged 30 when on *Titanic*; third-class passenger (ticket price: £7 5s); *Titanic* victim

Gustafsson, Mr Johan Birger – 15 June 1883 – aged 28 when on *Titanic*; third-class passenger (ticket price: £7 18s 6d); *Titanic* victim

Hargadon, Miss Kate – 15 June 1894 – aged 17 when on *Titanic*; cook to third-class passengers; *Titanic* victim

Mrs Elizabeth Lindsey Lines, mentioned in *Chicago American*, Tuesday, 16 April 1912

> 'Paris, 16 April…News Nearly 300 of the *Titanic*'s passengers made their bookings here and went aboard at Cherborg. Col. John Jacob Astor engaged passage here as did also Mr and Mrs George D. Widener and Mrs J. W. M. Cardeza and T. D. M. Cardeza.
>
> 'The latter engaged the most expensive suite on the vessel, paying $3,000. Among the other passengers who engaged passage here were Major Archibald Butt, Frank D. Millett, the well-known painter; J. Clinch Smith, Mrs Ernest H. Lines, Mr and Mrs Isador Straus and Washington Dodge.'

Titanic survivor quote

Mrs Emily Borie Ryerson, witness at United States Senate Inquiry (Day 16)

> 'After the *Titanic* sank we saw no lights, and no one seemed to know what direction to take. Lowe, the officer in charge of the boat, had called out earlier for all to tie together, so we now heard his whistle, and as soon as we could make out the other boats in the dark, five of us were tied together, and we drifted about without rowing, as the sea was calm, waiting for the dawn. It was very cold, and soon a breeze sprang up, and it was hard to keep our heavy boat bow on; but as the cries died down we could see dimly what seemed to be a raft with about 20 men standing on it, back to back.'

Day 168

16 June | On this day in *Titanic* history

The following people were born on this day:
Leather, Mrs Elizabeth Mary – 16 June 1861 – aged 50 when on *Titanic*; stewardess; rescued in lifeboat 16

Widener, Mr George Dunton – 16 June 1861 – aged 50 when on *Titanic*; first-class passenger (ticket price: £211 10s); *Titanic* victim

Perren, Mr William Charles – 16 June 1864 – aged 47 when on *Titanic*; boots steward in second class; *Titanic* victim

Rice, Mr John Reginald – 16 June 1886 – aged 25 when on *Titanic*; clerk in first class; *Titanic* victim

The following people died on this day:
Compton, Miss Sara Rebecca – 16 June 1952 – aged 39 when on *Titanic*; first-class passenger (ticket price: £83 3s 2d); rescued in lifeboat 14

Mock, Mr Philipp Edmund – 16 June 1951 – aged 30 when on *Titanic*; first-class passenger (ticket price: £57 15s); rescued in lifeboat 11

Hebb, Mr William Albert – 16 June 1932 – aged 22 when on *Titanic*; trimmer; rescued in collapsible lifeboat B

Mr Charles H. Lightoller, witness at United States Senate Inquiry (Day 5)

> 'The lookouts receive so much per month, the same as the men, and then they get 5 shillings per voyage of what we call "lookout money".'

Titanic survivor quote

Mrs Emily Borie Ryerson, witness at United States Senate Inquiry (Day 16)

> 'The ropes seemed to stick at one end and the boat tipped, some one called for a knife, but it was not needed until we got into the water, as it was but a short distance, and I then realized for the first time how far the ship had sunk. The deck we left was only about 20 feet from the sea. I could see all the portholes open and water washing in, and the decks still lighted. Then they called out, "How many seamen have you," and they answered one. "That is not enough," said the officer, "I will send you another," and he sent a sailor down the rope.'

Titanic fact

Carpathia dropped off the *Titanic*'s emptied lifeboats at Pier 59 in New York, back into the possession of the White Star Line.

Day 169

17 June | On this day in *Titanic* history

The following people were born on this day:
Jarvis, Mr Denzil John – 17 June 1865 – aged 46 when on *Titanic*; engineer; *Titanic* victim

Evans, Mr William Thomas – 17 June 1878 – aged 33 when on *Titanic*; trimmer; *Titanic* victim

Giglio, Mr Victor Gaitan Andrea – 17 June 1888 – aged 23 when on *Titanic*; servant to Mr Benjamin Guggenheim; *Titanic* victim

Dennis, Mr Samuel – 17 June 1889 – aged 22 when on *Titanic*; third-class passenger (ticket price: £7 5s); *Titanic* victim

Hickman, Mr Stanley George – 17 June 1890 – aged 21 when on *Titanic*; second-class passenger (ticket price: £73 10s); *Titanic* victim

Hagland, Mr Konrad Mathias Reiersen – 17 June 1892 – aged 19 when on *Titanic*; third-class passenger (ticket price: £6 19s 4d); *Titanic* victim

Richards, Master Sibley George – 17 June 1911 – aged 9 months when on *Titanic*; second-class passenger (ticket price: £18 15s); rescued in lifeboat 4

The following person died on this day:
Taussig, Mrs Tillie – 17 June 1957 – aged 39 when on *Titanic*; first-class passenger (ticket price: £79 13s); rescued in lifeboat 8

Mrs Tillie Taussig, mentioned in *New York Times*, Monday, 22 April 1912

> 'Mrs Emil Taussig, who was saved with her eighteen-year-old daughter, Ruth, while her husband stood by on the deck, also tells a story of half-empty boats, but she adds that incompetent stewards, unable to handle an oar, were assigned to man them.'

Titanic survivor quote

Mr Arthur J. Bright

> 'She broke in two. All at once she seemed to go up on end, you know, and come down about half way, and then the afterpart righted itself again, and the forepart had disappeared. A few seconds [later] the afterpart did the same thing and went down. I could distinctly see the propellers – everything – out of the water.'

Day 170

18 June | On this day in *Titanic* history

The following people were born on this day:

Lind, Mr Erik Gustaf – 18 June 1869 – aged 42 when on *Titanic*; first-class passenger (ticket price: £26 11s); *Titanic* victim

Moyes, Mr William Young – 18 June 1887 – aged 24 when on *Titanic*, senior sixth engineer, *Titanic* victim

Kelly, Mr William Patrick – 18 June 1888 – aged 23 when on *Titanic*; assistant electrician; *Titanic* victim

Canavan, Miss Mary – 18 June 1889 – aged 22 when on *Titanic*; third-class passenger (ticket price: £7 15s); *Titanic* victim

Dorking, Mr Edward Arthur – 18 June 1893 – aged 18 when on *Titanic*; third-class passenger (ticket price: £8 1s); rescued in collapsible lifeboat B

The following people died on this day:

Baxter, Mrs Hélène – 18 June 1923 – aged 50 when on *Titanic*; first-class passenger (ticket price: £247 10s 5d); rescued in lifeboat 6

Bliss, Mrs Emma – 18 June 1959 – aged 45 when on *Titanic*; stewardess; rescued in lifeboat 15

Portaluppi, Mr Emilio Ilario Giuseppe – 18 June 1974 – aged 30 when on *Titanic*; second-class passenger (ticket price: £12 14s 9d); rescued in lifeboat 14

Wheat, Mr Joseph Thomas – 18 June 1961 – aged 30 when on *Titanic*; assistant second steward; rescued in lifeboat 11

Olliver, Mr Alfred John – 18 June 1934 – aged 27 when on *Titanic*; quartermaster; rescued in lifeboat 5

Titanic survivor quote

Mr Alfred Olliver (quartermaster)

> 'There were so many people that when I got in the boat I went to put the plug in, and there were so many people around me I could not get near the plug to put the plug in. As he was lowering the boat I implored the passengers to move so that I could put the plug in, so that as soon as they put the boat in the water I let the tripper go and water came into the boat; but I then forced my way to the plug and put it in. If it was not for that the boat would have been swamped.'

Mr Joseph Bruce Ismay, witness at United States Senate Inquiry (Day 11)

> 'Because there was room in the boat. She was being lowered away. I felt the ship was going down, and I got into the boat.'

Day 171

19 June | On this day in *Titanic* history

The following people were born on this day:

Carter, Mr William Ernest – 19 June 1875 – aged 36 when on *Titanic*; first-class passenger (ticket price: £120); rescued in collapsible lifeboat C

Becker, Mrs Nellie E. – 19 June 1876 – aged 35 when on *Titanic*; second-class passenger (ticket price: £39); rescued in lifeboat 11

O'Dwyer, Miss Ellen – aged 22 when on *Titanic*; third-class passenger (ticket price: £7 17s 7d); rescued in lifeboat 10

Landergren, Miss Aurora Adelia – 19 June 1889 – aged 22 when on *Titanic*; third-class passenger (ticket price: £7 5s); rescued in lifeboat 13

Karlsson, Mr Einar Gervasius – 19 June 1890 – aged 21 when on *Titanic*; third-class passenger (ticket price: £7 15s 11d); rescued in lifeboat 13

Coelho, Mr Domingos Fernandeo – 19 June 1891 – aged 20 when on *Titanic*; third-class passenger (ticket price: £7 1s); *Titanic* victim

Astor, Mrs Madeleine Talmage – 19 June 1893 – aged 18 when on *Titanic*; first-class passenger (ticket price: £247 10s 6d); rescued in lifeboat 4

Pålsson, Miss Stina Viola – 19 June 1908 – aged 3 when on *Titanic*; third-class passenger (ticket price: £21 1s 6d); *Titanic* victim

The following people died on this day:

Yūsuf, Mrs Kātrīn – 19 June 1915 – aged 23 when on *Titanic*; third-class passenger (ticket price: £22 7s 2d); rescued in collapsible lifeboat C

Dymond, Mr Frank – 19 June 1947 – aged 36 when on *Titanic*; fireman; rescued in lifeboat 15

Mr George A. Hogg, witness at United States Senate Inquiry (Day 7)

Senator Perkins: 'There was quite a ripple on the water?'

Mr George A. Hogg: 'Not a ripple on the water, sir. It was as smooth as glass.'

Titanic survivor quote

Mr Lawrence Beesley

> 'The oarsman laid on their oars and all in the lifeboat were motionless as we watched her in absolute silence. Save some who would not look and buried their heads on each other's shoulders.'

Day 172

20 June | On this day in *Titanic* history

The following people were born on this day:

Coleman, Mr John – 20 June 1854 – aged 57 when on *Titanic*; engineers' mess steward; *Titanic* victim

Taussig, Mr Emil – 20 June 1857 – aged 54 when on *Titanic*; first-class passenger (ticket price: £79 13s); *Titanic* victim

Ray, Mr Frederick Dent – 20 June 1879 – aged 32 when on *Titanic*; saloon steward; rescued in lifeboat 13

Howell, Mr Arthur Albert – 20 June 1880 – aged 31 when on *Titanic*; saloon steward in first class; *Titanic* victim

Buley, Mr Edward John – 20 June 1885 – aged 26 when on *Titanic*, able seaman; rescued in lifeboat 10

Dropkin, Miss Jennie – 20 June 1887 – aged 24 when on *Titanic*; third-class passenger (ticket price: £8 1s); rescued in lifeboat unknown

Turja, Miss Anna Sofia – 20 June 1893 – aged 18 when on *Titanic*; third-class passenger (ticket price: £9 16s 10d); rescued in lifeboat 15

Titanic survivor quotes

Miss Anna Turja, quoted in *Cleveland Plain Dealer*, Friday, 19 April 1912

> 'There was some shooting. They would not allow those half crazed men to get into the boats. I was in one of the lifeboats. After we were lowered away the men in our boat started to row. I looked back to the *Titanic* and could see the big ship settling. She seemed already to be only half her former height. The officer in charge of our lifeboat kept urging the men at the oars to row harder. Some of the oarsmen fell exhausted. Then we women took turns at the oars, We must have been about a mile away from the *Titanic* when she went under. There was a big wave. The sea was calm, otherwise, and I asked a sailor what it was. He said, "The *Titanic* has sunk!"'

Mrs George D. Widener

> 'I was a passenger with my husband, George D. Widener, and my son, Harry Widener, on the steamship *Titanic* of the White Star Line on her voyage from Southampton on the 10th day of April, 1912. My husband and I gave a dinner at which Capt. Smith was present. Capt. Smith drank absolutely no wine or intoxicating liquor of any kind whatever at the dinner.'

Day 173

21 June | On this day in *Titanic* history

The following people were born on this day:
Ahlin, Mrs Johanna Persdotter – 21 June 1871 – aged 40 when on *Titanic*; third-class passenger (ticket price: £9 9s 6d); *Titanic* victim

Estanislau, Mr Manuel Gonçalves – 21 June 1874 – aged 37 when on *Titanic*; third-class passenger (ticket price: £7 1s); *Titanic* victim

Lang, Mr Fang – 21 June 1894 – aged 16 when on *Titanic*; third-class passenger (ticket price: £7); rescued in lifeboat 14 **Further information on the seven other Chinese passengers is shown in the Appendix.*

The following person died on this day:
Jalševac, Mr Ivan – 21 June 1945 – aged 30 when on *Titanic*; third-class passenger (ticket price: £7 17s 11d); rescued in lifeboat 15

Mr Ivan Jalševac, mentioned in *Galesburg Mail*, Thursday, 25 April 1912

> 'Ivan Jaalasovich, who was one of the passengers on the *Titanic*, and who in company with Frank Kurun of this city, escaped from the ill-fated steamer, started this morning on the return trip to his home in Agram, Austria, and before leaving today expressed to a Mail reporter his firm intention of remaining on the other side of the Atlantic the rest of his life. Jaalasovich had, prior to coming to this country, been an officer in the Austrian army, and had given up his commission in that body to come to America, having had the hope of bettering his fortunes here, built up by the glowing description of the country given by Kurun. He stated that when he found the ship was sinking, that he had almost given up any hope of being saved, but that one of the lifeboats was being lowered past the deck where he was standing [and] he decided to take a chance, and leaping over the railing, jumped into the boat. Those in the boat, which was already filled, made room for him and he was pressed into service at the oars…'

Titanic survivor quote

Mr Samuel Ernest Hemming (lamp trimmer)

> 'We went back in our bunks a few minutes. Then the joiner possibly (J. Hutchinson) came in and he said: "If I were you, I would turn out, you fellows. She is making water, one-two-three, and the racket court is getting filled up." Just as he went, the boatswain came, and he says, "Turn out, you fellows," he says; "you haven't half an hour to live." He said: "That is from Mr Andrews." He said: "Keep it to yourselves, and let no one know."'

Birth of a legend

Time to stretch her legs

Ship of dreams

The Queen of the Ocean

'We thought it quite possible that she could reach 24 knots.'

The finest steamer afloat

'Iceberg, right ahead.'

'Hard-a-starboard.'

'Come at once. We have struck a berg. It's a CQD, old man.'

A night to remember

'As a boat shoved off he threw me a kiss.'

Firing distress signals

'Well boys, it's every man for himself now.'

'Everything in the world was standing still.'

End of a dream

'She dived finally from human sight.'

Mowing the lawn

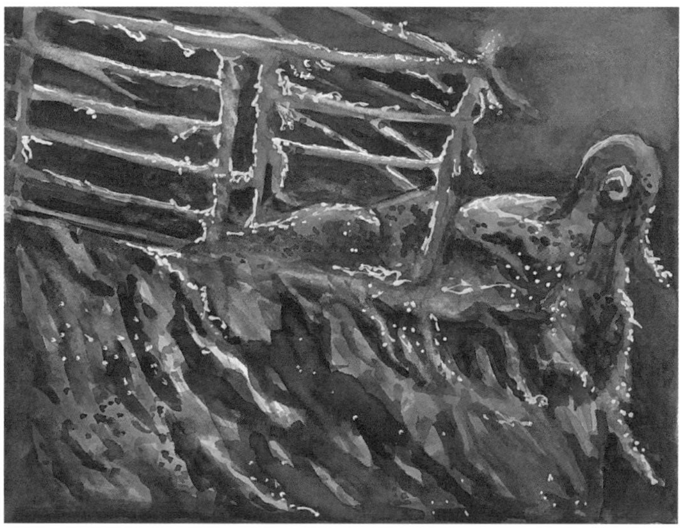

Beauty in the deep

3800m below

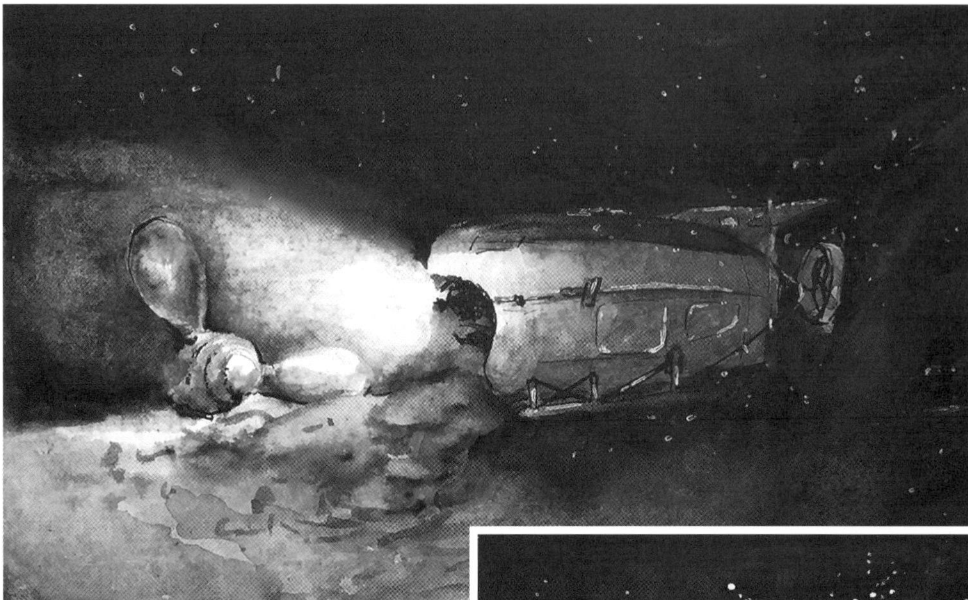

Taking a close up look

Splendour held
in time

Preserving history

Day 174

22 June | On this day in *Titanic* history

The following people were born on this day:
Somerton, Mr Francis William – 22 June 1881 – aged 30 when on *Titanic*; third-class passenger (ticket price: £8 1s); *Titanic* victim

Bailey, Mr Henry Joseph – 22 June 1865 – aged 46 when on *Titanic*; master-at-arms; rescued in lifeboat 16

Beux, Mr David – 22 June 1886 – aged 25 when on *Titanic*; assistant waiter; *Titanic* victim

The following people died on this day:
Aks, Mrs Leah – 22 June 1967 – aged 21 when on *Titanic*; third-class passenger (ticket price: £9 7s); rescued in lifeboat unknown

Gold, Mrs Jane Kate Coulson – 22 June 1948 – aged 45 when on *Titanic*; stewardess; rescued in lifeboat unknown

Titanic survivor quotes

Mrs Leah Aks, quoted in *Norfolk (Virginia Ledger – Dispatch)*, Wednesday, 24 April 1912

> 'With our boat but a short distance from the *Titanic*, I could see the big steamer distinctly as she went down, and the scenes are too horrifying to relate. The passengers and crew were praying, while many were crying and the groans could be heard distinctly. I saw Captain Smith step into the last boat, but he immediately stepped back on the ship. Then he stepped back into the lifeboat again, but turned, and as he saw the water going over his ship, stepped back on the deck and went down with her."

Mr James R. McGough

> 'They called for the women and children to board the boats first. Both women and men, however, hesitated, and did not feel inclined to get into the small boats, thinking the larger boat was the safer. I had my back turned looking in the opposite direction at that time and was caught by the shoulder by one of the officers, who gave me a push, saying, "Here, you are a big fellow; get into the boat." Our boat was launched with 28 people; we, however, transferred 5 from one of the other boats after we were out in the ocean, which was some time after the ship went down.'

Day 175

23 June | On this day in *Titanic* history

The following people were born on this day:
Witter, Mr James William Cheetham – 23 June 1880 – aged 31 when on *Titanic*; smoke room steward in second class; rescued in lifeboat unknown

Jackopson, Mr John Henry – 23 June 1881 – aged 30 when on *Titanic*; fireman; *Titanic* victim

Titanic survivor quotes

Mr Colonel Archibald Gracie

'After sinking with the ship, it appeared to me as if I was propelled by some great force through the water. This might have been occasioned by explosions under the water, and I remembered fearful stories of people being boiled to death. Again and again I prayed for deliverance, although I felt sure that the end had come. I had the greatest difficulty in holding my breath until I came to the surface. I knew that once I inhaled, the water would suffocate me. When I got under water I struck out with all my strength for the surface. I got to air again after a time, which seemed to me to be unending. There was nothing in sight save the ocean, dotted with ice and strewn with large masses of wreckage. Dying men and women all about me were groaning and crying piteously. By moving from one piece of wreckage to another, at last I reached a cork raft. Soon the raft became so full that it seemed as if she would sink if more came on board her. The crew for self-preservation had therefore to refuse to permit any others to climb aboard. This was the most pathetic and horrible scene of all. The piteous cries of those around us still ring in my ears, and I will remember them to my dying day. "Hold on to what you have, old boy!" we shouted to each man who tried to get on board. "One more of you would sink us all!" Many of those whom we refused answered as they went to their death, "Good luck – God bless you!"'

Mr George Brayton

'A number of us who were enjoying the crisp air were promenading about the deck. Captain Smith was on the bridge when the first cry from the lookout came that there was an iceberg ahead. It may have been 30 feet high when I saw it. It was possibly 200 yards away and dead ahead. Captain Smith shouted some orders… A number of us promenaders rushed to the bow of the ship. When we saw he could not fail to hit it, we rushed to the stern. Then came a crash, and the passengers were panic-stricken.'

Day 176

24 June | On this day in *Titanic* history

The following people were born on this day:
Wick, Mr George Dennick – 24 June 1854 – aged 57 when on *Titanic*; first-class passenger (ticket price: £164 17s 4d); *Titanic* victim

Mantle, Mr Roland Frederick – 24 June 1871 – aged 40 when on *Titanic*; steward in third class; *Titanic* victim

Tikkanen, Mr Juho – 24 June 1879 – aged 32 when on *Titanic*; third-class passenger (ticket price: £7 18s 6d); *Titanic* victim

The following person died on this day:
Chapman, Mr Joseph Charles – 24 June 1939 – aged 32 when on *Titanic*; boots steward in second class; rescued in lifeboat 9

Titanic survivor quotes

Col. George Dennick Wick, mentioned in *Daily Northwestern*, Wednesday, 17 April 1912

> 'William F. Bonnell, a cousin of the party, made the following statement as coming from Mrs Wick: When the collision occurred, Mrs Wick and her husband were in their stateroom. They thought that a boiler had exploded. When they reached the, deck, all was confusion, Miss Natalia and the Misses Bonnell were on deck. The four women entered the second lifeboat let down. Mrs Wick said the boat was not launched for an hour after the collision. Mr Wick stood at the rail as his wife and daughter were helped into the boat and waved his hand as the party left the *Titanic*. The last seen of him was standing on the deck waving a farewell. Mrs Wick said the party drifted about in the intense cold for five hours before they were picked up. I want to say that Mrs Wick told me that if the lifeboats had been launched as soon as the captain knew the extent of the damage, every one would have been saved.'

Mr Frank Herbert Morris (steward for first-class bathroom)

> 'The women and children were crying and we had great trouble to get the women into the boats.'

Mr Charles Donald Mackay (steward)

> 'They complained about being crushed up so much, and they could not sit down properly, and other people complained because they had to stand all night.'

Day 177

25 June | On this day in *Titanic* history

The following people were born on this day:
Keane, Mr Daniel – 25 June 1876 – aged 35 when on *Titanic*; second-class passenger (ticket price: £12 7s); *Titanic* victim

Godley, Mr George Auguste – 25 June 1873 – aged 38 when on *Titanic*; fireman; rescued in lifeboat unknown

Allum, Mr Owen George – 25 June 1894 – aged 17 when on *Titanic*; third-class passenger (ticket price: £8 6s); *Titanic* victim

The following people died on this day:
Appleton, Mrs Charlotte – 25 June 1924 – aged 53 when on *Titanic*; first-class passenger (ticket price: £51 9s 7d); rescued in lifeboat 2

Scheerlinck, Mr Jean – 25 June 1956 – aged 29 when on *Titanic*; third-class passenger (ticket price: £9 10s); rescued in lifeboat 11

Ryerson, Miss Emily Borie – 25 June 1960 – aged 18 when on *Titanic*; first-class passenger (ticket price: £262 7s 6d); rescued in lifeboat 4

Titanic survivor quotes

Mrs Charlotte Appleton, quoted in *Brooklyn Daily Eagle*, Friday, 19 April 1912

> 'Mr Ismay did everything possible to get the children and women off of the *Titanic*,' said Mrs Appleton. 'When our boat pulled away I heard the band playing distinctly and among the men standing on the deck was Mr Ismay. I heard him order the men to stay back and instruct the men in command of the boats to take all women and children off first. I cannot understand how he got aboard the *Carpathia* himself, for I was under the impression that I had seen him still on deck after the last boat had left the ship.'

Mr George A. Harder, witness at United States Senate Inquiry (Day 13)

> 'That was the second boat to leave on the starboard side, as far as I could see.
>
> As we were being lowered, they lowered one side quicker than the other, but we finally reached the water safely, after a few scares. When we got down into the water, somebody said the plug was not in; so they fished around to see if that was in, and I guess it was in. Then, they could not get the boat detached from the tackle, so they fussed around there for a while, and finally they asked if anybody had a knife, and nobody seemed to have a knife. Finally, one of the passengers had a knife in his possession, and they cut some rope; what it was I do not know.'

Day 178

26 June | On this day in *Titanic* history

The following people were born on this day:

Wittman, Mr Henry – 26 June 1873 – aged 38 when on *Titanic*; bathroom steward in first class; *Titanic* victim

Flynn, Mr John Irwin – 26 June 1875 – aged 36 when on *Titanic*; first-class passenger (ticket price: £26 5s 9d); rescued in lifeboat 5

McGann, Mr James – 26 June 1882 – aged 29 when on *Titanic*; trimmer; rescued in lifeboat unknown

Briant, Mr Albert – 26 June 1877 – aged 34 when on *Titanic*; greaser; *Titanic* victim

Becker, Master Richard F. – 26 June 1910 – aged 1 when on *Titanic*; second-class passenger (ticket price: £39); rescued in lifeboat 11

The following people died on this day:

Hartnell, Mr Fred – 26 June 1920 – aged 21 when on *Titanic*; saloon steward; rescued in lifeboat unknown

Brown, Mrs Caroline Lane – 26 June 1928 – aged 59 when on *Titanic*; first-class passenger (ticket price: £51 9s 7d); rescued in collapsible lifeboat D

Titanic survivor quotes

Mrs Caroline Lane Brown, mentioned in *Worcester Evening Gazette*, Saturday, 20 April 1912

> 'Mrs John Murray Brown of Acton, Mass, who with her sister, Mrs Robert C. Cornell and Mrs E.D. Appleton, was saved, was in the last lifeboat to get safely away from the *Titanic*. "The band played marching from deck to deck, and as the ship went under I could still hear the music," Mrs Brown said. "The musicians were up to their knees in water when last I saw them. We offered assistance to Capt. Smith, but he refused."

Mr Arthur J. Bright

> 'As soon as we got away from the ship, we were told to keep together, if possible, to keep as close to each other in the boats as possible. There was a light sighted away, I should say, possibly 4 or 5 miles away, off the port bow of the ship. It looked to me like a sailing ship – like a fishing boat. There were no lights to be seen about the hull of the ship, if it was a ship. We pulled toward that for a time.'

Day 179

27 June | On this day in *Titanic* history

The following people were born on this day:
Uruchurtu, Don. Manuel Ramirez – 27 June 1872 – aged 39 when on *Titanic*; first-class passenger (ticket price: £27 14s 5d); *Titanic* victim

Butt, Mr Robert Henry – 27 June 1890 – aged 21 when on *Titanic*; saloon steward; *Titanic* victim

The following people died on this day:
Bucknell, Mrs Emma Eliza – 27 June 1927 – aged 58 when on *Titanic*; first-class passenger (ticket price: £76 5s 10d); rescued in lifeboat 8

al-Ba'qlīnī, Miss Mārīyā – 27 June 1982 – aged 5 when on *Titanic*; third-class passenger (ticket price: £19 5s 2d); rescued in collapsible lifeboat C

Titanic survivor quotes

Mrs Emma Eliza Bucknell, mentioned in *Worcester Evening Gazette*, Saturday, 20 April 1912

> 'Mrs William Bucknell of Philadelphia, after telling of taking an oar in a lifeboat and rowing "till her hands were blistered" said:
>
> "After being taken aboard the *Carpathia*, J. Bruce Ismay went to a cabin and remained closeted until waited upon by a committee of the survivors, who demanded that they be permitted to see him. One of the questions that was put to him was as to what the White Star and the International Mercantile Marine Co.'s intended doing in the way of reparation. To this Mr Ismay replied that the company would do all in its power to make a partial repayment for the suffering of the survivors. Further than this he would say nothing."'

Mr George A. Harder

> 'Then we waited out there until the ship went down. We were out there until the ship went down. After it went down, we heard a lot of these cries and yells. You could not hear any shouts for help, or anything like that. It was a sort of continuous yelling or moaning. You could not distinguish any sounds. It was more like – what I thought it was – the steerage on rafts, and that they were all hysterical. That is the way it sounded in the distance. Then we stayed around there until daybreak, when we saw the *Carpathia*, and we rowed the distance; I do not know how far it was; probably 2 miles; it might have been less.'

Day 180

28 June | On this day in *Titanic* history

The following person was born on this day:
Lee, Mr Henry Reginald – 28 June 1882– aged 29 when on *Titanic*; scullion; rescued in lifeboat 13

The following person died on this day:
Tu'mah, Mrs Hinnah – 28 June 1976– aged 27 when on *Titanic*; third-class passenger (ticket price: £15 4s 11d); rescued in collapsible lifeboat C

Titanic survivor quote

Mr Robert W. Daniel

'Not until the last five minutes did the awful realization come that the end was at hand. The lights became dim and went out, but we could see. Slowly, ever so slowly, the surface of the water seemed to come towards us. So gradual was it that even after I had adjusted the life jacket about my body it seemed a dream. Deck after deck was submerged. There was no lurching or grinding or crunching. The *Titanic* simply settled. I was far up on one of the top decks when I jumped. About me were others in the water. My bathrobe floated away, and it was icily cold. I struck out at once. I turned my head, and my first glance took in the people swarming on the *Titanic*'s deck. Hundreds were standing there helpless to ward off approaching death. I saw Captain Smith on the bridge. My eyes seemingly clung to him. The deck from which I had leapt was immersed. The water had risen slowly, and was now to the floor of the bridge. Then it was to Captain Smith's waist. I saw him no more. He died a hero. The bows of the ship were far beneath the surface and to me only the four monster funnels and the two masts were now visible. It was all over in an instant.'

Mr James R. McGough, witness at United States Senate Inquiry (Day 18)

'When our lifeboats left the vessel, we were directed to row away a short distance from the large boat, feeling it would be but a short time until we would be taken back on the *Titanic*. We then rested on our oars; but after realizing that the *Titanic* was really sinking, we rowed away for about half a mile, being afraid that the suction would draw us down. Although there were several of us wanted drinking water, it was unknown to us that there was a tank of water and also some crackers in our boat, having no light on our boat; and we did not discover this fact – that is, as to the tank of water – until after reaching the *Carpathia*.'

Day 181

29 June | On this day in *Titanic* history

The following people were born on this day:

Ashby, Mr John – 29 June 1854 – aged 57 when on *Titanic*; second-class passenger (ticket price: £13); *Titanic* victim

Hilliard, Mr Herbert Henry – 29 June 1867 – aged 44 when on *Titanic*; first-class passenger (ticket price: £51 17s 3d); *Titanic* victim

McCoy, Miss Agnes – 29 June 1882 – aged 29 when on *Titanic*; third-class passenger (ticket price: £23 5s); rescued in lifeboat 16

The following people died on this day:

Leather, Mrs Elizabeth Mary – 29 June 1937 – aged 50 when on *Titanic*; stewardess; rescued in lifeboat 16

Brown, Mrs Elizabeth Catherine – 29 June 1925 – aged 39 when on *Titanic*; second-class passenger (ticket price: £39); rescued in lifeboat 14

Titanic survivor quote

Mrs J. Stuart White

'Nobody ever thought the ship was going down. I do not think there was a person that night, I do not think there was a man on the boat who thought the ship was going down. They speak of the bravery of the men. I do not think there was any particular bravery, because none of the men thought it was going down. If they had thought the ship was going down, they would not have frivoled as they did about it. Some of them said, "When you come back you will need a pass," and, "You can not get on tomorrow morning without a pass." They never would have said these things if anybody had had any idea that the ship was going to sink.

'In my opinion the ship when it went down was broken in two. I think very probably it broke in two. I heard four distinct explosions, which we supposed were the boilers. Of course, we did not know anything about it. We did what we were ordered to do. We went toward the light. That seemed to be the verdict of everybody in the boat. We had strict orders to do that from the officer or whoever started us off – land the passengers and come right back for the others. We all supposed that [the] boat was coming toward us, on account of all the rockets that we had sent up.'

Day 182

30 June | On this day in *Titanic* history

The following people were born on this day:
Gibson, Mrs Pauline Caroline – 30 June 1867 – aged 45 when on *Titanic*; first-class passenger (ticket price: £59 8s); rescued in lifeboat 7

Slight, Mr Harry John – 30 June 1878 – aged 33 when on *Titanic*; steward in third class; *Titanic* victim

Hinton, Mr Stephen William – 30 June 1881 – aged 30 when on *Titanic*; trimmer; *Titanic* victim

Dean, Mr Bertram Frank – 30 June 1886 – aged 25 when on *Titanic*; third-class passenger (ticket price: £20 11s 6d); *Titanic* victim

Donati, Sig. Italo Francesco – 30 June 1894 – aged 17 when on *Titanic*; assistant waiter; *Titanic* victim

The following people died on this day:
Dodge, Dr Washington – 30 June 1919 – aged 52 when on *Titanic*; first-class passenger (ticket price: £81 17s 2d); rescued in lifeboat 13

Backström, Mrs Maria Mathilda – 30 June 1947 – aged 33 when on *Titanic*; third-class passenger (ticket price: £15 17s); rescued in lifeboat unknown

Fredericks, Mr Walter Francis – 30 June 1960 – aged 21 when on *Titanic*; trimmer; rescued in lifeboat 15

Slade, Mr Alfred Edward – 30 June 1965 – aged 25 when on *Titanic*; fireman; left the *Titanic* in Southampton after being out drinking and did not make it back in time to sail and as a result, was listed as deserted

Brown, Miss Amelia Mary – 30 June 1976 – aged 18 when on *Titanic*; second-class passenger (ticket price: £13); rescued in lifeboat 11

Titanic survivor quote

Mr Harold S. Bride

> 'I now assisted in pushing off a collapsible lifeboat, which was on the port side of the forward funnel, onto the boat deck. Just as the boat fell I noticed Capt. Smith dive from the bridge into the sea. Then followed a general scramble down on the boat deck, but no sooner had we got there than the sea washed over. I managed to catch hold of the boat we had previously fixed up and was swept overboard with her.'

July

Day 183

1 July | On this day in *Titanic* history

The following people were born on this day:
Daly, Miss Margaret – 1 July 1878 – aged 33 when on *Titanic*; third-class passenger (ticket price: £6 19s); rescued in lifeboat 15

Thorneycroft, Mrs Florence Kate – 1 July 1879 – aged 32 when on *Titanic*; third-class passenger (ticket price: £16 2s); rescued in lifeboat 10

Robins, Mr Alexander A. – 1 July 1861 – aged 50 when on *Titanic*; third-class passenger (ticket price: £14 10s); *Titanic* victim

Burke, Mr Richard Edward – 1 July 1882 – aged 29 when on *Titanic*; lounge steward in first class; *Titanic* victim

The following person died on this day:
Murdock, Mr William John – 1 July 1941 – aged 37 when on *Titanic*; fireman; rescued in lifeboat unknown

Titanic survivor quotes

Mrs Florence Kate Thorneycroft, mentioned in *Utica Herald-Dispatch*, Saturday, 20 April 1912

> 'They were obliged to remain standing in the Despatched lifeboat for seven or eight hours until picked up. Among those in the boat were a Mrs Dean with a seven-week-old baby [and] a little two-year-old boy, who was clad only in his night dress. Although having to go through such an experience the little chap suffered no serious after results, but the mother who was compelled to stand for so long and hold the little babe, was practically exhausted... Mrs Thorneycroft spoke in the highest terms of the men who manned the boat in which she was... During the night and in the early morning many pieces of wreckage were seen in the water, but no bodies were seen by those in her boat.'

Mr John Collins

> 'We were all telling him not to get on. He said, "That is all right, boys, keep cool," he said; "God bless you," and he bid us goodbye and he swam along for about two minutes and we seen [*sic*] him, but did not see him moving off; we saw his head, but we did not see him moving his hands. Then we were washed out of his road. There were others that tried to get on, but we would not let them on. A big foreigner came up; I think he was a Dutchman. He came up to the stern and he hung on to me all the time.'

Day 184

2 July | On this day in *Titanic* history

The following people were born on this day:
Webber, Miss Susan – 2 July 1874 – aged 38 when on *Titanic*; second-class passenger (ticket price: £13); rescued in lifeboat 12

Ṭannūs Mu'awwad, Mr Ḥannā – 2 July 1877 – aged 34 when on *Titanic*; third-class passenger (ticket price: £6 8s 9d); *Titanic* victim

Shelley, Mrs Imanita Parrish – 2 July 1887 – aged 24 when on *Titanic*; second-class passenger (ticket price: £26); rescued in lifeboat 12

Sage, Mr George John – 2 July 1892 – aged 19 when on *Titanic*; third-class passenger (ticket price: £69 11s); *Titanic* victim

Reeves, Mr Frederick Vernon Hilton – 2 July 1891 – aged 20 when on *Titanic*; assistant pantryman steward; *Titanic* victim

Laroche, Miss Louise – 2 July 1910 – aged 1 when on *Titanic*; second-class passenger (ticket price: £41 11s 7d); rescued in lifeboat 14

The following person died on this day:
Vigott, Mr Philip Francis – 2 July 1941 – aged 32 when on *Titanic*; able seaman; rescued in lifeboat 13

Titanic survivor quotes

Miss Susan Webber, quoted in *Western Times*, Monday, 13 May 1912

> 'I feel that I cannot praise too highly the bravery of the many who were left on deck. We were floating in mid-ocean among the icebergs for six hours. The night was bitterly cold, but very calm. At last we saw the lights of the *Carpathia* coming to our aid; this was a welcome sight. We were taken on her and treated with every kindness, both passengers and crew doing everything they could for the survivors.'

Mrs J. Stuart White

> 'I never saw a finer body of men in my life than the men passengers on this trip – athletes and men of sense – and if they had been permitted to enter these lifeboats with their families the boats would have been appropriately manned and many more lives saved, instead of allowing the stewards to get in the boats and save their lives, under the pretence that they could row, when they knew nothing whatever about it.'

Day 185

3 July | On this day in *Titanic* history

The following person was born on this day:
Cunningham, Mr Andrew Orr – 3 July 1873 – aged 38 when on *Titanic*; bedroom steward in first class; rescued in lifeboat 4

The following person died on this day:
Connolly, Miss Kate – 3 July 1948 – aged 23 when on *Titanic*; third-class passenger (ticket price: £7 15s); rescued in lifeboat 13

Titanic survivor quote

Mr Arthur G. Peuchen

> 'I hardly got back in the grand staircase – I probably waited around there 10 minutes more – when I saw the ladies and gentlemen all coming in off of the deck looking very serious, and I caught up to Mr Beatty, and I said, "What is the matter?" He said, "Why the order is for lifebelts and boats." I could not believe it at first, it seemed so sudden. I said, "Will you tell Mr Ross?" He said, "Yes; I will go and see Mr Ross." I then went to my cabin and changed as quickly as I could from evening dress to heavy clothes. As soon as I got my overcoat on I got my life preserver and I came out of my cabin. In the hallway I met a great many people, ladies and gentlemen, with their lifebelts on, and the ladies were crying, principally, most of them. It was a very serious sight, and I commenced to realize how serious matters were. I then proceeded up to the boat deck, and I saw that they had cleared away.'

Mr Frederick Clench, witness at United States Senate Inquiry (Day 7)

> 'While Mr Lowe was gone I heard shouts. Of course I looked around, and I saw a boat in the way that appeared to be like a funnel. We started to back away then. We thought it was the top of the funnel. I put my head over the gunwale and looked along the water's edge and saw some men on a raft. Then I heard two whistles blown. I sang out "Aye, aye; I am coming over," and we pulled over and found it was a raft – not a raft, exactly, but an overturned boat – and Mr Lightoller was there on that boat, and I believe – I do not know whether I am right or not, but I think the wireless operator was on there, too. We took them on board the boat and we shared the amount of the room that was there.'

Titanic fact

One of the lifeboats was found in the ocean almost a month after the sinking.

Day 186

4 July | On this day in *Titanic* history

The following people were born on this day:
McGough, Mr George Francis 'Paddy' – 4 July 1875 – aged 36 when on *Titanic*; able seaman; rescued in lifeboat unknown

McGough, Mr James Robert – 4 July 1876 – aged 35 when on *Titanic*; first-class passenger (ticket price: £26 5s 9d); rescued in lifeboat 7

McCrie, Mr James Matthew – 4 July 1879 – aged 32 when on *Titanic*; second-class passenger (ticket price: £13); *Titanic* victim

Harder, Mrs Dorothy – 4 July 1890 – aged 21 when on *Titanic*; first-class passenger (ticket price: £55 8s); rescued in lifeboat 5

Johansson, Mr Nils – 4 July 1881 – aged 30 when on *Titanic*; third-class passenger (ticket price: £7 17s 1d); *Titanic* victim

The following people died on this day:
Chaffee, Mrs Carrie Constance – 4 July 1931 – aged 47 when on *Titanic*; first-class passenger (ticket price: £61 3s 6d); rescued in lifeboat 4

Moss, Mr Albert Johan – 4 July 1973 – aged 29 when on *Titanic*; third-class passenger (ticket price: £7 15s 6d); rescued in lifeboat unknown

Leroy, Miss Berthe – 4 July 1972 – aged 27 when on *Titanic*; first-class passenger (ticket price: £106 8s 6d); rescued in lifeboat 2

Wright, Miss Marion – 4 July 1965 – aged 26 when on *Titanic*; second-class passenger (ticket price: £13 10s); rescued in lifeboat 9

Roth, Miss Sarah A. – 4 July 1947 – aged 26 when on *Titanic*; third-class passenger (ticket price: £8 1s); rescued in collapsible lifeboat C

Svensson, Mr Johan Cervin – 4 July 1981 – aged 14 when on *Titanic*; third-class passenger (ticket price: £9 4s 6d); rescued in lifeboat 13

Quick, Miss Winifred Vera – 4 July 2002 – aged 8 when on *Titanic*; second-class passenger (ticket price: £26); rescued in lifeboat 11

Titanic survivor quote

Mr Edward Brown Steward, witness at British Wreck Commissioner's Inquiry (Day 9)

> 'The Captain came past us while we were trying to get this boat away with a megaphone in his hand, and he spoke to us. "Well, boys, do your best for the women and children, and look out for yourselves." He walked on the bridge.'

Day 187

5 July | On this day in *Titanic* history

The following people were born on this day:
Keeping, Mr Edwin Herbert – 5 July 1878 – aged 33 when on *Titanic*; first-class passenger (ticket price £211 10s); *Titanic* victim

Pond, Mr George – 5 July 1879 – aged 32 when on *Titanic*; fireman; *Titanic* victim

Stead, Mr William Thomas – 5 July 1849 – aged 62 when on *Titanic*; first-class passenger (ticket price: £26 11s); *Titanic* victim

Chevré, Mr Paul Romaine Marie Léonce – 5 July 1866 – aged 45 when on *Titanic*; first-class passenger (ticket price: £29 14s); rescued in lifeboat 7

Kenyon, Mrs Marion – 5 July 1880 – aged 40 when on *Titanic*; first-class passenger (ticket price: £51 17s 3d); rescued in lifeboat 8

Hedman, Mr Oskar Arvid – 5 July 1884 – aged 27 when on *Titanic*; third-class passenger (ticket price: £6 19s 6d); rescued in lifeboat 15

Cleaver, Miss Alice Catherine – 5 July 1889 – aged 22 when on *Titanic*; first-class passenger (ticket price: £151 16s); rescued in lifeboat 11

Read, Mr Joseph – 5 July 1891 – aged 20 when on *Titanic*; trimmer; *Titanic* victim

Fahlstrøm, Mr Arne Joma – 5 July 1893 – aged 18 when on *Titanic*; second-class passenger (ticket price: £13); *Titanic* victim

McGowan, Miss Anna 'Annie' – 5 July 1894 – aged 17 when on *Titanic*; third-class passenger (ticket price: £8 7s); rescued in lifeboat unknown

The following person died on this day:
Hoyt, Mr Frederick Maxfield – 5 July 1940 – aged 38 when on *Titanic*; first-class passenger (ticket price: £100); rescued in collapsible lifeboat D

Titanic survivor quote

Mr Charles E. Stengel

> 'I think between Sir Duff-Gordon and myself we decided which way to go. We followed a light that was to the bow of the boat, which looked like in the winter, in the dead of winter, when the windows are frosted with a light coming through them. It was in a haze. Most of the boats rowed toward that light, and after the green lights began to burn I suggested it was better to turn around and go toward the green lights, because I presumed there was an officer of the ship in that boat, and he evidently knew his business.'

Day 188

6 July | On this day in *Titanic* history

The following people were born on this day:
Kelly, Mr James – 6 July 1892 – aged 19 when on *Titanic*; third-class passenger (ticket price: £8 1s); *Titanic* victim

Sage, Mr Frederick – 6 July 1895 – aged 16 when on *Titanic*; third-class passenger (ticket price: £69 11s); *Titanic* victim

Nakid, Mr Sa'īd Antūn – 6 July 1891– aged 20 when on *Titanic*; third-class passenger (ticket price: £15 14s); rescued in lifeboat unknown

Van Den Steen, Mr Leo Peter – 6 July 1883 – aged 28 when on *Titanic*; third-class passenger (ticket price: £9 10s); *Titanic* victim

The following people died on this day:
Prichard, Mrs Alice Maud – 6 July 1948 – aged 36 when on *Titanic*; stewardess; rescued in lifeboat 11

Becker, Miss Ruth Elizabeth – 6 July 1990 – aged 12 when on *Titanic*; second-class passenger (ticket price: £39); rescued in lifeboat 13

Titanic survivor quotes

Miss Ruth Elizabeth Becker, mentioned in *Gettysburg Complier*, Wednesday, 15 May 1912

> 'She awakened the children and wanted to dress them but was told there was no time. She grasped shoes and stockings and in their nightclothes rushed them on deck, where the two smaller children were put into lifeboat No.18. She appealed to be placed in with them, and as the boat was being lowered, she was thrown in with the two small children. Ruth was left behind and placed into the next boat lowered and she found her on board the *Carpathia*. She says she saw the *Titanic* slowly sink bow first, and after the explosion disappear.'

Mr John Collins, witness at United States Senate Inquiry (Day 7)

> 'We rowed for the *Carpathia*, and whenever we got in amongst a lot of wreckage we rowed on ahead. The wind rose, and the waves were coming up, and we were rowing for all we were worth. Then the *Carpathia* blew her horn, and we all seen [*sic*] the *Carpathia*. She stopped in the one place. We were at this time within a mile of her, and she did not make any sign of coming over near to us. She stopped in the one place, and, I think, lowered two or three of her own boats, and her own boats were kept in the water when one of our boats, the sailboat, went up alongside of her.'

Day 189

7 July | On this day in *Titanic* history

The following people were born on this day:
Bengtsson, Mr Johan Viktor – 7 July 1885 – aged 26 when on *Titanic*; third-class passenger (ticket price: £7 15s 6d); *Titanic* victim

Lévy, Mr René Jacques – 7 July 1875 – aged 36 when on *Titanic*; second-class passenger (ticket price: £12 17s 6d); *Titanic* victim

Moran, Mr Daniel J. – 7 July 1883 – aged 28 when on *Titanic*; third-class passenger (ticket price: £24 3s); *Titanic* victim

Millar, Mr Robert – 7 July 1885 – aged 26 when on *Titanic*; extra fifth engineer; *Titanic* victim

The following people died on this day:
Christy, Mrs Alice Frances – 7 July 1939 – aged 52 when on *Titanic*; second-class passenger (ticket price: £30); rescued in lifeboat 12

Karun, Mr Franz – 7 July 1934 – aged 39 when on *Titanic*; third-class passenger (ticket price: £13 13s 4d); rescued in lifeboat 15

Navratil, Master Edmond Roger – 7 July 1953 – aged 2 when on *Titanic*; second-class passenger (ticket price: £26); rescued in lifeboat unknown

Crafter, Mr Frederick Horace – 7 July 1960 – aged 20 when on *Titanic*; saloon steward in first class; rescued in lifeboat 15

Titanic survivor quotes

Mr Franz Karun, mentioned in *Galesburg Evening Mail*, Tuesday, 23 April 1912

> 'When the *Titanic* sank, he heard the cries of the people and saw men struggling in the water after the boat went down. There was much ice floating about. Of the fifty-two persons on board the lifeboat, most of them were women and children, there being over a dozen children. It was four hours and a half before the *Carpathia* was sighted and almost five hours before the half-frozen survivors were taken aboard.'

Mr Ernest Archer

> 'Then when I got to No. 16 boat the officer told me to get into the boat and see that the plug was in; so I got in the boat. I seen [*sic*] that the plug was in tight; then they started to put passengers in, and I assisted to get them in."

Day 190

8 July | On this day in *Titanic* history

The following people were born on this day:
Brown, Mrs Caroline Lane – 8 July 1852 – aged 59 when on *Titanic*; first-class passenger (ticket price: £51 9s 7d); rescued in collapsible lifeboat D

Hewlett, Mrs Mary Dunbar – 8 July 1855 – aged 56 when on *Titanic*; second-class passenger (ticket price: £16); rescued in lifeboat 13

Barlow, Mr Charles Henry – 8 July 1881 – aged 30 when on *Titanic*; fireman; *Titanic* victim

Serreplaà, Miss Augusta Jeanne – 8 July 1881 – aged 30 when on *Titanic*; first-class passenger (ticket price: £31); rescued in lifeboat 4

The following person died on this day:
Rugg, Miss Emily – 8 July 1958 – aged 22 when on *Titanic*; second-class passenger (ticket price: £10 10s); rescued in lifeboat 12

Titanic survivor quotes

Mrs Mary Dunbar Hewlett, quoted in *Evanston Daily News*, Thursday, 25 April 1912

> 'Just as I was climbing the ladder a throng of steerage passengers fought their way up and started clambering for the stairs. When I reached the deck the men forced me into the lifeboat waiting to be lowered. It was boat No. 13 and next to the last lifeboat to leave the doomed vessel. When I got into the lifeboat it was very nearly filled and there were more men than women. Fifty were saved and there were not more than ten women in that number. The majority of the passengers in the boat I was in were steerage people. There were men with their wives in this boat while many first class men were separated from their wives. I cannot understand this'.

Mr Frederick Clench

> 'We was rowing up there, and up come the officer, after the ship was gone down, come up with us with his boat, and transferred some of his people he had in his boat into two boats of ours; I could not say the number of the other boats, but he transferred his people into ours so that he would have a clear boat to go around to look for the people who were floating in the water.'

Titanic fact

Pier 54 in New York City was the place where first- and second-class survivors disembarked; thousands of friends, relatives and sympathisers greeted their arrival.

Day 191

9 July | On this day in *Titanic* history

The following people were born on this day:
Knight, Mr Robert – 9 July 1869 – aged 42 when on *Titanic*; fitter; *Titanic* victim
Jermyn, Miss Annie Jane – 9 July 1885 – aged 26 when on *Titanic*; third-class passenger (ticket price: £7 15s); rescued in collapsible lifeboat D

The following people died on this day:
Renouf, Mrs Lillian – 9 July 1933 – aged 30 when on *Titanic*; second-class passenger (ticket price: £21); rescued in lifeboat unknown

Jacobsohn, Mrs Amy Frances Christy – 9 July 1947 – aged 24 when on *Titanic*; second-class passenger (ticket price: £27); rescued in lifeboat 12

Mrs Lillian Renouf, mentioned in *Newark Evening News*, Friday, 19 April 1912

> 'Mrs Renouf's husband, her two brothers, Clifford and Ernest Jeffries; a cousin, Charles Cann, and Herbert Denbouy, a friend of the family, formed a party. All were lost except Mrs Renouf. All of the party but the Renoufs were residents of England. Mrs Renouf is in a condition bordering on hysteria, and it was between sobs that she told her story to a News reporter who found her at the home of her brother, Frederick Jeffries, in this city. She is about thirty years old. Mrs Renouf said that she cannot realize the awful calamity and that it haunts her all the time. She will make her home with her brother for a while at least.'

Captain Herbert J. Haddock (captain of the steamship *Olympic*)

> 11.10 pm. *Titanic* replies and gives me his position. 41.46 N. 50.14 W., and says: "we have struck an iceberg." Reported this information to bridge immediately. Our distance from the *Titanic*, 505 miles.
>
> 11.20 pm. Signals with the *Titanic*. He says: "Tell captain get your boats ready and what is your position?"
>
> 11.35 pm. Sent message to *Titanic*: "Commander *Titanic*, 4.24 am GMT, 40.52 N. 61.18 W. Are you steering southly to meet us? – Haddock."
>
> 11.40 pm. *Titanic* says, "Tell captain we are putting the passengers off in small boats."
>
> 11.45 pm. Asked *Titanic* what weather he had had. He says, "Clear and calm."
> 11.50 pm. Message to *Titanic*: "Commander *Titanic*, am lighting up all possible boilers as fast as can. – Haddock."

Day 192

10 July | On this day in *Titanic* history

The following people were born on this day:
Archer, Mr Ernest Edward – 10 July 1876 – aged 35 when on *Titanic*; able seaman; rescued in lifeboat 16

O'Connor, Mr Maurice – 10 July 1896 – aged 15 when on *Titanic*; third-class passenger (ticket price: £7 15s); *Titanic* victim

The following person died on this day:
Wells, Miss Joan – 10 July 1933 – aged 4 when on *Titanic*; second-class passenger (ticket price: £23); rescued in lifeboat 14

Titanic survivor quotes

Mr Lawrence Beesley

> 'When arranging a tour around the United States I had decided to cross on the *Titanic*. It was rather a novelty to be on the largest ship yet launched. It was no exaggeration to say that it was quite easy to lose one's way on such a ship.'

Mr Charles H. Lightoller

> 'You could actually walk miles along the decks and passages covering different ground all the time. I was thoroughly familiar with pretty well every type of ship afloat but it took me 14 days before I could, with confidence, find my way from one part of that ship to another.'

Mr Frederick Clench, witness at United States Senate Inquiry (Day 7)

> 'It did not take much to wake me. I am a light sleeper. If anybody touches me, I will jump quick. Of course I put on my trousers and I went on deck on the starboard side of the well deck and I saw a lot of ice.'

Titanic fact

Some of the interior rooms of the *Titanic* were modelled after the Ritz Hotel, with first-class cabins finished in the Empire style.

Day 193

11 July | On this day in *Titanic* history

The following people were born on this day:
Romaine, Mr Charles Hallace – 11 July 1866 – aged 45 when on *Titanic*; first-class passenger (ticket price: £26 11s); rescued in lifeboat 9

O'Keeffe, Mr Patrick – 11 July 1890 – aged 21 when on *Titanic*; third-class passenger (ticket price: £7 15s); rescued in collapsible lifeboat B

Veström, Miss Hulda Amanda Adolfina – 11 July 1897 – aged 14 when on *Titanic*; third-class passenger (ticket price: £7 17s 1d); *Titanic* victim

Lenox-Conyngham, Miss Eileen – 11 July 1900 – aged 11 when on *Titanic*; first-class passenger (ticket price: £3); disembarked at Cherbourg

The following person died on this day:
Fitzpatrick, Mr Cecil William – 11 July 1964 – aged 21 when on *Titanic*; mess steward; rescued in collapsible lifeboat B

Titanic survivor quotes

Mr Charles Hallace Romaine, quoted in *Chicago Daily Journal*, Friday, 19 April 1912

> 'We had been crunching through ice all day,' said Mr Romaine, 'and I had been standing on the deck. I had become chilled and went inside for a warming drink before going to bed. Suddenly there came the shock, and my first thought was that we had struck a larger cake of ice than usual.
>
> 'The boat suddenly tilted, so sharply that my highball slid from the table. Then came a cry: "We're sinking," and the lights grew dimmer and dimmer and finally went out. Even then there was no panic of any kind, although there was a rush to the boats when they were first lowered. The officers in charge commanded, "women and children first," and the men stood back. Few of us, even then, thought there was any real danger.'

Mr Charles Burgess

> 'Never shall I forget the feeling of all on board our boat when the *Carpathia* hove in sight. We who were at the oars pulled with renewed spirits, and one by one the boats took up the hymn "Pull for the Shore, Sailor", as we put our backs to the work. The officers and crew of the *Carpathia* were kind and attentive beyond all praise, as were the American people on our arrival in New York. They fitted us all out with a double suit of everything. On our homeward journey by the *Lapland* we were treated most kindly...'

Day 194

12 July | On this day in *Titanic* history

The following people were born on this day:
Søholt, Mr Peter Andreas Lauritz Andersen – 12 July 1892 – aged 19 when on *Titanic*; third-class passenger (ticket price: £7 13s); *Titanic* victim

al-Ba'qlīnī, Miss Hilānah – 12 July 1909 – aged 2 when on *Titanic*; third-class passenger (ticket price: £19 5s 2d); rescued in collapsible lifeboat C

The following people died on this day:
Cornell, Mrs Malvina Helen – 12 July 1941 – aged 55 when on *Titanic*; first-class passenger (ticket price: £25 14s 10d); rescued in lifeboat 2

Buss, Miss Kate – 12 July 1972 – aged 20 when on *Titanic*; second-class passenger (ticket price: £36 15s); rescued in lifeboat 11

New York Times, Saturday, 12 July 1913

'LONDON, July 11. The King and Queen paid Liverpool a visit today in the course of their Lancashire tour, and inspected the shipping and the new Gladstone Dock.

'Aboard the tender *Galatea*, which took them around the mercantile fleet, their Majesties made the acquaintance of a stewardess, Mrs Robinson, one of the *Titanic* survivors. The King was immediately interested, and plied the stewardess with questions, but Mrs Robinson seemed disinclined to speak of the calamity.

"It is the sort of thing one doesn't like to talk much about afterward," she said to him. "It was too terrible."

"But do tell me this," persisted the King, "do you think more lives could have been saved?"

"Among the third-class passengers I certainly think so," was the reply. "But they appeared to think more of their belongings than they did of themselves."

'Their Majesties congratulated the stewardess on her escape.'

Titanic survivor quote

Mrs J. Stuart White

'Then we turned and went back, and lingered around there for a long time, trying to locate the other boats, but we could not locate them except by hearing them. The only way they could locate us was by my electric lamp. The lamp on the boat was absolutely worth nothing. They tinkered with it all along, but they could not get it in shape. I had an electric cane – a cane with an electric light in it – and that was the only light we had. We sat there for a long time, and we saw the ship go down, distinctly.'

Day 195

13 July | On this day in *Titanic* history

The following people were born on this day:

Astor, Colonel John Jacob – 13 July 1864 – aged 47 when on *Titanic*; first-class passenger (ticket price: £247 10s 6d); *Titanic* victim

Baxter, Mr Quigg Edmond – 13 July 1887 – aged 24 when on *Titanic*; first-class passenger (ticket price: £247 10s 5d); *Titanic* victim

Skoog, Master Karl Thorsten – 13 July 1900 – aged 11 when on *Titanic*; third-class passenger (ticket price: £27 18s); *Titanic* victim

The following people died on this day:

Widener, Mrs Eleanor – 13 July 1937 – aged 50 when on *Titanic*; first-class passenger (ticket price: £211 10s); rescued in lifeboat 4

Halford, Mr Richard – 13 July 1953 – aged 27 when on *Titanic*; steward in third class; rescued in lifeboat 15

Titanic survivor quotes

Mrs Eleanor Widener, mentioned in *Washington Times*, Friday, 19 April 1912

> 'Mrs Widener, said Daniels, did not want to go, and asked to be allowed to stand by her husband. However, Mr Widener told her to save herself and son, and forced her almost to seek the lifeboat. Mrs Widener kissed her husband goodbye. He told her not to worry, as it was possible that all would be saved, and the danger did not seem great. He stood by Colonel Astor and Major Butt, and was one of those who upheld Chief Officer Murdoch, when the latter at the point of a pistol drove back the foreigners who started the panic and stampede'.

Mr Paul Chevré, quoted in *The Times*, Saturday, 20 April 1912

> 'When our boat had rowed about half a mile from the vessel the spectacle was quite fairylike. The *Titanic*, which was illuminated from stern to stern, was perfectly stationary, like some fantastic piece of stage scenery. The night was clear and the sea perfectly smooth, but it was intensely cold. Presently the gigantic ship began to sink by the bows, and then those who had remained on board realised to the full the horror of their situation. Suddenly the lights went out, and an immense clamour filled the air. Little by little the *Titanic* settled down, and for three hours cries were heard. At moments the cries were lulled, and we thought it was all over, but the next instant they were renewed in still keener accents. As for us we did nothing but row, row, row to escape from the obsession of the heartrending cries. One by one the voices were stilled.'

Day 196

14 July | On this day in *Titanic* history

The following people were born on this day:

Artagaveytia, Mr Ramon – 14 July 1840 – aged 71 when on *Titanic*; first-class passenger (ticket price: £49 10s 1d); *Titanic* victim

Goodwin, Mrs Augusta – 14 July 1868 – aged 43 when on *Titanic*; third-class passenger (ticket price: £46 18s); *Titanic* victim

Sæther, Mr Simon Sivertsen – 14 July 1868 – aged 43 when on *Titanic*; third-class passenger (ticket price: £7 5s); *Titanic* victim

Lefebvre, Master Henry – 14 July 1906 – aged 5 when on *Titanic*; third-class passenger (ticket price: £25 9s 4d); *Titanic* victim

Master Henry Lefebvre, mentioned in *Semi Weekly Iowegian*, Friday, 19 April 1912

> 'Frank Lefebre, of Mystic, has almost given away to despair of ever seeing his wife and four children who were on board the ill-fated *Titanic*. They were coming from France to join him after separation of a year while he worked hard in the mines to accumulate enough money to pay their passage in the steerage.
>
> As soon as he got enough ahead he sent for the oldest son who came over to help him and the two were able to make enough to send for the wife and children about five weeks ago. The happy woman wrote that they would come on the *Titanic*.
>
> Her name appears among those missing. It was hoped something further might be heard of her this morning after the survivors landed, but no word has come and now the sorrowing husband and father has almost given up hope. He and his son work at one of the Lodwick mines.'

Titanic survivor quote

Mr James Witter

> 'Well I went down to the working out way where my cabin is, number seven glory hole, I was standing there talking to two or three fellows and the carpenter came along and I heard him say the bloody mail room was full of water. I said, what's that mail room full of water? He said yes. I said well what about those bulkhead doors forward he said they're not holding Jim. Of course then I walked into my cabin, number seven glory hole, and I opened my box. I called everybody. I said come along fellas get up she's going down so they opened my box, took up some matches [and] some cigarettes and I said come on fella get out... "What the hell are you talking about?" he said, "get out of here." Someone threw a boot at me. I said good night gentlemen just as easy as that.'

Day 197

15 July | On this day in *Titanic* history

The following people were born on this day:
Harris, Mrs Irene – 15 July 1876 – aged 35 when on *Titanic*; first-class passenger (ticket price: £83 9s 6d); rescued in collapsible lifeboat D

Fox, Mr Patrick – 15 July 1883 – aged 28 when on *Titanic*; third-class passenger (ticket price: £7 15s); *Titanic* victim

Abrahamsson, Mr August Abraham Johannes – 15 July 1892 – aged 19 when on *Titanic*; third-class passenger (ticket price: £7 18s 6d); rescued in lifeboat 15

Yazbeck, Mrs Selini – 15 July 1896 – aged 17 when on *Titanic*; third-class passenger (ticket price: £14 9s); rescued in lifeboat unknown

Moubarek, Master Halim Gonios ('William George') – 15 July 1907 – aged 4 when on *Titanic*; third-class passenger (ticket price: £15 4s 11d); rescued in collapsible lifeboat C

The following people died on this day:
Icard, Miss Rose Amélie – 15 July 1964 – aged 39 when on *Titanic*; first-class passenger (ticket price: £80); rescued in lifeboat 6

Aks, Master Frank Philip – 15 July 1991 – aged 10 months when on *Titanic*; third-class passenger (ticket price: £9 7s); rescued in lifeboat unknown

Mr Robert Hichens (quartermaster), witness at United States Senate Inquiry (Day 5)

> 'Mr Murdoch, the first officer, sir; the officer in charge. The sixth officer repeated the order, "The helm is hard astarboard, sir." But, during the time, she was crushing the ice, or we could hear the grinding noise along the ship's bottom. I heard the telegraph ring, sir. The skipper came rushing out of his room – Capt. Smith – and asked, "What is that?" Mr Murdoch said, "An iceberg." He said, "Close the emergency doors."

Titanic survivor quote

Mr Frederick Clench, witness at United States Senate Inquiry (Day 7)

> 'The lights went out after the second explosion. Then she gradually sank down into the water very slowly.'

Day 198

16 July | On this day in *Titanic* history

The following people were born on this day:
Parrish, Mrs Lutie Davis – 16 July 1852 – aged 59 when on *Titanic*; second-class passenger (ticket price: £26); rescued in lifeboat 12

Nichols, Mr Albert William Stanley – 16 July 1864 – aged 47 when on *Titanic*; boatswain; *Titanic* victim

The following people died on this day:
Warren, Mrs Anna Sophia – 16 July 1925 – aged 60 when on *Titanic*; first-class passenger (ticket price: £75 5s); rescued in lifeboat 5

Brereton, Mr George Andrew – 16 July 1942 – aged 37 when on *Titanic*; first-class passenger (ticket price: £26 11s); rescued in lifeboat 9

Frölicher, Miss Hedwig Margaritha – 16 July 1972 – aged 22 when on *Titanic*; first-class passenger (ticket price: £49 10s); rescued in lifeboat 5

Titanic survivor quotes

Mrs Anna Sophia Warren, quoted in *Portland Oregonian*, Saturday, 27 April 1912

> 'From the time of the accident until I left the ship there was nothing which in any way resembled a panic, and I believe that a panic would have been impossible owing to the immense size of the vessel, but there seemed to be a sort of aimless confusion and an utter lack of organised effort.
>
> 'Mr Lightoller, one of the officers of the *Titanic* who went down with the vessel, but afterwards was saved, told me that Colonel Gracie assisted him in cutting loose one of the life rafts and in pushing it overboard. Both were carried down with the vessel by the suction, but afterwards found safety on the same raft which they had previously cleared.'

Mr Ernest Archer

> 'We went to them, uncovered them, and got the falls ready for lowering. Then I went over to the starboard side and assisted in lowering about three boats. I could not mention the number of the boats I lowered. I never taken any notice. Then an officer came along – I could not mention his name – and he sang out that they wanted some seamen on the other side, on the port side, to assist over there. I went over then and assisted in getting Nos. 12, 14, and 15 out. I assisted in getting the falls and everything ready, and the passengers into No. 14 boat. Then I went to No. 16.'

Day 199

17 July | On this day in *Titanic* history

The following people were born on this day:
Daly, Mr Peter Dennis – 17 July 1860 – aged 51 when on *Titanic*; first-class passenger (ticket price: £26 11s); rescued in lifeboat unknown

Gale, Mr Shadrach – 17 July 1878 – aged 33 when on *Titanic*; second-class passenger (ticket price: £21); *Titanic* victim

Klasén, Mr Klas Albin – 17 July 1893 – aged 18 when on *Titanic*; third-class passenger (ticket price: £7 17s 1d); *Titanic* victim

Rogers, Mr Reginald Harry – 17 July 1893 – aged 18 when on *Titanic*; second-class passenger (ticket price: £10 10s); *Titanic* victim

The following people died on this day:
Hoyt, Mrs Jane Anne – 17 July 1932 – aged 33 when on *Titanic*; first-class passenger (ticket price: £100); rescued in collapsible lifeboat D

Badman, Miss Emily Louisa – 17 July 1946 – aged 18 when on *Titanic*; third-class passenger (ticket price: £8 1s); rescued in collapsible lifeboat C

Titanic survivor quotes

Mr Peter Dennis Daly, quoted in *New York Times*, Monday, 22 April 1912

> 'I had on my pyjamas and my slippers… and when I felt the shock, I slipped on my overcoat, and went on deck to see what had happened. I was told that the ship had struck an iceberg, and I then proceeded to the port side, but could see nothing but some ice which had fallen on the deck. I heard one young woman say to another: Oh, come and let's see the berg. We never have never seen one before.'

Mr Gus Cohen

> 'When the boat was listing and then I decided to find a lifebelt. I found it very quickly and then I went towards the lifeboats. I never had a chance to get in any one of them because the order was women and children first, so as I was a lad of nineteen I was out of it, so I decided to find my own salvation. I went across to the davit, climbed across the davit, which was a dangerous thing entirely, and went down the rope about 80 feet long and went into the water, into the sea rather.'

Day 200

18 July | On this day in *Titanic* history

The following people were born on this day:
Brown, Mrs Margaret – 18 July 1867 – aged 44 when on *Titanic*; first-class passenger (ticket price: £27 14s 5d); rescued in lifeboat 6

Windebank, Mr Alfred Edgar – 18 July 1873 – aged 38 when on *Titanic*; cook; rescued in lifeboat 13

Van Impe, Mr Jean Baptiste – 18 July 1875 – aged 36 when on *Titanic*; third-class passenger (ticket price: £24 3s); *Titanic* victim

Stone, Mr Edward Thomas – 18 July 1882 – aged 29 when on *Titanic*; bedroom steward in second class; *Titanic* victim

Peters, Miss Catherine – 18 July 1884 – aged 27 when on *Titanic*; third-class passenger (ticket price: £8 2s 9d); *Titanic* victim

Titanic survivor quotes

Mr James Widgery (bath steward)

> 'Several men up on the deck, quite a lot. There were no more women there. Then we were lowered down to the water, and just before we went away the chief officer called out to the boatswain to keep about 100 yards off. We got into the water, and I cut loose the oars – I was the only one that had a knife amongst us – and we stood off a little ways. Of course, we gradually got a little farther away from them all the time.'

Mrs Margaret Brown, quoted in *New York Times*, Saturday, 20 April 1912

> '"We did not wait for the *Carpathia* to come to us. We rowed toward it. We were picked up at 7 am. We were lifted up in a sort of nice little sling that was lowered down to us. After that it was all over. The passengers of the *Carpathia* were so afraid that we would not have room enough that they gave practically the whole ship to ourselves."
>
> 'When seen at the Ritz-Carlton Mrs Brown wore the same dress in which she had gone over the side of the liner in a lifeboat. The black and white silk facing of the black velvet outer coat was discolored, and her white topped shoes were spotted with salt water in which her feet had rested.
>
> "My hair? Oh, that I have not bothered with since Sunday," she said. "But this morning? Did you not rest last night?" "No," she said smilingly. "Why should I? There was work to do. I left the *Carpathia* at 3 o'clock this morning. There were many women there whom I had to look out for."'

Day 201

19 July | On this day in *Titanic* history

The following person was born on this day:
Pettersson, Miss Ellen Natalia – 19 July 1893 – aged 18 when on *Titanic*; third-class passenger (ticket price: £7 15s 6d); *Titanic* victim

The following people died on this day:
Ward, Mr William – 19 July 1941 – aged 37 when on *Titanic*; saloon steward; rescued in lifeboat 9

McCoy, Mr Bernard – 19 July 1945 – aged 24 when on *Titanic*; third-class passenger (ticket price: £23 5s); rescued in lifeboat 16

Mr Bernard McCoy, quoted in *Jersey Journal*, Tuesday, 23 April 1912

> 'When I reached the deck there were many excited persons there. I saw my cousin, Ernest McCoy of Union Hill. We were not long in finding out that the ship was sure to sink. We secured life preservers and strapped them securely on us. The ship was settling badly as we sprang into the water. Ernest jumped from one side of the *Titanic* while I jumped from the other. After being beaten severely by sailors with oars I managed to get into one of the lifeboats and my cousins, Alice and Kate McCoy of Union Hill sat on me and tried to cover me up.
>
> 'After a while one of the sailors saw my legs protruding, and seizing them asked me "what in _____" I was doing in the boat. He dragged me out and tried to throw me into the water. I grabbed him by the throat and said if I went overboard I would take him with me. When he saw that he could not throw me over he finally desisted and I was allowed to remain.'

Titanic survivor quote

Mr Charles Lightoller (second officer)

> 'Just then the ship took a slight but definite plunge – probably a bulkhead went – and the sea came rolling along up in a wave, over the steel fronted bridge, along the deck below us, washing the people back in a dreadful huddled mass. Those that didn't disappear under the water right away, instinctively started to clamber up that part of the deck still out of water, and work their way towards the stern, which was rising steadily out of the water as the bow went down.'

Day 202

20 July | On this day in *Titanic* history

The following people were born on this day:
Baxter, Mr Harry Ross – 20 July 1858 – aged 52 when on *Titanic*; steward in third class; *Titanic* victim

Louch, Mrs Alice Adelaide – 20 July 1869 – aged 42 when on *Titanic*; second-class passenger (ticket price: £26); rescued in lifeboat unknown

Leonard, Mr Matthew – 20 July 1886 – aged 25 when on *Titanic*; steward in third class; *Titanic* victim

The following person died on this day:
Marconi, Guglielmo – 20 July 1937 – aged 37 when *Titanic* set sail; founder of The Wireless Telegraph & Signal Company in the United Kingdom in 1897 (which became the Marconi Company).

Mr Guglielmo Marconi, quoted in *New York Times*, Thursday, 18 April 1912

> '"And just now all the world is thinking of this greatest of sea disasters, concerning which we know so little as yet. I feel that I must speak of it, but I do it reluctantly. I know that you will understand me if I say that all those who have been working with me entertain a true feeling of gratitude that wireless telegraphy has again in this instance helped to save human lives. I also want to take this opportunity to express my thanks to the press for the hearty approval it has given to the invention."
>
> 'At the conclusion of this modest mention of his own part in the saving of a part of the *Titanic*'s list of passengers and crew, there was another ovation, which lasted even longer than the others. The audience would not be satisfied, and the Chairman finally called for silence.'

Titanic survivor quote

Mrs Emily Richards

> 'One woman who spoke a tongue none of us could understand was picked up by the boat and believed that her children were lost. She was entirely mad. When her children were brought to her on the *Carpathia* she was wild with joy and lay down on the children on the floor trying to cover them with her body like a wild beast protecting its young.'

Day 203

21 July | On this day in *Titanic* history

The following people were born on this day:
Mayné, Mlle Berthe Antonine – 21 July 1887 – aged 24 when on *Titanic*; first-class passenger (ticket price: £49 10s 1d); rescued in lifeboat 6

Perry, Mr Edgar Lionel – 21 July 1892 – aged 19 when on *Titanic*; trimmer; rescued in lifeboat unknown

Andersson, Master Sigvard Harald Elias – 21 July 1907 – aged 4 when on *Titanic*; third-class passenger (ticket price: £31 5s 6d); *Titanic* victim

The following person died on this day:
Hendrickson, Mr Charles Osker – 21 July 1956 – aged 28 when on *Titanic*; leading fireman; rescued in lifeboat 1

One unusual story in *The Washington Herald*, Sunday, 21 July 1912

> 'Peter Pryal, a wealthy, retired mariner, of this city, who was a shipmate of Capt. Smith, of the *Titanic*, for seventeen years, declared today that he saw Capt Smith in Baltimore yesterday and talked with him. He said he met him on the street on Thursday and accosted him, but the commander brushed by him angrily.
>
> 'The next day, he declares, he saw him again and followed him around the city to a railroad station, where the captain bought a ticket for Washington. Just as he was about to pass through the gates, Pryal declares, the man turned to him and greeted him by name. "I am on business, don't worry me, Pryal," he said. "Be good to yourself, old shipmate, till we meet again." Pryal is under the care of a physician, suffering from nervous shock brought on by the experience. His physician declares he is absolutely sane. He is well known in Baltimore and an active church member.'

Titanic survivor quote

Mr Joseph Bruce Ismay, chairman and managing director of White Star Line

> 'When we left the ship, it would be on the starboard side. We rowed on, and we thought the light became more distinct, and then it seemed to draw away from us again.'

Titanic fact

Thirty people managed to survive by standing, sitting or kneeling on the upturned hull of collapsible lifeboat B, having failed in their attempts to right it.

Day 204

22 July | On this day in *Titanic* history

The following people were born on this day:
Duff-Gordon, Sir Cosmo Edmund – 22 July 1862 – aged 49 when on *Titanic*; first-class passenger (ticket price: £39 12s); rescued in lifeboat 1

Dahl, Mr Charles Edward – 22 July 1866 – aged 45 when on *Titanic*; third-class passenger (ticket price: £8 1s); rescued in lifeboat 15

Bazzi, Sig. Narciso – 22 July 1878 – aged 33 when on *Titanic*; waiter; *Titanic* victim

Van Impe, Miss Catharina – 22 July 1901 – aged 10 when on *Titanic*; third-class passenger (ticket price: £24 3s); *Titanic* victim

Skoog, Miss Mabel – 22 July 1902 – aged 9 when on *Titanic*; third-class passenger (ticket price: £27 18s); *Titanic* victim

The following person died on this day:
Snyder, Mr John Pillsbury – 22 July 1959 – aged 24 when on *Titanic*; first-class passenger (ticket price: £82 5s 4d); rescued in lifeboat 7

Titanic survivor quote

Sir Cosmo Edmund Duff-Gordon, mentioned in *Newark Star*, Friday, 19 April 1912

> 'I saw Sir Cosmo Duff-Gordon and the rest of his party seated in the boat and saw there was room for one more. I asked the mate if I might get in. Even then I did not think that I was saving my life.
>
> 'The boat was a high one and it was difficult for me to enter it. I finally managed to climb up, but was not very graceful in doing so. In fact, I rolled in and the mate jokingly said: "That is the funniest thing I have seen tonight." I have not seen that mate since. The action of this officer proves to me conclusively that not even the crew was aware of the danger.
>
> 'I wish you to emphatically say that Colonel John Jacob Astor was not threatened by any of the officers. The last seen of him he was calmly walking the upper deck.'

Titanic fact

The Renault Type CB Coupé de Ville owned by first-class passenger William Carter was the only automobile thought to be brought onto the ship.

Day 205

23 July | On this day in *Titanic* history

The following people were born on this day:
Bennett, Mr George Alfred – 23 July 1881 – aged 30 when on *Titanic*; fireman; *Titanic* victim

Caldwell, Mrs Sylvia Mae – 23 July 1883 – aged 28 when on *Titanic*; second-class passenger (ticket price: £29); rescued in lifeboat 13

Sutehall, Mr Henry Jr – 23 July 1886 – aged 25 when on *Titanic*; third-class passenger (ticket price: £7 1s); *Titanic* victim

Bentham, Miss Lillian W. – 23 July 1892 – aged 19 when on *Titanic*; second-class passenger (ticket price: £13); rescued in lifeboat 12

The following people died on this day:
Bessette, Miss Nellie Mayo – 23 July 1944 – aged 39 when on *Titanic*; first-class passenger (ticket price: £63 7s 2d); rescued in lifeboat 8

Mellors, Mr William John – 23 July 1948 – aged 19 when on *Titanic*; second-class passenger (ticket price: £10 10s); rescued in collapsible lifeboat A

Bassani, Mrs Albina – 23 July 1960 – aged 31 when on *Titanic*; first-class passenger (ticket price: £76 5s 10d); rescued in lifeboat 8

Ranger, Mr Thomas – 23 July 1964 – aged 29 when on *Titanic*; greaser; rescued in lifeboat 4

Titanic survivor quote

Mrs Sylvia Mae Caldwell, mentioned in *New York Herald*, Friday, 19 April 1912

> 'Albert F. Caldwell, his wife Sylvia, and their ten-month-old boy, A.G. Caldwell, who went to the Chelsea Hotel on their arrival at New York, according to a friend of the family, were asleep at the time of the accident and were awakened by the fearful shock and thrown from their beds. Hearing the cry, "All hands on deck!" they rushed out just as the lights were extinguished. There was a great confusion and Mrs Caldwell was suddenly taken and placed in one of the lifeboats.
>
> 'Her husband followed her to the side and was about to hand the baby to his wife when she called "Can't my husband come aboard with the baby?" Some one behind him shouted "Sure!" and he was shoved into the boat beside his wife, a couple of other men jumping in on top of him. As they rowed away one of the ropes caught against the vessel, causing the lifeboat to tip, and they had great difficulty in cutting loose. No officer seemed to be in charge, and one of the sailors finally succeeded in getting the boat free, and they drifted away. Several shots were fired, seemingly from some foreigners who stood on the deck of the vessel behind them.'

Day 206

24 July | On this day in *Titanic* history

The following people were born on this day:
Nicholls, Mr Joseph Charles – 24 July 1892 – aged 20 when on *Titanic*; second-class passenger (ticket price: £36 15s); *Titanic* victim

Cribb, Miss Laura Mae – 24 July 1895 – aged 16 when on *Titanic*; third-class passenger (ticket price: £16 2s); rescued in lifeboat 12

The following people died on this day:
McGough, Mr James Robert – 24 July 1937 – aged 35 when on *Titanic*; first-class passenger (ticket price: £26 5s 9d); rescued in lifeboat 7

Toms, Mr Fred – 24 July 1937 – aged 29 when on *Titanic*; saloon steward; rescued in lifeboat 15

Titanic survivor quote

Miss Laura Mae Cribb, quoted in *Newark Star*, Saturday, 20 April 1912

> "When we were taken aboard the *Carpathia* we were asked to what class we belonged. I, of course, told them I was of the third-class, as did others, but immediately we became aware that there was a difference between the accommodations of the two ships for the same classification of passengers. Such things as table linens and other homelike features were missing in our new habitations, we found. But eventually foreigners of a certain class were relegated to another part of the ship and we were given what was probably more attention than was usual for that grade of passengers. In fact, there was one of my father's old friends who saw that I and some of my new acquaintances got some of the things on the bill of fare that the second-class passengers got, and so we did not fare so badly after all, so far as eating was concerned."
>
> 'Miss Cribb is decidedly English in her speech and manner, but by birth she is an American, having been born in Newark seventeen years ago. The last eight years she has lived in Bornemouthe [*sic*], England, where her mother and two brothers and a sister still remain. She and her father had preceded the rest of the family on what was to have been their return to this country for the purpose of making their home.
>
> 'Mr Cribb was for several years assistant steward of the Essex Club of this city and he was later butler for Herbert Ballentine. In a similar capacity he was in the employ of Frank Gould on the yacht, Helenita, and until last September was butler at the home of E. S. Repello, 841 Madison avenue, New York. It was to the last mentioned place to which he had expected to return on his arrival upon the *Titanic*.'

Day 207

25 July | On this day in *Titanic* history

The following people were born on this day:
Williams, Mr Fletcher Fellowes Lambert – 25 July 1868 – aged 43 when on *Titanic*; first-class passenger (ticket price: £35); *Titanic* victim

Ashcroft, Mr Austin Aloysius – 25 July 1885 – aged 26 when on *Titanic*; clerk; *Titanic* victim

Kink, Mr Vincenz – 25 July 1885 – aged 26 when on *Titanic*; third-class passenger (ticket price: £8 13s 3d); *Titanic* victim

Salander, Mr Karl Johan – 25 July 1887 – aged 24 when on *Titanic*; third-class passenger (ticket price: £9 6s 6d); *Titanic* victim

The following person died on this day:
Dodge, Mrs Ruth – 25 July 1950 – aged 38 when on *Titanic*; first-class passenger (ticket price: £81 17s 2d); rescued in lifeboat 5

Titanic survivor quote

Mrs Ruth Dodge, quoted in *San Francisco Bulletin*, Tuesday, 30 April 1912

> 'The most terrible part of the experience was that awful crying after the ship went down. We were a mile away, but we heard it, oh how we heard it. It seemed to last about an hour, although it may have been only a short time, for some say a man could not have lived in that water over fifteen minutes. At last it died down.
>
> 'Our officer and the members of the crew wanted to go back and pick up those whom they could, but the women in the boat would not let them. They told them if they attempted to turn back their husbands would take the oars from them, and the other men outnumbered the crew. I told them I could not see how they could forbid turning back in the face of those awful cries. I will remember it until I die, as it is. I told them: "How do I know, you have your husbands with you, but my husband may be one of those who are crying."
>
> 'They argued that if we got back where the people were struggling, some of the steerage passengers, crazed with fear and the cold, might capsize the boat struggling to get in, or might force the officers to overload so we would all go down.
>
> 'After the crying died down, two or three of the women became hysterical about what I don't know; they were missing none of their people. I was trying to keep baby from realization of what was happening, but when these women shrieked he would begin crying and asking, "Where's papa?"'

Titanic fact

There were around 3,500 cork-filled life jackets on *Titanic*.

Day 208

26 July | On this day in *Titanic* history

The following people were born on this day:
Gilbert, Mr William – 26 July 1865 – aged 46 when on *Titanic*; second-class passenger (ticket price: £10 10s); *Titanic* victim

Johansson Palmquist, Mr Oskar Leander – 26 July 1885 – aged 26 when on *Titanic*; third-class passenger (ticket price: £7 15s6d); rescued in lifeboat 15

Hocking, Mr Richard George – 26 July 1888 – aged 23 when on *Titanic*; second-class passenger (ticket price: £11 10s); *Titanic* victim

The following people died on this day:
Ellis, Mr John Bertie – 26 July 1932 – aged 28 when on *Titanic*; assistant vegetable cook; rescued in lifeboat 2

McIntyre, Mr William – 26 July 1965 – aged 21 when on *Titanic*; trimmer; rescued in lifeboat unknown

Titanic survivor quote

Mr Richard George Hocking, mentioned in *Akron Beacon Journal*, Saturday, 20 April 1912

> 'Mrs Richards, or others who escaped, know nothing of the fate of George Hocking. They last saw him on Sunday evening before the crash came. He occupied a stateroom with two friends, Harry Cotterill and William Bailey. Until they reach New York members of the family hoped that George and his friends had been saved by some boat other than the *Carpathia*. Mrs Richards was seen at the Star Hotel at 57 Clarkson street and said: "I had put the children in bed and had gone to bed myself. We had been making good time all day, the ship was rushing through the sea at a tremendous rate and the air on deck was cold and crisp. I didn't hear the collision for I was asleep. But my mother came and shook me. 'There is surely danger,' she said to me, 'something has gone wrong.' So we put on our slippers and outside coats and got the children into theirs and went on deck. We had on our night gowns under our coats. As we went up the stairway some one was shouting down in a calm voice 'Everybody put on their life preservers before coming to deck.'
>
> "We went back and put them on assuring each other that it was nothing. When we got on deck we were told to pass through the dining room to a ladder that was placed against the side of the cabins and led to the upper deck. We were put through the port holes into the boats and the boat I was in had a foot of water in it. As soon as we were in we were told to sit down on the bottom. In that position we were so low we could not see over the gunwale."'

Day 209

27 July | On this day in *Titanic* history

The following people were born on this day:
Williams, Mr Walter John – 27 July 1883 – aged 28 when on *Titanic*; saloon steward in second class; rescued in lifeboat unknown

Doherty, Mr William John – 27 July 1889 – aged 21 when on *Titanic*; third-class passenger (ticket price: £8 9s 2d); *Titanic* victim

Jardim, Mr José Neto – 27 July 1890 – aged 21 when on *Titanic*; third-class passenger (ticket price: £7 1s); *Titanic* victim

Lines, Miss Mary Conover – 27 July 1895 – aged 16 when on *Titanic*; first-class passenger (ticket price: £39 8s); rescued in lifeboat 9

Quick, Miss Phyllis May – 27 July 1909 – aged 2 when on *Titanic*; second-class passenger (ticket price: £26); rescued in lifeboat 11

The following people died on this day:
Young, Miss Marie Grice – 27 July 1959 – aged 36 when on *Titanic*; first-class passenger (ticket price: £135 12s 8d); rescued in lifeboat 8

Abelseth, Miss Karen Marie – 27 July 1969 – aged 16 when on *Titanic*; third-class passenger (ticket price: £7 13s); rescued in lifeboat 16

Titanic survivor quote

Miss Phyllis May Quick, mentioned in *Detroit Journal*, Saturday, 20 April 1912

> 'Mrs Jane Quick, wife of Frederick Quick, 383 Brooklyn Avenue, with her two daughters, one aged eight and the other three years, all survivors of the *Titanic*, also arrived on the same train with Mrs Hamlin.
>
> "Jane!" cried a man's voice as Mrs Quick prepared to step from the coach with her children. A man rushed up, unconscious of the stares of curious travellers, and took his wife in his arms. The two children danced with glee at the pleasure of being home again with their father.
>
> "Oh Fred, it was terrible," sobbed Mrs Quick as she buried her head on her husband's shoulder, "I never expected to see my loved ones again."
>
> 'Mrs Quick and her children came from Plymouth, England, to join her husband who works for a Detroit contractor. With her children Mrs Quick was placed in lifeboat 11.'

Titanic fact

There were 48 lifebelts (rings) on *Titanic*.

Day 210

28 July | On this day in *Titanic* history

The following people were born on this day:
Backström, Mrs Maria Mathilda – 28 July 1878 – aged 33 when on *Titanic*; third-class passenger (ticket price: £15 17s); rescued in lifeboat unknown

Clarke, Mr John Frederick Preston – 28 July 1883 – aged 28 when on *Titanic*; musician in *Titanic*'s orchestra; *Titanic* victim

Williams, Mr Charles Eugene – 28 July 1888 – aged 23 when on *Titanic*; second-class passenger (ticket price: £13); rescued in lifeboat 14

Mullin, Miss Catherine – 28 July 1890 – aged 21 when on *Titanic*; third-class passenger (ticket price: £7 14s 8d); rescued in lifeboat 16

The following person died on this day:
Hedman, Mr Oskar Arvid – 28 July 1961– aged 27 when on *Titanic*; third-class passenger (ticket price: £6 19s 6d); rescued in lifeboat 15

Mr John Frederick Preston Clarke, mentioned in *Birkenhead News*, Saturday, 20 April 1912

> 'One of those on board the *Titanic*, and who it is feared is amongst those who have perished, is Mr J. F. P. Clarke, formerly a member of the orchestra of the Argyle Theatre of Varieties. Mr Clarke in his short time he played the Bass at the theatre.
>
> 'He then got employment with the White Star Line and was appointed as a member of the orchestra formed to render music on the ill-fated *Titanic*. He was in Birkenhead in the early part of last week, and said goodbye to friends preparatory to leaving for Southampton to join the *Titanic*. Mr Clarke is very well known in musical circles. Only two Saturdays ago he was in Birkenhead and was very well. He attended the performance at the Argyle Theatre and met a number of friends there.
>
> 'He appeared rather morose and he said had hopes of receiving great benefit from his voyage on the *Titanic*. After completing his services with the White Star Company, he hoped to embark on a musical season again. He was an enthusiastic musician, and has played in the Liverpool Philharmonic band. He had a large circle of friends in Birkenhead.
>
> 'The survivors from the *Titanic* are loud in their praise of the members of the orchestra. After the collision they brought their instruments on deck and played almost until the moment that the ship broke in two and foundered. The last item of a concert, such as surely has never taken place in the history of the sea, is reported to have been Nearer My God to Thee.'

Day 211

29 July | On this Day in *Titanic* history

The following people were born on this day:
Bonnell, Miss Elizabeth – 29 July 1850 – aged 61 when on *Titanic*; first-class passenger (ticket price: £26 11s); rescued in lifeboat 8

Bucknell, Mrs Emma Eliza – 29 July 1853 – aged 58 when on *Titanic*; first-class passenger (ticket price: £76 5s 10d); rescued in lifeboat 8

Ekström, Mr Johan – 29 July 1866 – aged 45 when on *Titanic*; third-class passenger (ticket price: £6 19s 6d); *Titanic* victim

Casey, Mr Thomas – 29 July 1867 – aged 44 when on *Titanic*; trimmer; *Titanic* victim

Perreault, Miss Mary Anne – 29 July 1878 – aged 33 when on *Titanic*; first-class passenger (ticket price: £93 10s); rescued in lifeboat 3

Dick, Mr Albert Adrian – 29 July 1880 – aged 31 when on *Titanic*; first-class passenger (ticket price: £57); rescued in lifeboat 3

Persson, Mr Ernst Ulrik – 29 July 1886 – aged 25 when on *Titanic*; third-class passenger (ticket price: £7 15s 6d); rescued in lifeboat 15

Parker, Mr Clifford Richard – 29 July 1894 – aged 20 when on *Titanic*; clerk; *Titanic* victim

Goodwin, Master Harold Victor – 29 July 1901 – aged 10 when on *Titanic*; third-class passenger (ticket price: £46 18s); *Titanic* victim

The following people died on this day:
Crosby, Mrs Catherine Elizabeth – 29 July 1920 – aged 64 when on *Titanic*; first-class passenger (ticket price: £71); rescued in lifeboat 7

Davis, Miss Mary – 29 July 1987 – aged 28 when on *Titanic*; second-class passenger (ticket price: £13); rescued in lifeboat unknown

Titanic survivor quote

Miss Mary Davis, *Perth Amboy Evening News*, 22 April 1912

> 'Miss Mary Davis of Tottenville, Staten Island, who is one of the survivors of the *Titanic* catastrophe... spoke particularly of the bravery of Col. Astor and Major Archibald Butt. She also told of seeing First Officer Murdock commit suicide by shooting. She said she was awakened when the liner struck the berg and ran out of her stateroom to the deck, clad only in a nightdress and a light jacket. She explained how she and others were placed in lifeboats and lowered into the water. She said that after she was placed in a lifeboat she was shoved overboard, but was quickly rescued by one of the sailors who manned the boat. She said she was unconscious when taken onboard the *Carpathia* and remained in that condition for many hours.'

Day 212

30 July | On this day in *Titanic* history

The following person were born on this day:
Bowen, Mr David John 'Dai' – 30 July 1891 – aged 20 when on *Titanic*; third-class passenger (ticket price: £16 2s); *Titanic* victim

The following people died on this day:
Slocombe, Mrs Maude Louise – 30 July 1967 – aged 30 when on *Titanic*; Turkish bath stewardess; rescued in lifeboat unknown

Nakid, Miss Maria – 30 July 1912 – aged 1 when on *Titanic*; third-class passenger (ticket price: £15 14s); rescued in lifeboat unknown

Titanic survivor quote

Miss Mariayam Nakid, mentioned in *Waterbury Republican*, Thursday, 25 April 1912

> 'My wife and I had been sleeping soundly and did not feel any jar. We could see by the slope of the floor that the boat was going down and not waiting to put on any more clothes we rushed up to the deck with the baby. We had to go up three storeys before we came to where the boats were. At that time the bow of the boat was way down in the water and the stern was sticking up at a slant, way out of the water… The boat we got into was one of the last to leave the ship and it left in such a hurry that there were only about twenty on board. Six of these were men to row the boat and there was one other man. The sailors saw my wife who had only her nightgown on, and me with the baby, and motioned for her to take a seat in the boat. She did so. I helped her over the side of the boat and was going to get in with her when one of the sailors pushed me back and motioned for me to stay behind. I pointed to the baby and he took it away from me and gave it to my wife. The baby started to cry and reached out her hands to me, but even then the sailors would not let me get in. Just then I saw another boat nearby being loaded preparatory to going in the water. It was nearly full of women when I saw a man try to get in. The sailors held him back but he managed to break thru [*sic*] them and jumped into the boat. When he stood up a sailor pulled out a revolver and shot him. The man's body tumbled over the side of the boat and that was the last I saw of him. I saw it would not be well to take a chance like that and waited. The baby pleaded with the sailors to let me get into the boat but they would not allow it. There was plenty of room there for more, and as there were no more women to get in, I determined to take a chance. As the boat was being lowered I jumped into it and fell flat on my face. The women covered me over with their skirts and I laid there. There had been so much confusion in casting off that the sailors did not see me jump and of course did not see me afterwards...'

Day 213

31 July | On this day in *Titanic* history

The following people were born on this day:
Self, Mr Alfred Henry – 31 July 1872 – aged 39 when on *Titanic*; greaser; *Titanic* victim

Mackie, Mr William Dickson – 31 July 1879 – aged 32 when on *Titanic*; junior fifth engineer; *Titanic* victim

Slocombe, Mrs Maude Louise – 31 July 1881 – aged 30 when on *Titanic*; Turkish bath stewardess; rescued in lifeboat unknown

Hamilton, Mr Ernest – 31 July 1886 – aged 25 when on *Titanic*; assistant smoke room steward in first class; *Titanic* victim

Davies, Master John Morgan Jr – 31 July 1903 – aged 8 when on *Titanic*; second-class passenger (ticket price: £36 15s); rescued in lifeboat 14

Titanic survivor quote

Master John Morgan Jr Davies, mentioned in *Akron Beacon Journal*, Saturday, 20 April 1912

> 'It is a thrilling story that Mrs Wells tells of her night ride to safety. Like so many others she did not realize her peril and had she not been literally forced into a lifeboat might have shared the fate of 1,600 others.
>
> "We were all in bed," Mrs Wells said, "Joan was asleep, but I was not. When the crash came I took the children and went on deck. I hadn't more than got there when some one grabbed me, saying: 'This way,' and hustled me and the children up to the lifeboat. An officer was shouting, 'Come on here, lively now, this way, women and children' and before I knew what was happening we were in a lifeboat, and the boat was going down the side while the men stood back serious and sober, watching us...
>
> "There were 40 or 50 in our boat and I couldn't get a chance to sit down, but stood up keeping the babies warm and dry in my skirts. The sailors pulled at the oars for all they were worth, but the boat kept drifting back against the ship. Finally we got away a hundred feet and we didn't have any more trouble. We spent the night in the boat and were picked up at daybreak."'

Titanic fact

Three actors in *A Night to Remember* (1958) would all later play 'Q' in the *James Bond* films: Desmond Llewelyn (Seaman at Steerage Gate), Geoffrey Bayldon (Wireless Operator, Cyril Evans) and Alec McCowen (Wireless Operator, Harold Thomas Cottam). Llewelyn played the 'Q' role in seventeen official *James Bond* films starting with *From Russia with Love* (1963) through to *The World Is Not Enough* (1999), Bayldon in *Casino Royale* (1967) and McCowen in *Never Say Never Again* (1983).

August

Day 214

1 August | On this day in *Titanic* history

The following people were born on this day:
Ward, Miss Annie Moore – 1 August 1874 – aged 37 when on *Titanic*; first-class passenger (ticket price: £512 6s 7d); rescued in lifeboat 3

Valvassori, Sig. Ettore Luigi – 1 August 1876 – aged 35 when on *Titanic*; waiter; *Titanic* victim

Turpin, Mr William John – 1 August 1882 – aged 29 when on *Titanic*; second-class passenger (ticket price: £21); *Titanic* victim

Jouannault, Mr Georges Jules – 1 August 1887 – aged 24 when on *Titanic*; assistant sauce cook; *Titanic* victim

The following person died on this day:
Cardeza, Mrs Charlotte Wardle – 1 August 1939 – aged 58 when on *Titanic*; first-class passenger (ticket price: £512 6s 7d); rescued in lifeboat 3

On 1 August 1980, the adventure film *Raise the Titanic* was released with a running time of 114 minutes. Directed by Jerry Jameson and written by Eric Hughes and Adam Kennedy, it was based on the book of the same name by Clive Cussler. It starred Jason Robards, Richard Jordan, David Selby, Anne Archer and Sir Alec Guinness. The film grossed around $7 million against an estimated $40 million budget. Producer Lew Grade remarked 'it would have been cheaper to lower the Atlantic'.

Titanic survivor quote

Mrs Charlotte Wardle Cardeza, quoted in *Washington Times*, Friday, 19 April 1912

> 'We crashed into the iceberg,' said Mrs Cardeza, 'at about 12 o'clock at night. I cannot bear to recall it. I can only thank God that fate was kind to us. With the sinking of the *Titanic* there were all kinds of terrible reports, among which was that Captain Smith, when he was pushed aboard a life raft, dove into the deep rather than suffer the mariner's shame of being saved. Four died, I understand, in the lifeboats before we were picked up. The terrible cold killed them.'

Titanic fact

On *Titanic*, there were five grand pianos, three of which were for first-class passengers. Among those three was Steinway's crown jewel, the Model B, which was specially modified for the ship. There was also a Model R, which was discontinued in 1942, and a Model K, which remains in production even to this day.

Day 215

2 August | On this day in *Titanic* history

The following people were born on this day:
Kirkham, Mr James – 2 August 1868 – aged 43 when on *Titanic*; greaser; *Titanic* victim

Bright, Mr Arthur John – 2 August 1869 – aged 42 when on *Titanic*; quartermaster; rescued in collapsible lifeboat D

Wick, Miss Mary Natalie – 2 August 1880 – aged 31 when on *Titanic*; first-class passenger (ticket price: £164 17s 4d); rescued in lifeboat 8

Crafter, Mr Frederick Horace – 2 August 1891 – aged 20 when on *Titanic*; saloon steward in first class; rescued in lifeboat 15

Ervine, Mr Albert George – 2 August 1893 – aged 18 when on *Titanic*; assistant electrician, *Titanic* victim

The following person died on this day:
Landergren, Miss Aurora Adelia – 2 August 1947 – aged 22 when on *Titanic*; third-class passenger (ticket price: £7 5s); rescued in lifeboat 13

Titanic survivor quote

Miss Mary Natalie Wick, quoted in *Washington Times*, Friday, 19 April 1912

> 'The boat we were in was the second to let down over the side, but the first to strike the water. In it, though it would have held more, were but twenty women, two sailors, and a steward. The latter were to do the rowing. As we took to the oars the officer shouted to us to row over to a distant light and to land there, sending the boat back for others.
>
> 'We watched the other boats being lowered as we got under way. And then, in a few minutes, we noticed that the *Titanic* began to list more heavily. After a while, when we were a considerable distance away, a whole deck of lights, the lowest deck, was suddenly snuffed out. At the same time the mast lights dropped a little farther down in the star-pointed sky.
>
> 'After this the tragedy moved with a relentless swiftness. Deck by deck we watched the lights go out, as the boat dropped lower and lower into the sea. At last but four rows of lights were left. Then the water reached the port holes, and as it rushed in here, there was one great explosion, and another, and then the ship left the horizon unbroken. And those that were in the lifeboats which were close to the vessel say that the orchestra played till the very last, and that the men went down into the sea singing "Nearer, My God, To Thee".'

Day 216

3 August | On this day in *Titanic* history

The following people were born on this day:
Hayter, Mr Arthur – 3 August 1867 – aged 44 when on *Titanic*; bedroom steward; *Titanic* victim

Sage, Mrs Annie Elizabeth – 3 August 1867 – aged 44 when on *Titanic*; third-class passenger (ticket price: £69 11s); *Titanic* victim

Joughin, Mr Charles John – 3 August 1879 – aged 33 when on *Titanic*; chief baker; rescued in collapsible lifeboat B

Pålsson, Mrs Alma Cornelia – 3 August 1882 – aged 29 when on *Titanic*; third-class passenger (ticket price: £21 1s 6d); *Titanic* victim

Ström, Mrs Elna Matilda – 3 August 1882 – aged 29 when on *Titanic*; third-class passenger (ticket price: £10 9s 3d); *Titanic* victim

Ryerson, Miss Susan Parker 'Suzette' – 3 August 1890 – aged 21 when on *Titanic*; first-class passenger (ticket price: £262 7s 6d); rescued in lifeboat 4

Smyth, Miss Julia – 3 August 1893 – aged 18 when on *Titanic*; third-class passenger (ticket price: £7 14s 8d); rescued in lifeboat 13

The following people died on this day:
Harrison, Mr Aragõa Drummond – 3 August 1947 – aged 40 when on *Titanic*; saloon steward in first class; rescued in lifeboat 9

Vartanian, Mr David – 3 August 1966 – aged 22 when on *Titanic*; third-class passenger (ticket price: £7 4s 6d); rescued in lifeboat unknown

Stähelin-Maeglin, Dr Max – 3 August 1968 – aged 32 when on *Titanic*; first-class passenger (ticket price: £30 10s); rescued in lifeboat 3

Mr David Vartanian, mentioned in *The Times*, Monday, 13 May 1912

> 'In the disaster to the *Titanic* Armenia has furnished more than her quota of unfortunate passengers. Six Armenians from Keghi, travelling to Marseilles, had proceeded to Cherbourg and taken passage in the *Titanic*. They were: Haroutioun Der Zakarian aged 35, Mamprè Der Zakarian (22), Neshan Krikorian (22), Sarkis Mardirosian (21), David Vartanian (20) and Arsen Siraganian (20). The first five were married men. Of these Neshan Krikorian and David Vartanian were among the saved, and are reported to be undergoing medical treatment at a hospital in New York.'

Titanic survivor quote

Mr Charles H. Lightoller, witness at United States Senate Inquiry (Day 5)

> 'They were not at all eager to get into the boat, anyway, any of them. I had to sing out. Naturally, no one looked on it as serious and they were not in any hurry to go down to the sea in a boat.'

Day 217

4 August | On this day in *Titanic* history

The following people were born on this day:

Van De Velde, Mr Johannes Josef – 4 August 1876 – aged 35 when on *Titanic*; third-class passenger (ticket price: £9 10s); *Titanic* victim

Midtsjø, Mr Karl Albert – 4 August 1890 – aged 21 when on *Titanic*; third-class passenger (ticket price: £7 15s 6d); rescued in lifeboat 15

Burrage, Mr Arthur Victor Edwards – 4 August 1891 – aged 20 when on *Titanic*; plate steward; rescued in lifeboat 13

Glynn, Miss Mary Agatha – 4 August 1893 – aged 18 when on *Titanic*; third-class passenger (ticket price: £7 15s); rescued in lifeboat 13

The following people died on this day:

Davies, Mrs Elizabeth Agnes Mary – 4 August 1933 – aged 48 when on *Titanic*; second-class passenger (ticket price: £36 15s); rescued in lifeboat 14

Perkis, Mr Walter John – 4 August 1954 – aged 37 when on *Titanic*; quartermaster; rescued in lifeboat 4

Bryhl, Miss Dagmar Jenny Ingeborg – 4 August 1969 – aged 20 when on *Titanic*; second-class passenger (ticket price: £26s); rescued in lifeboat 12

Cohen, Mr Gurshon 'Gus' – 4 August 1978 – aged 19 when on *Titanic*; third-class passenger (ticket price: £8 1s); rescued in lifeboat 12

Mr Karl Albert Midtsjø, quoted in *Chicago Evening Post*, Tuesday, 23 April 1912

> 'We than started out, feeling that everything was all right. After we had gone a few steps a young lady of our party came back from the upper deck and we asked her what was going on up there. She said the order had been given to put on the lifebelts. We returned to our staterooms, which were only a few feet away, got our lifebelts and notified all the women in the corridor to do the same and come with us. None of us knew how to put on the belts, but I saw an officer in the companionway and he showed us how to put them on, telling us that there was no danger and that everything would be all right.
>
> 'When we arrived upon the deck only a few people were there. As it was about seventy-five feet from the boat deck to the water, the officers were having great difficulty in getting the people to go into the lifeboats, assuring them at the same time that it would not be a long while before they probably would be back on the big boat.'

Day 218

5 August | On this day in *Titanic* history

The following people were born on this day:
Sloan, Miss Mary Jane – 5 August 1866 – aged 45 when on *Titanic*; stewardess; rescued in lifeboat unknown

Robert, Mrs Elisabeth Walton (née McMillan) – 5 August 1868 – aged 43 when on *Titanic*; first-class passenger (ticket price: £211 6s 9d); rescued in lifeboat 2

Miss Mary Jane Sloan, mentioned in *Bucks Free Press*, Friday, 26 April 1912

> 'Amongst the crew on the ill-fated *Titanic* was Miss M Sloan, a sister of Mr Thomas Sloan of West-street yard superintendent at Messrs Wethered's brewery. Miss Sloan was chief stewardess on the great liner, having been transferred from the *Olympic* when the latter met with an accident. The news of the disaster caused grave anxiety to the relatives, and it was not until Friday that the welcome intelligence arrived that Miss Sloan was amongst the survivors, a cablegram being sent from New York to Mr Sloan by his sister.'

Titanic survivor quote

Miss Edith Eileen Brown

> 'I could see nothing, but the shrieks and the screams that filled the air were terrible. I presume that it was just as she went under water that the most horrible moan as from a thousand lips came over the water as that many men and I believe some women were hurled into the water and struggled about trying to find something on which to save themselves. After the awful moan it was quiet for a moment and the screams broke out again, but I heard no cries for help.'

Mrs Catherine Crosby, witness at United States Senate Inquiry (Day 18)

> 'I was suffering from the cold while I was drifting around, and one of the officers put a sail around me and over my head to keep me warm, and I was hindered from seeing any of the other lifeboats, drifting in the vicinity or observe anything that took place while we were drifting around until the *Carpathia* took us on board: we received very good treatment on the *Carpathia*, and finally arrived to New York; it was reported on the *Carpathia* by passengers, whose names I do not recollect, that the lookout who was on duty at the time the *Titanic* struck the iceberg had said: "I know they will blame me for it, because I was on duty, but it was not my fault; I had warned the officers three or four times before striking the iceberg that we were in the vicinity of icebergs, but the officer on the bridge paid no attention to my signals."'

Day 219

6 August | On this day in *Titanic* history

The following people were born on this day:
Mallet, Mr Albert Denis Pierre – 6 August 1866 – aged 45 when on *Titanic*; second-class passenger (ticket price: £37 1s); *Titanic* victim

Corbett, Mrs Irene – 6 August 1881 – aged 30 when on *Titanic*; second-class passenger (ticket price: £13); *Titanic* victim

Robinson, Mr James William – 6 August 1881 – aged 30 when on *Titanic*; saloon steward in first class; *Titanic* victim

Port, Mr Frank – 6 August 1889 – aged 22 when on *Titanic*; steward; rescued in lifeboat unknown

Parsons, Mr Richard Henry – 6 August 1893 – aged 18 when on *Titanic*; saloon steward in second class; *Titanic* victim

The following people died on this day:
Lee, Mr Reginald Robinson – 6 August 1913 – aged 41 when on *Titanic*; lookout; rescued in lifeboat 13

Theissinger, Mr Alfred – 6 August 1949 – aged 46 when on *Titanic*; bedroom steward; rescued in lifeboat unknown

Noel, Master William Henry Middleton – 6 August 1954 – aged 13 when on *Titanic*; first-class passenger (ticket price: £3); disembarked at Cherbourg

Titanic survivor quote

Mr Alfred Theissinger, quoted in *Washington Herald*, Friday, 19 April 1912

> 'As we stood there the *Titanic* suddenly sank. I say suddenly, but it seemed to be a gradual sinking. I remember the band. Those brave men were playing the immortal hymn, Nearer, My God, to Thee. Then an awful cry arose out of the depths of the sea. The vessel had disappeared. The unfortunates who found no means of escape dotted the ocean. All was silent.
>
> 'Before daylight a lifeboat came near to us on the raft. We were rescued and taken aboard. Selbert was dead. Then the shout: A steamer!
>
> 'And finally we were rescued by the *Carpathia*. The *Carpathia* had provided for our comfort and each of us was given coffee and sandwiches. I am through with the sea; but for the heroes that stuck to their posts I have only words of praise. For the captain I feel sorrowful. He was well named, The grand old man of the sea.'

Titanic fact

Cosmo Duff-Gordon gave £5 to each of the seven crew members who had shared his lifeboat.

Day 220

7 August | On this day in *Titanic* history

The following people were born on this day:
Compton, Mrs Mary Eliza – 7 August 1847 – aged 64 when on *Titanic*; first-class passenger (ticket price: £83 3s 2d); rescued in lifeboat 14

Smith, Mrs Mary Eloise – 7 August 1893 – aged 18 when on *Titanic*; first-class passenger (ticket price: £60); rescued in lifeboat 6

The following people died on this day:
Allison, Master Hudson Trevor – 7 August 1929 – aged 11 months when on *Titanic*; second-class passenger (ticket price: £151 16s); rescued in lifeboat 11

Parrish, Mrs Lutie Davis – 7 August 1930 – aged 59 when on *Titanic*; second-class passenger (ticket price: £26); rescued in lifeboat 12

Mrs Mary Eliza Compton, quoted in *Newark Evening News*, Friday, 19 April 1912

> 'Mrs Alexander T. Compton and her daughter, Miss Alice Compton, of Lakewood, N. J., and New Orleans, two of the *Titanic*s rescued, reached here completely prostrated over the loss of Mrs Compton's son Alexander, who went down with the big liner. Mr Compton was a large stockholder in the Laurel House, Lakewood, and was also financially interested in the Waumbek, Jefferson, N. H. The family spent considerable time in Lakewood.'

Titanic survivor quote

Mr Frederick Clench, witness at United States Senate Inquiry (Day 7)

> 'Mr Lightoller took charge of us and sighted the *Carpathia*'s lights. Then we started heading for that. We had to row a tidy distance to the *Carpathia*, because there was boats ahead of us, you see, and we had a boat in tow with us, besides all the people we had aboard.'

Titanic fact

If all the lifeboats were loaded to full capacity during an emergency evacuation, there were available seats for only one-third of all those on board. *Titanic*'s supply of lifeboats actually exceeded the British Board of Trade's requirements.

Day 221

8 August | On this day in *Titanic* history

The following people were born on this day:
Brown, Mr Thomas William Solomon – 8 August 1851 – aged 60 when on *Titanic*; second-class passenger (ticket price: £39); *Titanic* victim

Walker, Mr William Anderson – 8 August 1863 – aged 48 when on *Titanic*; first-class passenger (ticket price: £34 5s); *Titanic* victim

Bessant, Mr William Edward Lowe – 8 August 1871 – aged 40 when on *Titanic*; fireman; *Titanic* victim

Grønnestad, Mr Daniel Danielsen – 8 August 1879 – aged 32 when on *Titanic*; third-class passenger (ticket price: £8 6s 15d); *Titanic* victim

Ellis, Mr John Bertie – 8 August 1883 – aged 28 when on *Titanic*; assistant vegetable cook; rescued in lifeboat 2

Jones, Mr Albert Hugh Brabner – 8 August 1894 – aged 17 when on *Titanic*; saloon steward in second class; *Titanic* victim

The following people died on this day:
Spedden, Master Robert Douglas – 8 August 1915 – aged 6 when on *Titanic*; first-class passenger (ticket price: £134 10s); rescued in lifeboat 3

Laroche, Miss Simonne Marie Anne Andrée – 8 August 1973 – aged 3 when on *Titanic*; second-class passenger (ticket price: £41 11s 7d); rescued in lifeboat 10

Titanic survivor quote

Mme Juliette Marie Louise Laroche, quoted in *Le Matin*, Friday, 3 May 1912

'Three weeks ago, Mme Laroche, her two little girls and her husband boarded a huge liner at Cherbourg. Only three of them returned… Mme Laroche recounted the dramatic hours they had lived through.

"After the impact, the panic was terrible... people were pushing and shoving, we were completely surrounded. I could feel that someone had snatched my eldest daughter from my arms, my little Simone... I saw her being thrown into a lifeboat which was suspended above the abyss. 'My little girl,' I cried, 'my little girl! They've taken my little girl away!' But at the same moment, I also felt... hands were lifting me... I found myself in the lifeboat, beside my little Simone... There, on the bridge, in the midst of the crush, I caught a glimpse of my husband, who, both arms outstretched above the crowd, was holding our younger daughter. He was trying to shield her from all the pushing... he was struggling with the crewmen, and was showing them the little girl, to try to convince them that she had been separated from me, her mother... Finally, someone grasped my little Louise from my husband's arms, she would soon be in my own arms… I hardly had time to shout a final farewell to my husband. I heard his voice, above the hubub, shouting to me: 'I'll see you soon, my darling! There'll be room for everyone, go with the lifeboats... Look after our little girls... See you soon!'

Day 222

9 August | On this day in *Titanic* history

The following people were born on this day:
Kirkaldy, Mr Thomas Benjamin – 9 August 1876 – aged 35 when on *Titanic*; bedroom steward; *Titanic* victim

Keen, Mr Percy Edward – 9 August 1881 – aged 30 when on *Titanic*; saloon steward in first class; rescued in lifeboat 15

Hume, Mr John Law – 9 August 1890 – aged 21 when on *Titanic*; musician in *Titanic*'s orchestra; *Titanic* victim

Sandström, Miss Beatrice Irene – 9 August 1910 – aged 1 when on *Titanic*; third-class passenger (ticket price: £16 14s); rescued in lifeboat 13

The following person died on this day:
Dropkin, Miss Jennie – 9 August 1951 – aged 24 when on *Titanic*; third-class passenger (ticket price: £8 1s); rescued in lifeboat unknown

Titanic survivor quotes

Miss Beatrice Irene Sandström, mentioned in *Chicago Daily News*, Wednesday, 24 April 1912

> 'Mrs Agnes Sandström and her two little daughters were exhausted with their long trip. She told of their escape from the sinking ship.
>
> "We were placed in the boat and they swung us down into the water," said Mrs Sandström. "We didn't want to leave the ship. They told us before we started that nothing could happen to us. That was why so many people were drowned. They thought the ship could not sink, but they did not trust the lifeboats. After we got in the water the men rowed and rowed and rowed. That's all I know. I was scared for my babies."

Mr Jack B. Thayer

> 'About this time people began jumping from the stern, my friend Milton Long and myself stood beside each other and jumped on the rail. We did not give each other any messages for back home cause neither thought we would ever get back.'

Mr Frederick Clench, witness at United States Senate Inquiry (Day 7)

> 'Mr Lowe was in No. 14 boat, and he sings out, "Anybody attempting to get into these boats while we are lowering them, I will shoot them," and he shot three shots.'

Day 223

10 August | On this day in *Titanic* history

The following people were born on this day:

Baxter, Mr Thomas Ferguson – 10 August 1856 – aged 55 when on *Titanic*; linen steward in first class; *Titanic* victim

Lundahl, Mr Johan Svensson – 10 August 1861 – aged 50 when on *Titanic*; third-class passenger (ticket price: £7 1s 1d); *Titanic* victim

Ryerson, Mrs Emily Maria – 10 August 1863 – aged 48 when on *Titanic*; first-class passenger (ticket price: £262 7s 6d); rescued in lifeboat 4

Hardy, Mr John – 10 August 1871 – aged 40 when on *Titanic*; chief steward in second class; rescued in collapsible lifeboat D

Leroy, Miss Berthe – 10 August 1884 – aged 27 when on *Titanic*; first-class passenger (ticket price: £106 8s 6d); rescued in lifeboat 2

Terrell, Mr Frank – 10 August 1884 – aged 27 when on *Titanic*; assistant saloon steward; rescued in lifeboat 11

Sirota, Mr Morris – 10 August 1892 – aged 19 when on *Titanic*; third-class passenger (ticket price: £8 1s); *Titanic* victim

Titanic survivor quotes

Mrs Eva Hart

> 'There was nothing, just deathly, terrible silence in the dark night with the stars overhead.'

Mr Lawrence Beesley

> 'The history of the RMS *Titanic* of the White Star Line, is one of the most tragically short it is possible to conceive. The world had waited expectantly for its launching and again for its sailing; had read accounts of its tremendous size and its unexampled completeness and luxury; had felt it a matter of the greatest satisfaction that such a comfortable and above all such a safe boat had been designed and built – the "unsinkable lifeboat" – and then in a moment to hear that it had gone to the bottom as if it had been the veriest tramp steamer of a few hundred tons; and with it fifteen hundred passengers, some of them known all the world over! The improbability of such a thing ever happening was what staggered humanity.'

Titanic fact

It took approximately four hours to unload all of the lifeboats after *Carpathia* arrived.

Day 224

11 August | On this day in *Titanic* history

The following people were born on this day:
Williams, Mr Charles Duane – 11 August 1860 – aged 51 when on *Titanic*; first-class passenger (ticket price: £61 7s 7d); *Titanic* victim

Perkis, Mr Walter John – 11 August 1874 – aged 37 when on *Titanic*; quartermaster; rescued in lifeboat 4

Millar, Mr Thomas – 11 August 1879 – aged 32 when on *Titanic*; deck engineer; *Titanic* victim

Couch, Mr Frank – 11 August 1884 – aged 27 when on *Titanic*; able seaman; *Titanic* victim

Madigan, Miss Margaret – 11 August 1890 – aged 21 when on *Titanic*; third-class passenger (ticket price: £7 15s); rescued in lifeboat 15

The following people died on this day:
Shulver, Mr Christopher Arthur (John Dilley) – 11 August 1922 – aged 28 when on *Titanic*; fireman; rescued in lifeboat unknown

Longley, Miss Gretchen Fiske – 11 August 1965 – aged 21 when on *Titanic*; first-class passenger (ticket price: £77 19s 2d); rescued in lifeboat 10

Titanic survivor quotes

Mr Christopher Arthur Shulver, quoted in *The Stevens Point Journal*, Saturday, 27 April 1912

> 'The stokers were beginning to get alarmed over it, but the officers told us to keep our mouths shut. They didn't want to alarm the passengers. Another fireman said that because of the fire the ship sank more rapidly than otherwise would have been the case. It had been necessary to take the coal out of sections two and three on the starboard side forward, he said. And when the water came rushing in after the collision the bulkheads would not hold because they did not have the supporting weight of the coal. Somebody reported to Chief Engineer Bell that the forward bulkhead had given way and he replied: My God, we are lost.'

Titanic fact

The swimming pool on F Deck was 30ft long and 14ft wide.

Day 225

12 August | On this day in *Titanic* history

The following people were born on this day:
Andrews, Miss Kornelia Theodosia – 12 August 1849 – aged 62 when on *Titanic*; first-class passenger (ticket price: £77 19s 2d); rescued in lifeboat 10

Wiklund, Mr Karl Johan – 12 August 1893 – aged 18 when on *Titanic*; third-class passenger (ticket price: £6 9s 11d); *Titanic* victim

Goodwin, Master William Frederick – 12 August 1898 – aged 13 when on *Titanic*; third-class passenger (ticket price: £46 18s); rescued in lifeboat 14

The following person died on this day:
Touma, Miss Maria Youssef ('Mary Thomas') – 12 August 1953 – aged 9 when on *Titanic*; third-class passenger (ticket price: £15 4s 11d); rescued in collapsible lifeboat C

Titanic survivor quote

Mrs Nellie Becker

'One steward carried Marion, another Richard, and Ruth and I followed. When we got to the lifeboat they were just lowering it. They put Marion and Richard in and then shouted. "That's all for this boat." But I screamed, "Those are my children. Oh! please, let me go with them," and the kind officer, who stood near, picked me up and almost threw me in as it was being lowered.

'Arriving at the *Carpathia*, some of that crew came into our boat, and putting the children into sacks, they were pulled up. The women were put into swings – tied in – for we were so numb from cold that we could not move or help ourselves.

'Looking back over it all, it seems nothing short of a miracle that we were all saved. Surely the Lord walked by our side the whole way, and saved us or we never should have been saved. How I am praising Him for His loving kindness, and thanking Him with every breath that my dear husband was not with us. I left him ill in India, and Dr Kugler said to me before I left, "I think Mr Becker should go home with you, but we will try the mountains first." How often I have thought of that I am convinced that the Lord directed her decision. It all seems like a horrible nightmare to me yet – something that cannot be true. But it has driven me closer to my Saviour and made me feel that nothing is too much to do for Him who has been so wonderfully good to me.'

Day 226

13 August | On this day in *Titanic* history

The following people were born on this day:
Boothby, Mr Walter Thomas – 13 August 1874 – aged 37 when on *Titanic*; bedroom steward in second class; *Titanic* victim

Navratil, Mr Michel – 13 August 1880 – aged 31 when on *Titanic*; second-class passenger (ticket price: £26); *Titanic* victim

Frölicher, Miss Hedwig Margaritha – 13 August 1889 – aged 22 when on *Titanic*; first-class passenger (ticket price: £49 10s); rescued in lifeboat 5

West, Miss Constance Mirium – 13 August 1907 – aged 4 when on *Titanic*; second-class passenger (ticket price: £27 15s); rescued in lifeboat 11

The following people died on this day:
Niskänen, Mr Juha – 13 August 1927 – aged 45 when on *Titanic*; third-class passenger (ticket price: £7 18s); rescued in lifeboat 9

Chaudanson, Miss Victorine – 13 August 1962 – aged 36 when on *Titanic*; first-class passenger (ticket price: £262 7s 6d); rescued in lifeboat 4

Combes, Mr George – 13 August 1916 – aged 34 when on *Titanic*; fireman; rescued in lifeboat 3

Titanic survivor quotes

Mr Charles H. Lightoller (second officer)

> 'Striking the water was like a thousand knives being driven into one's body. The temperature was 28 degrees, four degrees below freezing.'

Mr Lawrence Beesley

> 'Then creeping over the edge of the sea we saw a single light and presently a second below it. It seemed almost too good to be true and I think everyone's eyes were filled with tears, men's as well as women's. All around us we heard shouts and cheers.'

Mrs Millvina Dean

> 'Nobody knew about me and the *Titanic*, to be honest, nobody took any interest, so I took no interest either. But then they found the wreck, and after they found the wreck, they found me.'

Day 227

14 August | On this day in *Titanic* history

The following person was born on this day:
Smith, Mr Lucian Philip – 14 August 1887 – aged 24 when on *Titanic*; first-class passenger (ticket price: £60); *Titanic* victim

The following person died on this day:
Asplund, Mr Johan Charles – 14 August 1943 – aged 23 when on *Titanic*; third-class passenger (ticket price: £7 15s 11d); rescued in lifeboat 13

Titanic survivor quotes

Mr J. Bruce Ismay, chairman and managing of the White Star Line

> 'What do you think I am? Do you believe that I'm the sort that would have left that ship as long as there were any women and children on board? That's the thing that hurts, and it hurts all the more because it is so false and baseless. I have searched my mind with deepest care, I have thought long over each single incident that I could recall of that wreck. I'm sure that nothing wrong was done; that I did nothing that I should not have done. My conscience is clear and I have not been a lenient judge of my own acts.'

Mr Reginald Robinson Lee (lookout)

> 'My mate happened to pass the remark to me. He said, "Well; if we can see through that we will be lucky." That was when we began to notice there was a haze on the water. There was nothing in sight.'

Mr Colonel Archibald Gracie

> 'My friend Clinch Smith made the proposition that we should leave and go toward the stern. But there arose before us from the decks below a mass of humanity several lines deep converging on the Boat Deck facing us and completely blocking our passage to the stern. There were women in the crowd as well as men and these seemed to be steerage passengers who had just come up from the decks below. Even among these people there was no hysterical cry, no evidence of panic. Oh, the agony of it.'

Titanic fact

The captain ordered firearms to be given to some officers to stop passengers rushing lifeboats.

Day 228

15 August | On this day in *Titanic* history

The following people were born on this day:
Potter, Mrs Lily Alexenia – 15 August 1855 – aged 56 when on *Titanic*; first-class passenger (ticket price: £83 3s 2d); rescued in lifeboat 7

Humphreys, Mr Sidney – 15 August 1859 – aged 52 when on *Titanic*; quartermaster; rescued in lifeboat 11

Kearl, Mr George Edward – 15 August 1886 – aged 25 when on *Titanic*; trimmer; *Titanic* victim

Proudfoot, Mr Richard Royston – 15 August 1890 – aged 21 when on *Titanic*; trimmer; *Titanic* victim

Pedrini, Sig. Alessandro – 15 August 1890 – aged 21 when on *Titanic*; assistant waiter; *Titanic* victim

Perotti, Sig. Alfonso – 15 August 1891 – aged 20 when on *Titanic*; assistant waiter; *Titanic* victim

The following person died on this day:
Sandström, Miss Marguerite Rut – 15 August 1963 – aged 4 when on *Titanic*; third-class passenger (ticket price: £16 14s); rescued in lifeboat 13

Titanic survivor quote

Mr Sidney Humphreys, mentioned in *Chicago Inter Ocean*, Sunday, 21 April 1912

> 'Humphreys said the boys were called to their regular posts in the main cabin entry and taken in charge by their captain, a steward. They were ordered to remain in the cabin and not get in the way. Throughout the first hour of confusion and terror these lads sat quietly on their benches in various parts of the first cabin.
>
> 'Then, just toward the end, when the order was passed around that the ship was going down and every man was free to save himself if he kept away from the lifeboats in which the women were taken, the bellboys scattered to all parts of the ship.
>
> 'Humphreys said he saw numbers of them smoking cigarettes and joking with the passengers. They seemed to think that their violation of the rule against smoking while on duty was a sufficient breach of discipline. Not one of them attempted to enter a lifeboat. Not one of them was saved.'

Titanic fact

The cost of using the heated swimming pool was 1 shilling; it was reserved for first-class passengers only.

Day 229

16 August | On this day in *Titanic* history

The following people were born on this day:
Mangiavacchi, Mr Emilio Giovanni Andrea Onorato – 16 August 1864 – aged 47 when on *Titanic*; clerk; *Titanic* victim

Byrne, Mr James Edward – 16 August 1867 – aged 44 when on *Titanic*; bedroom steward in second class; *Titanic* victim

Compton, Mr Alexander Taylor Jr – 16 August 1874 – aged 37 when on *Titanic*; first-class passenger (ticket price: £83 3s 2d); *Titanic* victim

Mock, Mr Philipp Edmund – 16 August 1881 – aged 30 when on *Titanic*; first-class passenger (ticket price: £57 15s); rescued in lifeboat 11

Sägesser, Mlle Emma – 16 August 1887 – aged 24 when on *Titanic*; first-class passenger (ticket price: £69 6s); rescued in lifeboat 9

The following person died on this day:
Abrahamsson, Mr August Abraham Johannes – 16 August 1961 – aged 19 when on *Titanic*; third-class passenger (ticket price: £7 18s 6d); rescued in lifeboat 15

Titanic survivor quotes

Mr Philipp Edmund Mock, quoted in *Worcester Telegram*, Saturday, 20 April 1912

> 'Many men were hanging on to rafts in the sea,' said Mr Mock. 'William T. Stead, the author, and Col. John Jacob Astor clung to a raft. Their feet became frozen and they were compelled to release their hold. Both were drowned.'

Mr John Collins (assistant cook)

> 'I ran… to the port side on the saloon deck with another steward and a woman and two children, and the steward had one of the children in his arms and the woman was crying. I took the child off of the woman and made for one of the boats. Then the word came around…there was a collapsible boat getting launched on the starboard side and that all women and children were to make for it. So me and another steward and the two children and the woman came around…the starboard side… We saw the collapsible boat taken off of the saloon deck, and then the sailors and the firemen that were forward seen the ship's bow in the water and seen that she was intending to sink her bow. They shouted out for all they were worth we were to go aft… and we were just turning around and making for the stern end when the wave washed us off the deck – washed us clear of it – and the child was washed out of my arms; and the wreckage and the people that was around me, they kept me down for at least two or three minutes under the water.'

Day 230

17 August | On this day in *Titanic* history

The following person was born on this day:
Shorney, Mr Charles Joseph – 17 August 1889 – aged 22 when on *Titanic*; second-class passenger (ticket price: £8 1s); *Titanic* victim

Titanic survivor quotes

Miss Kornelia Theodosia Andrews, mentioned in *Daily Home News*, Saturday, 20 April 1912

'Miss Andrews told yesterday how they waited for the fourth lifeboat, because there was not room for the three together in the first three boats. When they got out on the water they found that their men companions, who had said they could row, had done so only for the purpose of saving themselves. Miss Longley had to take an oar with the one able bodied seaman in the boat.

'In a boat alongside of them a sailor lighted a cigarette. He flung the match carelessly among the women, who screamed their protests. "Ah, we're all going to hell, anyway," replied the sailor, "and we might as well be cremated now as then."'

Mr Charles H. Lightoller

'Just then the ship took a slight but definite plunge – probably a bulkhead went – and the sea came rolling along up in a wave, over the steel fronted bridge, along the deck below us, washing the people back in a dreadful huddled mass. Those that didn't disappear under the water right away, instinctively started to clamber up that part of the deck still out of water, and work their way towards the stern, which was rising steadily out of the water as the bow went down. It was a sight that doesn't bear dwelling on – to stand there, above the wheelhouse, and on our quarters, watching the frantic struggles to climb up the sloping deck, utterly unable to even hold out a helping hand.'

Mr Jack B. Thayer

'The partly filled lifeboat standing by about 100 yards away never came back. Why on Earth they never came back is a mystery. How could any human being fail to heed those cries.'

Titanic fact

If Captain Smith was sending a message to another ship, it would go free of charge because the Marconi Company allowed the captain and the officers of the ship a grant of so many words free of charge.

Day 231

18 August | On this day in *Titanic* history

The following people were born on this day:
Brewe, Dr Arthur Jackson – 18 August 1866 – aged 45 when on *Titanic*; first-class passenger (ticket price: £39 12s); *Titanic* victim

Ward, Mr William – 18 August 1874 – aged 37 when on *Titanic*; saloon steward; rescued in lifeboat 9

McCoy, Miss Alicia – 18 August 1885 – aged 26 when on *Titanic*; third-class passenger (ticket price: £23 5s); rescued in lifeboat 16

Brown, Miss Amelia Mary – 18 August 1893 – aged 18 when on *Titanic*; second-class passenger (ticket price: £13); rescued in lifeboat 11

Odell, Master Jack – 18 August 1900 – aged 11 when on *Titanic*; first-class passenger (ticket price: £24); disembarked at Queenstown

Rice, Master Arthur – 18 August 1907 – aged 4 when on *Titanic*; third-class passenger (ticket price: £29 2s 6d); *Titanic* victim

Titanic survivor quotes

Mrs Edith Brown Haisman

> 'I was in lifeboat 13. I always remembered that. My father was waving to us and talking to a clergyman, the Rev. Carter. The *Titanic* went in the ice and I heard three bangs. Before we hit, there had been terrific vibrations from the engines during the night as the ship was really racing over the sea. As the lifeboat pulled away we heard cries from people left on the *Titanic* and in the water and explosions in the ship. There were lots of bodies floating... We were in the lifeboat nine hours. I kept looking in the water for my father and when we reached New York we went to the hospitals to see if he had been picked up.'

Mrs Lucian P. Smith

> 'There was some delay in getting lifeboats down: in fact, we had plenty of time to sit in the gymnasium and chat with another gentleman and his wife. I kept asking my husband if I could remain with him rather than go in a lifeboat. He promised me I could. There was no commotion, no panic, and no one seemed to be particularly frightened; in fact, most of the people seemed interested in the unusual occurrence, many having crossed 50 and 60 times. However, I noticed my husband was busy talking to any officer he came in contact with; still I had not the least suspicion of the scarcity of lifeboats, or I never should have left my husband.'

Titanic fact

For the voyage to New York, *Titanic* had 250 barrels of flour, 10,000lb cereal and 1,000 loaves of bread.

Day 232

19 August | On this day in *Titanic* history

The following people were born on this day:
Jenner, Mr Thomas Henry (Harry) – 19 August 1856 – aged 55 when on *Titanic*; saloon steward; *Titanic* victim

Karlsson, Mr Nils August – 19 August 1889 – aged 22 when on *Titanic*; third-class passenger (ticket price: £7 10s 5d); *Titanic* victim

Prior, Mr Harold John – 19 August 1890 – aged 20 when on *Titanic*; steward in third class; rescued in lifeboat unknown

The following person died on this day:
Scarrott, Mr Joseph George – 19 August 1938 – aged 33 when on *Titanic*; able seaman; rescued in lifeboat 14

Titanic survivor quotes

Mr Joseph George Scarrott, quoted in *The Sphere*, April 1912

> 'All our boats then proceeded towards the *Carpathia*. She had stopped right over where our ship had gone down. She had got our wireless message for assistance. When we got alongside we were got aboard as soon as possible. We found some survivors had already been picked up. Everything was in readiness for us: dry clothes, blankets, beds, hot coffee, spirits, etc.; everything to comfort us. The last of the survivors were got aboard about 8.30 am. The dead bodies that were in some of the boats were taken aboard and after identification were given a proper burial. They were two male passengers, one fireman, and one able seaman. We steamed about in the vicinity for a few hours in the hope of finding some more survivors, but did not find any.'

Mr Alfred White (greaser in the engine room)

> 'I was on the whale deck in the bow calling the watch that was to relieve when the ice first came aboard. The collision opened the seams below the waterline but did not even scratch the paint above the line. I know that because I was one of those who helped to make an examination over the side with a lantern. I went down into the engine room at 12.40 am. We even made coffee, so there was not much thought of danger. An hour later I was still working at the light engines. I heard the chief engineer tell one of his subordinates that number six bulkhead had given way.'

Day 233

20 August | On this day in *Titanic* history

The following people were born on this day:
Vioni, Sig. Roberto – 20 August 1886 – aged 25 when on *Titanic*; waiter; *Titanic* victim

McMicken, Mr Arthur – 20 August 1885 – aged 26 when on *Titanic*; saloon steward in first class; rescued in lifeboat 11

Bunnell, Mr Wilfred James – 20 August 1891 – aged 20 when on *Titanic*; plate steward; *Titanic* victim

Desvernine, Mr Louis Gabriel – 20 August 1891 – aged 20 when on *Titanic*; assistant pastry cook; *Titanic* victim

Ford, Mr William Neal Thomas – 20 August 1897 – aged 14 when on *Titanic*; third-class passenger (ticket price: £34 7s 6d); *Titanic* victim

Van Billiard, Master James William – 20 August 1901 – aged 10 when on *Titanic*; third-class passenger (ticket price: £14 10s); *Titanic* victim

The following people died on this day:
Ball, Mr Percy – 20 August 1933 – aged 18 when on *Titanic*; plate steward; rescued in lifeboat 13

Angle, Mrs Florence 'Mary' Agnes – 20 August 1969 – aged 36 when on *Titanic*; second-class passenger (ticket price: £26); rescued in lifeboat 11

Titanic survivor quotes

Mr Charles Burgess

> 'As we dropped we pushed away from the ship's side with our oars, and the rush of water miraculously caught the bow and forced us away just in time as the other boat dropped alongside. We pulled away from the ship for about ten minutes and then laid on our oars waiting, as we expected to get orders to return to the ship again. We had no idea that she would sink or that the damage done was so great. It was then we noticed that she was sinking by the head. Slowly the lights from the portholes became extinguished as the water rose up deck after deck.'

Mr Lawrence Beesley

> 'Each night the sun sank right in our eyes along the sea, making an undulating glittering pathway, a golden track charted on the surface of the ocean which our ship followed unswervingly until the sun dipped below the edge of the horizon, and the pathway ran ahead of us faster than we could steam and slipped over the edge of the skyline – as if the sun had been a golden ball and had wound up its thread of gold too quickly for us to follow.'

Day 234

21 August | On this day in *Titanic* history

The following people were born on this day:
Dodd, Mr George Charles – 21 August 1867 – aged 44 when on *Titanic*; second steward in first class; *Titanic* victim

Sawyer, Mr Robert James – 21 August 1880 – aged 31 when on *Titanic*; window cleaner; *Titanic* victim

Moody, Mr James Paul – 21 August 1887 – aged 24 when on *Titanic*; sixth officer; *Titanic* victim

Peter / Joseph, Miss Anna ('Mary Joseph') – 21 August 1909 – aged 2 when on *Titanic*; third-class passenger (ticket price: £22 7s 2d); rescued in collapsible lifeboat C

The following people died on this day:
Widgery, Mr Isaac George – 21 August 1942 – aged 37 when on *Titanic*; bath steward; rescued in lifeboat unknown

Hays, Miss Margaret Bechstein – 21 August 1956 – aged 24 when on *Titanic*; first-class passenger (ticket price: £83 3s 2d); rescued in lifeboat 7

Sunderland, Mr Victor Francis – 21 August 1973 – aged 20 when on *Titanic*; third-class passenger (ticket price: £8 1s); rescued in collapsible lifeboat B

Miss Margaret Bechstein Hays, mentioned in *The Daily Banner*, Thursday, 9 May 1912

> 'When the steamship *Carpathia*, which picked up the survivors of the wrecked *Titanic*, came into the port of New York she carried two small French boys who nobody knew. They were cared for by a Miss Hays of New York. The children were mere babies. One four, the other a little over two years of age, and answered to the names Louis and Lolo. They knew no other names. The babies were found wrapped in a blanket in one of the *Titanic*'s lifeboats. According to the stories told by survivors, a man passed the children into the boat just as it was leaving the side of the ship. Officers stepped forward to prevent his taking a place in the boat, but he declared he did not wish to go, asking that the children be taken. They proved to be the sons of Mme Navratil of Nice, France. They had been taken away by their father, who probably perished on the *Titanic*.'

Titanic fact

On board *Titanic*, there were 57,600 pieces of crockery including pots and pans; 29,000 pieces of glassware; and 44,000 pieces of cutlery.

Day 235

22 August | On this day in *Titanic* history

The following people were born on this day:
Rutter, Mr Sidney Frank – 22 August 1885 – aged 26 when on *Titanic*; fireman; *Titanic* victim

Skoog, Master Harald – 22 August 1906 – aged 5 when on *Titanic*; third-class passenger (ticket price: £27 18s); *Titanic* victim

Titanic survivor quotes

Mr Joseph George Scarrott

> 'The men stood back to allow the women to pass, except in one or two cases where men tried to rush, but they were very soon stopped. This occurred at the boat I was in charge of (No. 14). About half a dozen foreigners tried to jump in before I had my complement of women and children, but I drove them back with the boat's tiller. Shortly afterwards the fifth officer, Mr Lowe, came and took charge of the boat. I told him what had happened. He drew his revolver and fired two shots between the boat and ship's side into the water as a warning to any further attempts of that sort.'

Mr Spencer Victor Silverthorne

> 'I was in the smoking room reading near a bridge whist game at one of the tables. When the crash came I said: "We've hit something," and went out on the starboard side to look. None of us was alarmed. It occurred to me that we might have bumped some small craft.
>
> 'I went back in the smoking room with the others. One of the bridge players had not left the smoking room at all and was waiting impatiently for the others to come back and resume the game. They returned and took up their hands and we were all about to settle down when an officer ordered us on deck and told us to get into the boats.
>
> 'There not being enough women on deck to fill the first ones we did not like the idea of leaving the ship then but did as we were told. Had we been in our rooms we would have had to stand aside as other men did then.'

Titanic fact

The first-class dining saloon could cater for 550 diners at any one time; this was eighteen more than the RMS *Olympic*'s 532.

Day 236

23 August | On this day in *Titanic* history

The following people were born on this day:
Wilton, Mr William Edward – 23 August 1858 – aged 53 when on *Titanic*; trimmer; *Titanic* victim

Johnson, Miss Eleanor Ileen – 23 August 1910 – aged 1 when on *Titanic*; third-class passenger (ticket price: £11 2s 8d); rescued in lifeboat 15

The following people died on this day:
Port, Mr Frank – 23 August 1943 – aged 22 when on *Titanic*; steward; rescued in lifeboat unknown

Candee, Mrs Helen Churchill – 23 August 1949 – aged 22 when on *Titanic*; second-class passenger (ticket price: £36 15s); rescued in lifeboat unknown

Mr Frank Port, mentioned in *Western Morning News*, Monday, 29 April 1912

> 'A representative called at the residence of Mr J. A. Pascoe, Crownhill, whose brother, Mr C. H. Pascoe, is believed to be the only West country member of the crew saved in the *Titanic* disaster, and had the pleasure of an interesting conversation with two of the survivors, Mr W. L. Halford, of Bournemouth, and Mr F. Port, Southampton, third class stewards. Mr C. H. Pascoe himself arrived with Messrs. Halford and Port, but while the latter two were detained in Plymouth in connection with the enquiry today, Mr Pascoe, after spending a few hours in the company of his brother and other relatives, entrained for Southampton. During that time however, he made some interesting reference to the disaster. When Mr Pascoe left the *Titanic* the water had already reached the forecastle. There was room for forty more in the boat in which he had left had there been time to take them. However, his boat was only 200 yards away when the *Titanic* sank. Mr Pascoe, who is 43 years of age, is a native of Cadgwith, Cornwall, but resides in Southampton. Messrs. Port and Halford spoke highly of the manner they were treated in America, prior to being transferred to the *Lapland*.'

Mr Charles H. Lightoller, witness at United States Senate Inquiry (Day 5)

> 'I told the sixth officer, Mr Moody, to ring up the crow's nest and tell them to keep a sharp lookout for ice, particularly small ice and growlers. That was received and replied to – and also to pass the word along.'

Day 237

24 August | On this day in *Titanic* history

The following people were born on this day:
Birnbaum, Mr Jakob – 24 August 1887 – aged 24 when on *Titanic*; first-class passenger (ticket price: £26); *Titanic* victim

Pain, Dr Alfred – 24 August 1888 – aged 23 when on *Titanic*; second-class passenger (ticket price: £10 10s); *Titanic* victim

Richard, Mr Emile Phillippe – 24 August 1888 – aged 23 when on *Titanic*; second-class passenger (ticket price: £15 11s); *Titanic* victim

Orešković, Miss Marija – 24 August 1892 – aged 19 when on *Titanic*; third-class passenger (ticket price: £8 13s); *Titanic* victim

Rice, Master Eric – 24 August 1905 – aged 6 when on *Titanic*; third-class passenger (ticket price: £29 2s); rescued in lifeboat 14

The following people died on this day:
Fleming, Miss Margaret – 24 August 1941 – aged 42 when on *Titanic*; first-class passenger (ticket price: £110 17s 8d); rescued in lifeboat 4

McGovern, Ms Mary – 24 August 1957 – aged 20 when on *Titanic*; third-class passenger (ticket price: £7 17s 7d); rescued in lifeboat 13

Ware, Mrs Florence Louise – 24 August 1973 – aged 32 when on *Titanic*; second-class passenger (ticket price: £21); rescued in lifeboat 10

Mrs Florence Louise Ware, quoted in *Bristol Times and Mirror*, Saturday, 11 May 1912

> 'I sailed with my husband, J. J. Ware. He was a builder in Bristol and expected to take up the same business with Charles Ware of number 186, South Main Street, New Britain. We had 250 dollars for our expenses, and he carried 1,000 dollars cash in a belt. Everything is gone. We sailed second cabin. My husband who is lost, was always talking about our chances in America and I was happy.
>
> 'I heard the noise when the steamer hit the iceberg, but did not pay any attention to it. The next I knew a man was pounding on the door of our stateroom and saying, "Men and women put on lifebelts and get upon deck." We did not put on our lifebelts, but ran on deck. As soon as we arrived there some men led me towards a boat. I did not wish to leave my husband, but the men said I must. My husband just shook my hand and said I should go and that he would see me soon.'

Day 238

25 August | On this day in *Titanic* history

The following people were born on this day:
Connolly, Miss Kate – 25 August 1870 – aged 41 when on *Titanic*; third-class passenger (ticket price: £7 12s 7d); *Titanic* victim

Lindell, Mrs Elin Gerda – 25 August 1881 – aged 30 when on *Titanic*; third-class passenger (ticket price: £15 11s); *Titanic* victim

Van Impe, Mrs Rosalie Paula – 25 August 1881 – aged 30 when on *Titanic*; third-class passenger (ticket price: £24 3s); *Titanic* victim

Strilić, Mr Ivan – 25 August 1885 – aged 26 when on *Titanic*; third-class passenger (ticket price: £8 13s 3d); *Titanic* victim

The following person died on this day:
Kink-Heilmann, Miss Luise Gretchen – 25 August 1992 – aged 4 when on *Titanic*; third-class passenger (ticket price: £22 6s); rescued in lifeboat 2

Titanic survivor quotes

Mr Hugh Woolner, witness at United States Senate Inquiry (Day 10)

> 'We felt it under the smoking room. We felt a sort of stopping, a sort of, not exactly shock, but a sort of slowing down; and then we sort of felt a rip that gave a sort of a slight twist to the whole room.'

Mrs Emily Esther Louisa Hart

> 'So that was the last I saw of my poor lost dear, no farewell kiss, no fond word, but in a moment he had gone and we were hanging over the sea, fifty or sixty feet above it, and then there were two or three horrible jerks as the boat was lowered from the davits and we were in the water, so crowded that we could scarcely move.'

Mrs Daniel Warner Marvin (on her honeymoon)

> 'As I was put into the boat, he cried to me, "It's all right, little girl. You go. I will stay." As our boat shoved off he threw me a kiss, and that was the last I saw of him.'

Titanic fact

40,000 fresh eggs, 1,500 gallons of fresh milk, 600 gallons of condensed milk, 1,200 quarts of ice cream and 6,000 pounds of butter were on board *Titanic*.

Day 239

26 August | On this day in *Titanic* history

The following people were born on this day:
Lesueur, Mr Louis Gustave Joseph – 26 August 1876 – aged 35 when on *Titanic*; first-class passenger (ticket price: £512 6s 7d); rescued in lifeboat 3

Gieger, Miss Amalie Henriette – 26 August 1876 – aged 35 when on *Titanic*; first-class passenger (ticket price: £211 10s); rescued in lifeboat 4

Goldsmith, Mrs Emily Alice – 26 August 1880 – aged 31 when on *Titanic*; third-class passenger (ticket price: £20 10s 6d); rescued in collapsible lifeboat C

Mullin, Mr Thomas – 26 August 1891 – aged 20 when on *Titanic*; steward in third class; *Titanic* victim

Burgess, Mr Charles Reginald – 26 August 1893 – aged 18 when on *Titanic*; extra third baker; rescued in lifeboat unknown

Mrs Emily Alice Goldsmith, mentioned in *New York Times*, Sunday, 21 April 1912

> 'Representatives from different houses that had opened their doors to survivors would appear with the names and conditions of those who were ready to move on. A priest from the Swedish Home was there to arrange for clothing and money for thirteen charges. A big man from the Salvation Army arrived with the list of those under his care. He had been down to arrange for the transportation of Mrs Emily Goldsmith and her little son, who must move on to Detroit without the husband and father who sailed with them from Southampton.'

Titanic survivor quotes

Mr Charles Burgess, quoted in *The Daily Banner*, Thursday, 16 May 1912

> 'I went down to our quarters and told them to get up and come up with lifebelts. They simply ridiculed me and told me when the ship was sinking to give them another call. I went down again later, but they took no notice and only abused me for disturbing their sleep. I never saw them again. On returning we were ordered to get in the boat and lower to A deck and take in women and children. We took in about 40 women and six children, and as there were no more about ten male passengers were told to get in. We were then lowered down, and we totalled about 70.'

Mrs Daisy Minahan, witness at United States Senate Inquiry (Day 16)

> 'Officers were yelling and cursing at men to stand back and let the women get into the boats. In going from one lifeboat to another we stumbled over huge piles of bread lying on the deck.'

Day 240

27 August | On this day in *Titanic* history

The following people were born on this day:
Coxon, Mr Daniel – 27 August 1853 – aged 58 when on *Titanic*; third-class passenger (ticket price: £7 5s); *Titanic* victim

Cherry, Miss Gladys – 27 August 1881 – aged 30 when on *Titanic*; first-class passenger (ticket price: £86 10s); rescued in lifeboat 8

PIRRIE TO REBUILD YACHT
The Sun (New York), Tuesday, 27 August 1912

> 'Chief of *Titanic*'s Builders Greatly Impressed by Disaster
> Special Cable Despatch to THE SUN LONDON, Aug. 26
> It is stated that Lord Pirrie, the chairman of Harland & Wolff, the builders of the *Titanic*, has been so impressed by the disaster to that vessel that he has decided to reconstruct his own yacht *Valiant* on lines suggested by the foundering of the White Star vessel. Meanwhile he will not allow the *Valiant* to be used.'

Titanic survivor quote

Miss Gladys Cherry, mentioned in *New York Herald*, Sunday, 21 April 1912

> '"The pitiful sadness of our rowing, rowing toward the lights of a ship that disappeared. We in boat No. 8 saw some tramp steamer's mast head lights, and then we saw the glow of red as she swung toward us for a few minutes. Then darkness and despair.' Lady Rothes yesterday, at the Ritz-Carlton, told of her experience on board the *Titanic*. "I went to bed at half-past seven," she said, "and my cousin, Miss Gladys Cherry, who shared my room – No. 77 on deck B – also retired. It was bitterly cold. I was awakened by a slight jar and then a grating noise. I turned on the light and saw that it was 11.46, and I wondered at the sudden quiet. Gladys had not been awakened and I called her and asked did she not think it strange that the engines had stopped. As I opened our cabin door I saw a steward. He said we had struck some ice. Our fur coats over our night gowns were all the clothes we had. My cousin asked the chief steward if there was any danger and he answered, 'Oh no, we have just grazed some ice and it does not amount to anything."

Titanic fact

On board *Titanic*, there were 2,200 pounds of coffee; 800 pounds of tea and 10,000 pounds of sugar.

Day 241

28 August | On this day in *Titanic* history

The following person was born on this day:
Chaffee, Mrs Carrie Constance – 28 August 1864 – aged 47 when on *Titanic*; first-class passenger (ticket price: £61 3s 6d); rescued in lifeboat 4

The following person died on this day:
Strandén, Mr Juho Niilosson – 28 August 1943 – aged 30 when on *Titanic*; third-class passenger (ticket price: £7 18s 6d); rescued in lifeboat 15

Mr Robert Hichens (quartermaster), witness at British Wreck Commissioner's Inquiry (Day 3)

> 'I heard Mr Lightoller speak to Mr Moody and tell him to speak through the telephone to the crow's-nest to keep a sharp lookout for small ice and growlers until daylight and pass the word along to the lookout man.'

Titanic survivor quote

Mrs Carrie Constance Chaffee, quoted in *Evening Tribune*, Tuesday, 23 April 1912

> '"When my husband and I, having been instructed to put on our lifebelts, were coming up the stairs to 'A' deck, we were met in the landing by an officer, the only one I had seen since the ship struck. I had said to Mr Chaffee: 'We had better stay together,' but the officer said something about the women going to one side of the deck. I passed out on the deck, leaving Mr Chaffee there.
>
> 'On the deck I was told, with other women, to go up to the boat deck and a moment later we were ordered down to 'A' deck again. I did not see my husband again. The boat in which I was taken from the ship was, I think, the last of those on one side to get away. There were a good many babies in our lifeboat. When I took my seat a baby was thrust into my arms and I was given a sea sick woman to hold up on the other side, and so we were lowered, first one end of the boat going down at a dangerous angle, and then the other...it seemed to writhe, breaking into the three parts in which it was divided. First the middle seemed to go down, lifting bow and stern into the air. Then it twisted the other way, throwing the middle up. Finally the bow went under, and it plunged, stern last..." From 1 o'clock until 4 in the morning, Mrs Chaffee said, the men in their lifeboat rowed steadily. Toward 4 o'clock a stewardess continually declared she could see the radiance of a searchlight behind them, and that they were going in the wrong direction…'

Day 242

29 August | On this day in *Titanic* history

The following people were born on this day:
Wake, Mr Percy – 29 August 1874 – aged 37 when on *Titanic*; assistant baker; *Titanic* victim

Hickman, Mr Lewis – 29 August 1881 – aged 30 when on *Titanic*; second-class passenger (ticket price: £73 10s); *Titanic* victim

McNamee, Mr Neal – 29 August 1884 – aged 27 when on *Titanic*; third-class passenger (ticket price: £16 2s); *Titanic* victim

Mattmann, Sig. Adolf – 29 August 1891 – aged 20 when on *Titanic*; ice man; *Titanic* victim

De Plesmaeker, Mr Alfons – 29 August 1895 – aged 16 when on *Titanic*; third-class passenger (ticket price: £9 10s); *Titanic* victim

The following person died on this day:
Moubarek, Master Halim Gonios ('William George') – 29 August 1975 – aged 4 when on *Titanic*; third-class passenger (ticket price: £15 4s 11d); rescued in collapsible lifeboat C

Titanic survivor quotes

Mr Charles H. Lightoller (second officer)

> 'One young couple walked steadily up and down the boat deck throughout pretty well the whole of the proceedings. Once or twice the young chap asked if he could help. He was a tall, clean-bred Britisher, on his honeymoon I should say. The girl – she was little more – never made the slightest attempt to come towards the boats, much less be taken on board, although I looked towards her several times with a look of silent invitation, but no, she was not going to be parted from her man…'

Miss Elizabeth Dowdell

> 'No sooner were we off than the *Titanic* began to go down rapidly. The bow disappeared first. There was no playing by the bands, and only the cries and sobs of those aboard and in the boats was to be heard above the wash of the sea.'

Titanic fact

The cost of using the Turkish and electric baths on F Deck was 4 shillings.

Day 243

30 August | On this day in *Titanic* history

The following people were born on this day:
Slight, Mr William Henry James – 30 August 1875 – aged 36 when on *Titanic*; larder cook; *Titanic* victim

Pryce, Mr Charles William – 30 August 1889 – aged 22 when on *Titanic*; saloon steward in first class; *Titanic* victim

Biggs, Mr Edward Charles – 30 August 1891 – aged 20 when on *Titanic*; fireman; *Titanic* victim

Carraú-Esteves, Mr José Pedro – 30 August 1894 – aged 17 when on *Titanic*; first-class passenger (ticket price: £47 2s); *Titanic* victim

The following people died on this day:
al-Ba'qlīnī, Miss Ūwjīnīyā – 30 August 1912 – aged 3 when on *Titanic*; third-class passenger (ticket price: £19 5s 2d); rescued in collapsible lifeboat C

Marsden, Miss Evelyn – 30 August 1938 – aged 28 when on *Titanic*; stewardess; rescued in lifeboat 16

Miss Evelyn Marsden, mentioned in *The Daily Herald*, 4 November 1912.

> 'Memories of the *Titanic* disaster were recalled on Saturday by the arrival at the Outer Harbour of the White Star steamer *Irishman*, which had on board two survivors from the ill-fated liner. Miss Evelyn Marsden, now the wife of Dr James (surgeon on the *Irishman*), is travelling on the vessel with her husband. Mrs James, who at the time of the great maritime mishap was a stewardess on the *Titanic*, is the daughter of Mr and Mrs W. H. Marsden, of Hoyleton (South Australia). When the *Irishman* reached the wharf Mrs James went ashore and renewed her acquaintances with many of her old South Australian friends. The second officer of the *Irishman* (Mr J. Boxhall) was also one of the fortunate ones to be saved from the *Titanic*, on which vessel he acted as fourth officer.

Titanic survivor quote

Mr Harry Senior (fireman)

> 'The ship was pretty near sinking then, and the captain shouted, "Each man for himself." I had noticed him on the bridge before that. He was pacing up and down, sending up rockets and giving orders. It is a dirty lie to say that such a man as he would [have] shot himself.'

Day 244

31 August | On this day in *Titanic* history

The following people were born on this day:
Hendy, Mr Edward Martin – 31 August 1872 – aged 39 when on *Titanic*; saloon steward; *Titanic* victim

Priest, Mr Arthur John – 31 August 1887 – aged 24 when on *Titanic*; fireman; rescued in lifeboat 15

Quick, Mrs Jane – 31 August 1878 – aged 33 when on *Titanic*; second-class passenger (ticket price: £26); rescued in lifeboat 11

Mullin, Miss Mary – 31 August 1893 – aged 18 when on *Titanic*; third-class passenger (ticket price: £15 10s); *Titanic* victim

Barber, Miss Ellen Mary – 31 August 1885 – aged 26 when on *Titanic*; first-class passenger (ticket price: £78 17s); rescued in lifeboat 6

The following person died on this day:
Senior, Mr Harry – 31 August 1937 – aged 31 when on *Titanic*; fireman; rescued in lifeboat unknown

Mr Arthur John Priest, mentioned in *The Times*, Monday, 23 April 1917

> 'A Fireman's Adventures'
>
> 'The remarkable adventures of a young fireman have been brought to light through the sinking of the *Donegal*. John Priest, who lives in Southampton, is only 29 years of age. He has been on the sea since his youth, and has served in many waters. He has seen four of the vessels in which he was serving sink, and three others have been damaged.
>
> 'Several years ago he was in the *Asturias* when she met with an accident on her maiden voyage. He was one of the survivors of the disaster which overtook the *Titanic* on her maiden voyage five years ago, escaping with frost-bitten toes and an injured leg.
>
> 'He was also in the *Olympic* when that vessel and the *Hawke* came into collision near the Isle of Wight some years ago, and he served in the *Alcantra* during her gallant fight with the German raider *Greif* in the North Sea.
>
> 'The *Alcantra* was sunk by a torpedo and Priest was wounded by shrapnel during the fight. Later he joined the *Britannic*, and was serving in that ship when she sank last November. His last ship was the *Donegal* from which he escaped with a rather serious injury to his head.'

September

Day 245

1 September | On this day in *Titanic* history

The following people were born on this day:
McReynolds, Mr William Thomas Carson – 1 September 1889 – aged 22 when on *Titanic*; junior sixth engineer; *Titanic* victim

Pears, Mrs Edith – 1 September 1889 – aged 22 when on *Titanic*; first-class passenger (ticket price: £66 12s); rescued in lifeboat 8

Longley, Miss Gretchen Fiske – 1 September 1890 – aged 21 when on *Titanic*; first-class passenger (ticket price: £78 19s 2d); rescued in lifeboat 10

Panula, Master Juha Niilo – 1 September 1904 – aged 7 when on *Titanic*; third-class passenger (ticket price: £39 13s 9d); *Titanic* victim

The following people died on this day:
Cunningham, Mr Andrew Orr – 1 September 1932 – aged 38 when on *Titanic*; bedroom steward in first class; rescued in lifeboat 4

Lavington, Miss Elizabeth – 1 September 1949 – aged 40 when on *Titanic*; stewardess; rescued in lifeboat 11

Today is an important day in *Titanic* history. In 1985, the *Titanic* was discovered at coordinates 41.7325° N, 49.9469° W, by the team led by Jean-Louis Michel and Robert Ballard.

Titanic survivor quotes

Mr George Alfred Hogg (lookout)

> 'I have always had night glasses in the White Star boats. I asked for the glasses, and I did not see why I should not have them. I had them from Belfast to Southampton; but from Southampton to where the accident occurred we never had them.'

Mr Paul Maugé (secretary to the chef of the à la carte restaurant)

> 'I did ask the chef to jump many times, but the chef was too fat I must say – too big, you know. He could not jump.'

Titanic fact

All the sinks in the first-class accommodation were made of marble.

Day 246

2 September | On this day in *Titanic* history

The following people were born on this day:
Lefebvre, Mr Paul Georges – 2 September 1876 – aged 35 when on *Titanic*; saloon steward; *Titanic* victim

Bryhl, Miss Dagmar Jenny Ingeborg – 2 September 1891 – aged 20 when on *Titanic*; second-class passenger (ticket price: £26); rescued in lifeboat 12

The following people died on this day:
Harris, Mrs Irene – 2 September 1969 – aged 35 when on *Titanic*; first-class passenger (ticket price: £83 9s 6d); rescued in collapsible lifeboat D

Cumings, Mrs Florence Briggs – 2 September 1949 – aged 35 when on *Titanic*; first-class passenger (ticket price: £71 5s 8d); rescued in lifeboat 4

Titanic survivor quote

Miss Dagmar Jenny Ingeborg Bryhl, mentioned in *Rockford Daily Register Gazette*, 25 April 1912

> 'For seven hours the boat in which Miss Bryhl was saved floated before the *Carpathia* came in sight. The air was keen and the women and the men were nearly frozen before they were picked up. Miss Bryhl wore nothing but a coat over her nightgown, a covering altogether insufficient to keep out the cold of the frosty night, and it was with a sigh of relief that the approach of the Cunard liner was seen.
>
> 'On board the *Carpathia* the *Titanic* survivors were given a drink of brandy and hot coffee and were allowed to retire. Miss Bryhl says that there was nothing extraordinary in the treatment accorded the *Titanic* passengers on the *Carpathia*. The food was plentiful but it did not seem of the best kind and the young woman subsided [*sic*] mainly on a diet of oranges until she arrived in New York.
>
> 'Of all of Miss Bryhl's belongings she saved only her watch, which was a present from her sweetheart. She chanced to wear the watch suspended from a chain, hung about her neck, which accounts for the fact that she bore it in safety from the liner. The only clothing she possessed when she arrived on the *Carpathia* was a nightgown and a coat, and a complete new outfit was necessary for her before she was able to leave the hospital to which she was taken from the boat.'

Titanic fact

The first leg of the journey, from Southampton to Cherbourg, France, was a distance of 84 miles.

Day 247

3 September | On this day in *Titanic* history

The following people were born on this day:

Elsbury, Mr William James – 3 September 1863 – aged 48 when on *Titanic*; third-class passenger (ticket price: £7 5s); *Titanic* victim

Goshawk, Mr Alfred James – 3 September 1871 – aged 40 when on *Titanic*; saloon steward in first class; *Titanic* victim

Cheverton, Mr William Frederick – 3 September 1884 – aged 27 when on *Titanic*; saloon steward in first class; *Titanic* victim

Aspeslagh, Mr Georges – 3 September 1885 – aged 26 when on *Titanic*; assistant plateman; *Titanic* victim

Hansen, Mr Jens Henrik Marius Juul – 3 September 1886 – aged 25 when on *Titanic*; third-class passenger (ticket price: £7 17s 1d); *Titanic* victim

Peñasco Y Castellana, Mrs Maria Josefa Perezde Soto y Vallejo – 3 September 1889 – aged 22 when on *Titanic*; first-class passenger (ticket price: £108 18s); rescued in lifeboat 8

Avery, Mr James Albert – 3 September 1891 – aged 20 when on *Titanic*; trimmer; rescued in lifeboat 15

The following people died on this day:

Humphreys, Mr Sidney – 3 September 1919 – aged 52 when on *Titanic*; quartermaster, rescued in lifeboat 11

Sandström, Miss Beatrice Irene – 3 September 1995 – aged 1 when on *Titanic*; third-class passenger (ticket price: £16 14s); rescued in lifeboat 13

Titanic survivor quote

Lucy Noël Martha Leslie, Countess of Rothes, quoted in *New York Herald*, Sunday, 21 April 1912

> 'For three hours we pulled steadily for the two masthead lights that showed brilliantly in the darkness. For a few minutes we saw the ship's port light, then it vanished, and the masthead lights got dimmer on the horizon until they, too, disappeared. A Mrs Smith did yeoman service. She rowed for five hours with Tom Jones without taking a rest. Really, she was magnificent, not only in her attitude, but in the whole-souled way in which she worked. Mrs Pearson also rowed, and my maid, Roberta Maioni, rowed the last half of the night... There was no excitement of any kind, save that once the third-class passengers became obstreperous, but it was instantly put down. When the awful end came, I tried my best to keep the Spanish woman from hearing the agonising sound of distress. They seemed to continue forever, although it could not have been more than ten minutes until the silence of a lonely sea dropped down.'

Day 248

4 September | On this day in *Titanic* history

The following people were born on this day:
Lyons, Mr William Henry – 4 September 1886 – aged 25 when on *Titanic*; able seaman; *Titanic* victim

Roberton, Mr George Edward – 4 September 1892 – aged 19 when on *Titanic*; assistant saloon steward in second class; rescued in lifeboat 6

The following people died on this day:
Wotton, Mr Henry Swaffin – 4 September 1925 – aged 54 when on *Titanic*; first-class passenger (ticket price: £1 10s); disembarked at Cherbourg

Wallcroft, Miss Ellen 'Nellie' – 4 September 1949 – aged 36 when on *Titanic*; second-class passenger (ticket price: £21); rescued in lifeboat 14

Titanic survivor quote

Miss Ellen 'Nellie' Wallcroft, quoted in *Maidenhead Advertiser*, Monday, 29 April 1912

> 'It was a lovely starlight night, but not light enough for us to see who were in the boats. The officer told the men to lay on their oars, so as to be handy later on. We did not seem to be long on the water. We could see the ship gradually going down, but all the lights were on, when suddenly two terrible explosions took place. The ship seemed to go forward and then split in the middle and then there were two more explosions that seemed from underneath the water. No more could be seen of that grand ship; all was silent for a moment and then the cries of 1600 men. All were crying for help; it was terrible. I should think the cries must have lasted two hours, or even more and then the day dawned and we could see about six large icebergs. Each looked as large as a house and all the time the cries of the drowning were getting fainter.
>
> 'Then Officer Lowe wanted to go back to the rescue, but the women begged him not to go. He got about four boats together and distributed his passengers amongst them as many as he possibly could and then went back to the rescue. I believe they rescued six alive from that raft; the others had all died from exposure, as it was intensely cold. The boat I got in was No. 10. There had been six picked up but one man was mad. He shook the boat and we were afraid it would capsize. Two men revived but they were terribly frozen and two were dead and fell into the water at the bottom of the boat.'

Titanic fact

On 12 February 1867, Edward Smith began his first journey to sea: a return trip to Hong Kong.

Day 249

5 September | On this day in *Titanic* history

The following person died on this day:
Pearce, Mr Alfred Ernest – 5 September 1887 – aged 24 when on *Titanic*; steward; *Titanic* victim

MISS RYERSON TO WED *New York Times*, Sunday, 5 September 1915

'*Titanic* Survivor Is Engaged to George Hyde Clark

Cooperstown, N. Y., Sept. 4. Mrs Arthur Ryerson of Chicago and Philadelphia announced today the engagement of her daughter, Miss Emily Borie Ryerson, to George Hyde Clarke, a son of the late George Hyde Clarke and a brother of Mrs Arthur Osgood Choate of New York. Miss Ryerson, together with her mother, her sister, Miss Susan Ryerson, and her brother, John A. Ryerson, are all survivors of the *Titanic* disaster, in which Mr Ryerson perished. Mr Clarke was graduated from Harvard. After taking the course in farming at Cornell University, he will take the occupation of a gentleman farmer, settling at Hyde Hall, which he inherited from his father. No date has been mentioned for the marriage, but it is expected to take place this Autumn.'

Titanic survivor quote

Mr William H. Taylor (fireman)

'I was asleep when the accident occurred, sir. The alarm bell for accidents rang outside of our door. I went up on deck, and could not see anything. I went down in our room again. I stayed in the room about 10 minutes, and somebody reported that there was water in No. 1 hatch. Then we packed our bags, took them in the mess room, in the alleyway, to wait for orders. The officer was coming along the alleyway, and ordered us to put on lifebelts. The lifebelts were on the deck, on the boat deck. They shoved out No. 15 boat and I was ordered into it. The boat was pretty full. We pulled a distance away from the *Titanic*, because we were afraid of the suction, and we kept on pulling toward the light, with the other rowboats. About half-past 7 the next morning we boarded the *Carpathia*.'

Titanic facts

There were six lookouts on board (all six survived).

There were also six quartermasters on board (all six survived).

Day 250

6 September | On this day in *Titanic* history

The following people were born on this day:
Blackwell, Mr Stephen Weart – 6 September 1866 – aged 45 when on *Titanic*; first-class passenger (ticket price: £35 10s); *Titanic* victim

Dowdell, Miss Elizabeth – 6 September 1880 – aged 31 when on *Titanic*; third-class passenger (ticket price: £12 9s 6d); rescued in lifeboat 13

Alexander, Mr William Albert – 6 September 1887 – aged 24 when on *Titanic*; third-class passenger (ticket price: £7 17s 9d); *Titanic* victim

Dore, Mr Albert James – 6 September 1889 – aged 22 when on *Titanic*; trimmer; rescued in lifeboat unknown

The following person died on this day:
Becker, Master Richard F. – 6 September 1975 – aged 1 when on *Titanic*; second-class passenger (ticket price: £39); rescued in lifeboat 11

Titanic survivor quote

Miss Elizabeth Dowdell, quoted in *Jersey Journal*, Saturday, 20 April 1912

> 'Many aboard the lifeboats, when they saw their dear ones on deck doomed, threw themselves overboard. Some had to be forcibly restrained. The last thing I heard was what I believed to have been the captain's voice crying "every man for himself."
>
> 'While we were rowing about many came alongside and were pulled aboard. We had seventy in our boat by the time the *Carpathia* picked us up. I do not know how many we took on board at the start.
>
> 'All during this time rockets were being sent up from the doomed vessel. Revolver shots added to the din and the dying voices. Then there was one great explosion. I guessed it was the boilers. The *Titanic* did not stay up long after that, but tilted, bow downward, with a great part of the stern in the air. She stayed for a moment, then plunged under. Her lights were burning to the last. One woman from a capsized boat came near to us. She was swimming. "Man, let go of me," she pleaded to some one who was hanging upon her. "I will not," responded the masculine voice. "If I do I will drown." He did let go, however, and the woman was hauled aboard. She said she had been swimming for an hour, and supporting this unknown man for half of that time.

Titanic fact

The intended distance of the longest leg of the voyage, from Queenstown, Ireland, to New York, USA, was 2,825 miles.

Day 251

7 September | On this day in *Titanic* history

The following people were born on this day:
Mills, Mr Christopher – 7 September 1860 – aged 51 when on *Titanic*; butcher, rescued in collapsible lifeboat C

McCarthy, Mr William – 7 September 1863 – aged 48 when on *Titanic*; able seaman; rescued in lifeboat 4

Pernot, Mr René – 7 September 1872 – aged 39 when on *Titanic*; second-class passenger (ticket price: £15 1s); *Titanic* victim

Jones, Mr Arthur Ernest – 7 September 1874 – aged 37 when on *Titanic*; plate steward in second class; *Titanic* victim

Combes, Mr George – 7 September 1877 – aged 34 when on *Titanic*; fireman; rescued in lifeboat 3

Judd, Mr Charles Edward – 7 September 1880 – aged 31 when on *Titanic*; fireman; rescued in collapsible lifeboat B

The following people died on this day:
Hart, Mrs Esther Ada – 7 September 1928 – aged 48 when on *Titanic*; second-class passenger (ticket price: £26 5s); rescued in lifeboat 14

Newsom, Miss Helen Monypeny – 7 September 1965 – aged 19 when on *Titanic*; first-class passenger (ticket price: £26 5s 8d); rescued in lifeboat 5

Titanic survivor quote

Mr Charles Edward Judd, quoted in *Bucks Free Press*, Friday, 3 May 1912

> 'Oh, dear, I shall never forget that night as long as I live; it was awful. I was in the water for 5½ hours before I was picked up, and when they picked me up I was helpless. I have been in hospital (on board with frost-bitten feet and legs); the doctor was rubbing my legs until they nearly bled to try to get life into them. I was afraid I should lose both feet. Oh, dear, what an awful thing it is to be frost-bitten, but never mind, thank God, I am alive. I am the luckiest man in the world to be alive, I can tell you. Poor Jack Hurst is gone to the bottom. He leaves a wife and six children. It is very sad. Out of a crew of 905 there are only 205 left to tell the tale.'

Miss Helen Monypeny Newsom, mentioned in *New York Times*, Sunday, 2 March 1913

> 'Both Mr Behr and his bride are survivors of the *Titanic* disaster. Mr Behr is a well-known amateur tennis player.'

Day 252

8 September | On this day in *Titanic* history

The following people were born on this day:
Goldenberg, Mr Samuel L. – 8 September 1864 – aged 47 when on *Titanic*; first-class passenger (ticket price: £89 2s 1d); rescued in lifeboat 5

Johansson, Mr Oscar Wilhelm – 8 September 1879 – aged 32 when on *Titanic*; third-class passenger (ticket price: £7 15s 6d); rescued in lifeboat unknown

Mangan, Miss Mary – 8 September 1879 – aged 32 when on *Titanic*; third-class passenger (ticket price: £7 15s); *Titanic* victim

Bracken, Mr James Hollen – 8 September 1881 – aged 30 when on *Titanic*; second-class passenger (ticket price: £13); *Titanic* victim

Cardwell, Mr John – 8 September 1885 – aged 26 when on *Titanic*; second-class passenger (ticket price: £29); rescued in lifeboat 13

Saccaggi, Sig. Giovanni Giuseppe Emilio – 8 September 1887 – aged 24 when on *Titanic*; assistant waiter; *Titanic* victim

Thomas, Mr Joseph Wakefield – 8 September 1888 – aged 23 when on *Titanic*; fireman; *Titanic* victim

Peacock, Master Albert Edward – 8 September 1911 – aged 7 months when on *Titanic*; third-class passenger (ticket price: £13 15s 6d); *Titanic* victim

Mr Samuel Levi Goldenberg, mentioned in *New York Herald*, Sunday, 21 April 1912

> 'Mrs Goldenberg ran out on deck. Mrs Goldenberg, who had not waited to dress, but had only pulled a skirt over her night dress, was helped into one of the first boats by Mr Goldenberg, Mr Ismay and one of the *Titanic*'s officers. She begged her husband to follow her, but he refused. The boat was unfilled when the crew started to lower it, according to Mrs Goldenberg, and she urged her husband to take a place, as other men had done. "When I saw that he was not coming," said Mrs Goldenberg, "I called, 'For God's sake say goodbye to me then,' and suddenly Mr Ismay and one of the crew seized Mr Goldenberg and threw him over the side. He managed to catch the boat ropes and I and others pulled him in..."'

Mr Samuel Levi Goldenberg, mentioned in *New York Times*, Wednesday, 24 April 1912

> 'All that I know about it,' said Mr Smyth, 'is that the baggage belonged to Mr Goldenberg and that it was the only piece saved from the wrecked liner. I do not know what was in the carry-all, as the customs regulations were waived in favor of the *Titanic* survivors, and as a result the Goldenberg piece was not inspected. That is all that I know about it.'

Day 253

9 September | On this day in *Titanic* history

The following people were born on this day:

Sjöstedt, Mr Ernst Adolf – 9 September 1852 – aged 59 when on *Titanic*; second-class passenger (ticket price: £13 10s); *Titanic* victim

Molson, Mr Harry Markland – 9 September 1856 – aged 55 when on *Titanic*; first-class passenger (ticket price: £30 10s); *Titanic* victim

Doling, Mrs Ada Julia Elizabeth – 9 September 1877 – aged 34 when on *Titanic*; second-class passenger (ticket price: £23); rescued in lifeboat unknown

Johansson, Mr Karl Johan – 9 September 1880 – aged 31 when on *Titanic*; third-class passenger (ticket price: £7 15s 6d); *Titanic* victim

Pearce, Mr Ernest – 9 September 1880 – aged 31 when on *Titanic*; third-class passenger (ticket price: £7); *Titanic* victim

Rogers, Mr Edward James William – 9 September 1880 – aged 31 when on *Titanic*; assistant storekeeper; *Titanic* victim

Flynn, Mr James – 9 September 1883 – aged 28 when on *Titanic*; third-class passenger (ticket price: £7 15s); *Titanic* victim

Moran, Miss Bertha Bridget – 9 September 1883 – aged 28 when on *Titanic*; third-class passenger (ticket price: £24 3s); rescued in lifeboat 15

Giles, Mr Ralph – 9 September 1887 – aged 24 when on *Titanic*; second-class passenger (ticket price: £13 10s); *Titanic* victim

Earnshaw, Mrs Olive – 9 September 1888 – aged 23 when on *Titanic*; first-class passenger (ticket price: £83 3s 2d); rescued in lifeboat 7

Goodwin, Master Sidney Leslie – 9 September 1910 – aged 1 when on *Titanic*; third-class passenger (ticket price: £46 18s); *Titanic* victim

The following person died on this day:

Heikkinen, Miss Laina – 9 September 1943 – aged 16 when on *Titanic*; third-class passenger (ticket price: £7 18s 6d); rescued in lifeboat unknown

Titanic survivor quote

Mrs Helen Bishop

> 'The conduct of the crew, as far as I could see, was absolutely beyond criticism. It was perfect. The men in our boat were wonderful. One man lost his brother. When the *Titanic* was going down I remember he just put his hand over his face; and immediately after she sank he did the best he could to keep the women feeling cheerful all the rest of the time. We all thought a great deal of that man.'

Day 254

10 September | On this day in *Titanic* history

The following people were born on this day:
Kinsella, Mr Louis – 10 September 1881 – aged 30 when on *Titanic*; fireman; *Titanic* victim

Mulvihill, Miss Bridget Elizabeth – 10 September 1886 – aged 25 when on *Titanic*; third-class passenger (ticket price: £7 15s); rescued in lifeboat 15

The following person died on this day:
Major, Mr William James – 10 September 1960 – aged 32 when on *Titanic*; fireman; rescued in lifeboat 13

Titanic survivor quotes

Mrs D. H. Bishop

> 'As the ship sank we could hear the screaming a mile away. Gradually it became fainter and fainter and died away. Some of the lifeboats that had room for more might have gone to their rescue, but it would have meant that those who were in the water would have swarmed aboard and sunk her.'

Miss Hilda Slater

> 'It was terrible, from the moment the vessel struck, or as soon as the members of the orchestra could be collected, there was a steady round of lively airs. It did much to keep up the spirits of everyone and probably served as much as the efforts of the officers trying to prevent panic. When I got there, I found an indescribable scene. A number of the steerage men passengers had attempted to seize one of the boats and there was a brisk revolver fire: many men fell under it. The prompt, and drastic action of the officers restored order.
>
> 'Of all the heroes who went to their death when the *Titanic* dived to its ocean grave, none deserved greater credit than the members of the vessel's orchestra. The orchestra played until the last. When the vessel took its final plunge the strains of a lively air, mingled gruesomely with the cries of those who realised that they were face to face with death.'

Titanic fact

As *Titanic* was sinking Charles John Joughin, chief baker, was busy drinking all the whisky he could find from the ship's storage to prepare himself for the cold water. As the ship went down, he said he 'rode it down as if it were an elevator.' He then spent several hours in the freezing cold water and lived to tell the tale.

Day 255

11 September | On this day in *Titanic* history

The following people were born on this day:

Gill, Mr Joseph Stanley – 11 September 1873 – aged 38 when on *Titanic*; bedroom steward; *Titanic* victim

Stroud, Mr Harry John – 11 September 1876 – aged 35 when on *Titanic*; saloon steward; *Titanic* victim

Woodward, Mr John Wesley – 11 September 1879 – aged 32 when on *Titanic*; musician in *Titanic*'s orchestra; *Titanic* victim

Daniel, Mr Robert Williams – 11 September 1884 – aged 27 when on *Titanic*; first-class passenger (ticket price: £30 10s); rescued in lifeboat 3

Mabey, Mr John Charles – 11 September 1888 – aged 26 when on *Titanic*; steward in third class; *Titanic* victim

Noel, Master William Henry Middleton – 11 September 1898 – aged 13 when on *Titanic*; first-class passenger (ticket price: £3); disembarked at Cherbourg

The following person died on this day:

Bird, Miss Ellen – 11 September 1949 – aged 31 when on *Titanic*; first-class passenger (ticket price: £221 15s 7d); rescued in lifeboat 8

Mr Robert Hichens (quartermaster), witness at British Wreck Commissioner's Inquiry (Day 3)

> 'I took the log which was part of my duty at half a minute to ten, as near as I can tell, and the vessel was going 45 knots by the Cherub log every two hours.'

Titanic survivor quote

Mr Robert Williams Daniel, quoted in *The Times*, Saturday, 20 April 1912

> 'Mr Robert E. Daniel, a young cotton broker, of Philadelphia, said: "I was in my cabin dictating to the stenograph when the ship struck the berg. The shock was not violent. The officers who survived told me afterwards the *Titanic* slipped up on the iceberg and tore her bottom out. I went on dictating until somebody knocked at my door and cried out that the ship was sinking.
>
> "I grabbed a life preserver and went to the deck. The 16 boats were filled with passengers, most of them women. Twelve of the boats pulled away from the port side and four from the starboard. There was no panic. Had there been sufficient lifeboats it is my opinion that practically all the passengers would have been saved."
>
> 'Mr Daniel leaped overboard when he discovered that the ship was sinking, and was picked up by one of the boats.'

Day 256

12 September | On this day in *Titanic* history

The following person was born on this day:
Canavan, Mr Patrick – 12 September 1890 – aged 21 when on *Titanic*; third-class passenger (ticket price: £7 15s); *Titanic* victim

The following people died on this day:
Rothes, Lucy Noël Martha, Countess of – 12 September 1956 – aged 33 when on *Titanic*; first-class passenger (ticket price: £86 10s); rescued in lifeboat 8

Cook, Mrs Selena – 12 September 1964 – aged 22 when on *Titanic*; second-class passenger (ticket price: £10 10s); rescued in lifeboat 14

West, Miss Constance Mirium – 12 September 1963 – aged 4 when on *Titanic*; second-class passenger (ticket price: £27 15s); rescued in lifeboat 11

Titanic survivor quotes

Mrs Lucy Noël Martha, Countess of Rothes, mentioned in *Worcester Gazette*, Friday, 19 April 1912

> 'Miss Alice Farnam Leader, a New York physician escaped from the *Titanic* on the same boat which carried the Countess Rothes. "The countess is an expert oarswoman," said Dr Leader, "and thoroughly at home on the water. She practically took command of our boat when it was found that the seamen who had been placed at the oars could not row skilfully. Several of the women took their place with the countess at the oars and rowed in turns while the weak and unskilled stewards sat quietly in one end of the boat."'

Mr Harold Godfrey Lowe (fifth officer)

> 'I was half awakened by hearing voices in our quarters, because it is an unusual thing, and it woke me up. I suppose I lay down there for a little while until I fully realised, and then I jumped out of bed and opened my door a bit and looked out, and I saw ladies in our quarters with lifebelts on.'

Mrs Charlotte Collyer (from a letter to her mother-in-law, written a week after the disaster when safe in New York)

> 'My dear Mother, I don't know how to write to you or what to say. I feel I shall go mad sometimes but dear, as much as my heart aches, it aches for you too for he is your son and the best that ever lived. Oh, mother, how can I live without him, he was so calm. The agony of that night can never be told. I haven't a thing in the world that was his, only his rings. Everything we had went down.'

Titanic fact

Cargo nets were used to help board the surviving children onto the rescue ship *Carpathia*.

Day 257

13 September | On this day in *Titanic* history

The following people were born on this day:
Wareham, Mr Robert Arthur – 13 September 1874 – aged 37 when on *Titanic*; bedroom steward; *Titanic* victim

Kinchenten, Mr Frederick Charles – 13 September 1875 – aged 36 when on *Titanic*; greaser; *Titanic* victim

Sage, Miss Dorothy Florence – 13 September 1897 – aged 14 when on *Titanic*; third-class passenger (ticket price: £69 11s); *Titanic* victim

Rice, Master Francis – 13 September 1909 – aged 2 when on *Titanic*; third-class passenger (ticket price: £29 2s 6d); *Titanic* victim

The following people died on this day:
Dean, Mrs Eva Georgetta – 13 September 1975 – aged 32 when on *Titanic*; third-class passenger (ticket price: £20 11s 6d); rescued in lifeboat 10

Törnquist, Mr William Henry – 13 September 1946 – aged 26 when on *Titanic*; seaman; rescued in lifeboat unknown

The Daily New Era, 18 April 1912

> 'Mr Milton Hershey of Hershey, Pa. the chocolate manufacturer, was in Lancaster on Wednesday and told his friends of the narrow escape he made from having taken passage on the *Titanic*. It was his intention to sail on the great liner originally, but finding that it would reach New York later than he desired to land he booked on the German liner *Amerika*, which arrived several days ago. It was this steamer, the *Amerika*, that sent a warning to Captain Smith, of the *Titanic* on the day prior to the collision that sent her to the bottom, advising him of the presence of huge icebergs in the very locality in which the great steamer met her doom. Captain Smith received the *Amerika*'s wire for he replied thanking the captain of the German liner for his advice.'

Titanic survivor quote

Mr Robert Hichens (quartermaster)

> 'Just about a minute, I suppose, after the collision, the Captain rushed out of his room and asked Mr Murdoch what was that, and he said, "An iceberg, Sir," and he said, "Close the watertight door."

Titanic fact

The average cost of a berth in standard (second class) was $60 (£12). The average cost of a steerage (third class) berth was $15-$40 (£3-£8).

Day 258

14 September | On this day in *Titanic* history

The following people were born on this day:
Foley, Mr John – 14 September 1865 – aged 46 when on *Titanic*; deck storekeeper, rescued in lifeboat unknown

Abelson, Mrs Anna – 14 September 1883 – aged 28 when on *Titanic*; second-class passenger (ticket price: £24); rescued in lifeboat 10

Norman, Mr Robert Douglas – 14 September 1884 – aged 27 when on *Titanic*; electrical engineer; *Titanic* victim

Podesta, Mr Alfred John Alexander – 14 September 1887 – aged 24 when on *Titanic*; fireman; rescued in lifeboat unknown

Abelseth, Miss Karen Marie – 14 September 1895 – aged 16 when on *Titanic*; third-class passenger (ticket price: £7 13s); rescued in lifeboat 16

Carter, Master William Thornton II – 14 September 1900 – aged 11 when on *Titanic*; first-class passenger (ticket price: £120); rescued in lifeboat 4

Johnston, Master William Andrew – 14 September 1903 – aged 8 when on *Titanic*; first-class passenger (ticket price: £23 9s); *Titanic* victim

The following person died on this day:
Jerwan, Mrs Marie Marthe – 14 September 1974 – aged 23 when on *Titanic*; second-class passenger (ticket price: £13 15s 10d); rescued in lifeboat 11

Titanic survivor quote

Miss Barbara Joyce West

> 'We were all asleep when the collision took place, but were only jolted in our berths – my husband and children not even being awakened, and it was only the hurrying of passengers outside the cabin that caused alarm. The steward bade us all get up and dress thoroughly with plenty of warm things. Arthur placed lifebelts upon the children and then carried them to the boat deck. I followed carrying my handbag. After seeing us safely into the lifeboat Arthur returned to the cabin for a thermos of hot milk, and, finding the lifeboat let down he reached it by means of a rope, gave the flask to me, and, with a farewell, returned to the deck of the ship.'

Titanic fact

There were two lifeboats that were not launched and simply floated away: collapsible lifeboats A and B.

Day 259

15 September | On this day in *Titanic* history

The following people were born on this day:

Harrison, Mr Norman E. – 15 September 1873 – aged 38 when on *Titanic*; junior second engineer; *Titanic* victim

Hoyt, Mr Frederick Maxfield – 15 September 1873 – aged 38 when on *Titanic*; first-class passenger (ticket price: £100); rescued in collapsible lifeboat D

De Mulder, Mr Theodoor – 15 September 1881 – aged 30 when on *Titanic*; third-class passenger (ticket price: £9 10s); rescued in lifeboat 11

Wiklund, Mr Jakob Alfred – 15 September 1894 – aged 17 when on *Titanic*; third-class passenger (ticket price: £6 9s 11d); *Titanic* victim

Burns, Miss Mary Delia – 15 September 1896 – aged 15 when on *Titanic*; third-class passenger (ticket price: £7 17s 7d); *Titanic* victim

The following person died on this day:

Blake, Mr Percival Albert – 15 September 1968 – aged 22 when on *Titanic*; trimmer; rescued in lifeboat 15

Titanic survivor quote

Mr Frederick Maxfield Hoyt, mentioned in *Paterson Morning Call*, Tuesday, 23 April 1912

> 'Then Mr Hoyt began to consider his own situation. To him there seemed but one chance. He must swim for his life, and that would be impossible heavily and encumbrously clad as he was. So he went below to his stateroom and carefully stripped to his underclothing in preparation for the final struggle. Then he ascended to the upper deck and climbed the bridge where the gallant Captain Smith was still on duty, and whom he knew very well. They briefly exchanged views of the outcome which was inevitable to their trained eyes.
>
> 'But, said Mr Hoyt, I feel like taking a drink before I take the plunge, don't you, captain? The captain said he agreed with him and repairing to the captain's stateroom each took a big drink, to fortify him against the bitter cold.
>
> 'On reascending to the deck, Captain Smith said: You will have to jump and you had better do it soon.
>
> 'Yes, I know it, said Mr Hoyt, but I won't take the plunge from this deck, but will go to one of the lower decks.
>
> 'That will be better, said Captain Smith.'

Titanic fact

The length of the next stage, from Cherbourg to Queenstown, Ireland, was 307 miles.

Day 260

16 September | On this day in *Titanic* history

The following people were born on this day:
Nutbean, Mr William – 16 September 1881 – aged 30 when on *Titanic*; fireman; rescued in lifeboat unknown

Hichens, Mr Robert – 16 September 1882 – aged 29 when on *Titanic*; quartermaster; rescued in lifeboat 6

Mr Robert Hichens (quartermaster)

> 'On the return to the bridge, I had been on the bridge about a couple of minutes when the carpenter came back and reported the duty carried out. Standing by waiting for another message – it is the duty of the quartermaster to strike the bell every half hour – as the stand-by quartermaster, sir, I heard the second officer repeat to Mr Moody, the sixth officer, to speak through the telephone, warning the lookout men in the crow's nest to keep a sharp lookout for small ice until daylight and pass the word along to the other lookout men. The next order I received from the second officer was to go and find the deck engineer and bring him up with a key to open the heaters up in the corridor of the officers' quarters, also the wheelhouse and the chart room, on account of the intense cold. At a quarter to 10 I called the first officer, Mr Murdoch, to let him know it was one bell, which is part of our duty; also took the thermometer and barometer, the temperature of the water, and the log. At 10 o'clock I went to the wheel, sir. Mr Murdoch come up to relieve Mr Lightoller. I had the course given me from the other quartermaster, north 71 west, which I repeated to him, and he went and reported it to the first officer or the second officer in charge, which he repeated back – the course, sir. All went along very well until 20 minutes to 12, when three gongs came from the lookout, and immediately afterwards a report on the telephone, "Iceberg right ahead." The chief officer rushed from the wing to the bridge, or I imagine so, sir. Certainly I am inclosed in the wheelhouse, and I can not see, only my compass. He rushed to the engines. I heard the telegraph bell ring; also give the order "Hard astarboard," with the sixth officer standing by me to see the duty carried out and the quartermaster standing by my left side. Repeated the order, "Hard astarboard. The helm is hard over, sir."'

Titanic fact

The iceberg that the *Titanic* collided with is thought to have been anywhere between 50 to 100 feet above the water. The entire iceberg is believed to have been between 200 and 400 feet long.

Day 261

17 September | On this day in *Titanic* history

The following people were born on this day:
Nysveen, Mr Johan Hansen – 17 September 1851 – aged 60 when on *Titanic*; third-class passenger (ticket price: £6 4s 9d); *Titanic* victim

Blank, Mr Henry – 17 September 1872 – aged 39 when on *Titanic*; first-class passenger (ticket price: £31); rescued in lifeboat 7

Murdock, Mr William John – 17 September 1874 – aged 37 when on *Titanic*; fireman; rescued in lifeboat unknown

Bendell, Mr Frank – 17 September 1888 – aged 23 when on *Titanic*; fireman; *Titanic* victim

Hampe, Mr Léon Jérome – 17 September 1892 – aged 19 when on *Titanic*; third-class passenger (ticket price: £9 10s); *Titanic* victim

Arnold-Franchi, Mrs Josefine – 17 September 1893 – aged 18 when on *Titanic*; third-class passenger (ticket price: £17 16s); *Titanic* victim

Vanderplancke, Mr Leo Edmondus – 17 September 1896 – aged 15 when on *Titanic*; third-class passenger (ticket price: £18); *Titanic* victim

Asplund, Master Clarence Gustaf Hugo – 17 September 1902 – aged 9 when on *Titanic*; third-class passenger (ticket price: £31 7s 9d); *Titanic* victim

The following people died on this day:
Beane, Mrs Ethel – 17 September 1983 – aged 22 when on *Titanic*; second-class passenger (ticket price: £26); rescued in lifeboat 9

Karun, Miss Manca – 17 September 1971 – aged 5 when on *Titanic*; third-class passenger (ticket price: £13 8s 4d); rescued in lifeboat 15

Titanic survivor quote

Mr Henry Blank, quoted in *Newark Evening News*, Friday, 19 April 1912

> 'The minute the boat struck the water we began to pull away...We were afraid that if the ship went down she might draw us under with her. We had gone about a mile and a half when we saw her plunge forward and then down out of sight. Before she sank we heard the explosion of her boilers.
>
> 'There was no confusion whatever on the *Titanic* up to the time our boat was lowered. Everything was quiet and orderly. Many of the passengers had not yet left their rooms.'

Titanic fact

In August 2000, explorer Rory Golden became the first Irish diver to visit the site of RMS *Titanic*, as part of an official expedition that recovered 800 artefacts from nearly 4,000 metres deep.

Day 262

18 September | On this day in *Titanic* history

The following people were born on this day:
Hickman, Mr Leonard Mark – 18 September 1887 – aged 24 when on *Titanic*; second-class passenger (ticket price: £73 10s); *Titanic* victim

Saundercock, Mr William Henry – 18 September 1892 – aged 19 when on *Titanic*; third-class passenger (ticket price: £8 1s); *Titanic* victim

The following people died on this day:
Foley, Mr John – 18 September 1934 – aged 46 when on *Titanic*; deck storekeeper, rescued in lifeboat unknown

Littlejohn, Mr Alexander James – 18 September 1949 – aged 40 when on *Titanic*; saloon steward in first class; rescued in lifeboat 13

Watt, Mrs Elizabeth – 18 September 1951 – aged 40 when on *Titanic*; second-class passenger (ticket price: £15 15s); rescued in lifeboat 9

Titanic survivor quote

Mr John Foley, quoted in *Cook County Herald*, Friday, 26 April 1912

'Jack Foley, a member of the crew of the *Titanic*, who was in the lifeboat in which Mrs Astor was placed, praised her courage during the arduous trip in the open boat.

"Mrs Astor is the bravest little woman I ever met," said he. "Sitting next to her was a Swedish woman with a little three-year-old girl. The little girl was whimpering with the cold. Mrs Astor took her shawl and threw it about the shoulders of this woman.

"When the explosion occurred aboard the *Titanic* Mrs Astor made some kind of a sound, but I couldn't understand whether she said anything or merely sobbed. She turned her head away from the direction of the vessel. When the ship went down, and that awful moan was heard, she didn't say anything, but she covered her face with her hands."

Mr Alexander James Littlejohn, quoted in *The Daily Telegraph*, April 1912

'I got about 35 women and children into boat 13. We shouted for more women but there were none forthcoming. We had a few first-class male passengers in. An officer ordered two of us to get in and help row the boat, and I happened to be one of the fortunate ones... We could see the *Titanic* sinking by the head. Her forward "E" deck ports were under the water and we could see the lights gradually go out on the "E" deck as she settled down.'

Day 263

19 September | On this day in *Titanic* history

The following person was born on this day:
Crovella, Sig. Paolo Luigi – 19 September 1895 – aged 19 when on *Titanic*; assistant waiter; *Titanic* victim

Titanic survivor quotes

Mrs Catherine Evers

> 'My brother, who is 19 years old, went to Ireland six months ago to visit relatives. With his cousins, John and Philip Kiernan of Grove and Second Streets, Jersey City, he took passage on the *Titanic* from Queenstown. The Kiernan boys were drowned.
>
> 'My brother is suffering from wounds he says were inflicted on his head and arms by sailors of the *Titanic*. He was asleep when the ship hit the iceberg and awoke and dressed. He said he had to fight his way past members of the crew to the upper decks.
>
> 'When he saw the condition of the ship he put on a lifebelt and leaped overboard. He tried to get into one lifeboat which was only partly filled and the sailors beat him off with their oars. He tried to enter another partly filled boat and was again beaten off, being partly stunned this time.
>
> 'Then two young women in the boat, Kate and Mary Murphy, reached into the water and grasped him. They pleaded with the sailors that there was plenty of room in the boat, and at last got him aboard.'

Miss Anna Sofia Nysten

> 'We had gone to bed but I'm not sure I was asleep. The first thing I distinctly remember was a terrific jar and a sound such as one hears when the bottom of an automobile scrapes the ground as you drive in a deep-rutted road. That was the ship striking the iceberg.'

Titanic facts

The *Belfast Morning News* reported on 1 June 1911: 'The captain can, by simply moving an electric switch, instantly close the doors throughout, practically making the vessel unsinkable.'

Fourth Officer Joseph Boxhall was technical advisor for the film *A Night to Remember*.

Day 264

20 September | On this day in *Titanic* history

The following people were born on this day:
Ricaldone, Sig. Rinaldo Renato – 20 September 1889 – aged 22 when on *Titanic*; assistant waiter; *Titanic* victim

Sap, Mr Julius ('Jules') – 20 September 1890 – aged 21 when on *Titanic*; third-class passenger (ticket price: £9 10s); rescued in lifeboat 11

Kelland, Mr Thomas – 20 September 1892 – aged 19 when on *Titanic*; library steward in second class; *Titanic* victim

The following people died on this day:
Lulić, Mr Nikola – 20 September 1962 – aged 29 when on *Titanic*; third-class passenger (ticket price: £8 13s 3d); rescued in lifeboat 15

Thayer, Mr John Borland Jr – 20 September 1945 – aged 17 when on *Titanic*; first-class passenger (ticket price: £110 17s 8d); rescued in collapsible lifeboat B

Titanic survivor quote

Mr John Borland Thayer, mentioned in *Washington Times*, Friday, 19 April 1912

> 'Colonel Gracie says that soon after he was swept into the sea and was struggling to keep above water, he encountered J. B. Thayer, Jr., and together they fought for their lives.
>
> "Too much praise can not be given to the valour of the men and women who were the actors in the terrible tragedy, as greater deeds of heroism were never done by man," he said. "Only in rare instances was it necessary for the officers to use force to prevent frenzied men from pushing aside women.
>
> "No fiction tells of heroines who acted as did the brave women, wives of the men left behind. Officers were compelled to drag many of them away because they refused to leave their loved ones."
>
> 'The Washingtonian had given up all hope of being saved and was resigned to his fate, he says, when the huge ship sank. He had seen the last lifeboat lowered and was standing on the topmost deck witnessing the wild scenes amid the awful wailing of women and the crying of children when the sinking ship with hardly a moment's warning fairly plunged into the maw of the ocean.
>
> "Emotions never before experienced by man thrilled me, as I stood there and felt the great ship trembling, as the mammoth waves struck her side, and looked down into the black waters at the lifeboats being tossed about with their burden of agonised humanity," narrated Colonel Gracie to a representative of *The Washington Times*.'

Day 265

21 September | On this day in *Titanic* history

The following people were born on this day:

Widener, Mrs Eleanor – 21 September 1861 – aged 50 when on *Titanic*; first-class passenger (ticket price: £211 10s); rescued in lifeboat 4

Beckwith, Mrs Sallie – 21 September 1865 – aged 46 when on *Titanic*; first-class passenger (ticket price: £52 11s 1d); rescued in lifeboat 5

Watt, Mrs Elizabeth – 21 September 1871 – aged 40 when on *Titanic*; second-class passenger (ticket price: £15 15s); rescued in lifeboat 9

Wilde, Mr Henry Tingle – 21 September 1872 – aged 39 when on *Titanic*; chief officer, *Titanic* victim

Evans, Miss Edith Corse – 21 September 1875 – aged 36 when on *Titanic*; first-class passenger (ticket price: £31 13s 7d); *Titanic* victim

Holverson, Mrs Mary Aline – 21 September 1876 – aged 35 when on *Titanic*; first-class passenger (ticket price: £52); rescued in lifeboat 8

Oreškovič, Miss Jelka – 21 September 1881 – aged 30 when on *Titanic*; third-class passenger (ticket price: £8 13s 3d); *Titanic* victim

al-Ba'qlīnī, Miss Mārīyā – 21 September 1906 – aged 5 when on *Titanic*; third-class passenger (ticket price: £19 5s 2d); rescued in collapsible lifeboat C

Titanic survivor quote

Mr Charles H. Lightoller (second officer)

> 'Unfortunately whilst in Southampton, we had a reshuffle amongst the Senior Officers. Owing to the *Olympic* being laid up, the ruling lights of the White Star Line thought it would be a good plan to send the Chief Officer of the *Olympic*, just for the one voyage, as Chief Officer of the *Titanic*, to help, with his experience of her sister ship.
>
> 'This doubtful policy threw both Murdoch and me out of our stride; and, apart from the disappointment of having to step back in our rank, caused quite a little confusion. Murdoch from Chief, took over my duties as First. I stepped back on Blair's toes, as Second, and picked up the many threads of his job, whilst he – luckily for him as it turned out – was left behind. The other officers remained the same. However, a couple of days in Southampton saw each of us settled in our new positions and familiar with our duties.'

Titanic fact

Of the twenty-three female crew members, only three did not survive the sinking of *Titanic*. Tragically, of the 885 male crew members, 693 went down with the ship.

Day 266

22 September | On this day in *Titanic* history

The following people were born on this day:
McInerney, Mr Thomas – 22 September 1873 – aged 38 when on *Titanic*; greaser; *Titanic* victim

Bennett, Mrs Mabel – 22 September 1878 – aged 33 when on *Titanic*; stewardess; rescued in lifeboat 5

Fortune, Miss Ethel Flora – 22 September 1883 – aged 28 when on *Titanic*; first-class passenger (ticket price: £263); rescued in lifeboat 10

Campbell, Mr William – 22 September 1891 – aged 20 when on *Titanic*; carpenter/joiner; *Titanic* victim

The following people died on this day:
Hippach, Mrs Ida Sophia – 22 September 1940 – aged 44 when on *Titanic*; first-class passenger (ticket price: £57 19s 7d); rescued in lifeboat 4

Faulkner, Mr William Stephen – 22 September 1949 – aged 37 when on *Titanic*; bedroom steward in first class; rescued in lifeboat 11

Goldsmith, Mrs Emily Alice – 22 September 1955 – aged 31 when on *Titanic*; third-class passenger (ticket price: £20 10s 6d); rescued in collapsible lifeboat C

Lucas, Mr William Arthur – 22 September 1921 – aged 25 when on *Titanic*; able seaman; rescued in collapsible lifeboat D

Ettlinger, Mr Peter – 22 September 1962 – aged 31 when on *Titanic*; signed on to the *Titanic* as a saloon steward in second class on 4 April 1912; discharged from the ship prior to its departure on 10 April

Mallet, Master André Clement – 22 September 1973 – aged 1 when on *Titanic*; second-class passenger (ticket price: £37 1d); rescued in lifeboat 10

Bennett, Mrs Mabel Kate – 22 September 1974 – aged 33 when on *Titanic*; stewardess; rescued in lifeboat 5

Titanic survivor quote

Miss Ethel Flora Fortune, quoted in *Washington Times*, Monday, 22 April 1912

> 'The lifeboat, said Mrs Fortune, was greatly overcrowded. Four of the survivors were in the boat and the rest were supposed to be women, with the exception of one stoker and a Chinaman. There was a figure forward dressed in a brown mackintosh with a shawl like that of a steerage passenger over its head. The face was completely hidden. Miss Alice Fortune sat directly beside the supposed woman.'

Day 267

23 September | On this day in *Titanic* history

The following people were born on this day:
Voegelin-Dubach, Sig. Johannes – 23 September 1876 – aged 35 when on *Titanic*; coffee man; *Titanic* victim

Siebert, Mr Sidney Conrad – 23 September 1882 – aged 29 when on *Titanic*; bedroom steward; *Titanic* victim

Vandewalle, Mr Nestor Cyriel – 23 September 1883 – aged 28 when on *Titanic*; third-class passenger (ticket price: £9 10s); *Titanic* victim

Snow, Mr Eustace Philip – 23 September 1890 – aged 21 when on *Titanic*; trimmer; rescued in lifeboat unknown

Dodge, Master Washington – 23 September 1907 – aged 4 when on *Titanic*; first-class passenger (ticket price: £81 17s 2d); rescued in lifeboat 5

The following person died on this day:
Hichens, Mr Robert – 23 September 1940 – aged 29 when on *Titanic*; quartermaster; rescued in lifeboat 6

Mr Robert Hichens, mentioned in *Chicago Examiner*, Sunday, 21 April 1912

> 'Senator William Alden Smith, chairman of the committee, discovered new facts which made it imperative that Hichens testify. The wireless was used, and the *Lapland* held until a tug arrived. Hichens is said to have been at the *Titanic*'s wheel when she struck the berg.'

Titanic survivor quote

Mr Robert Hichens (quartermaster), witness at United States Senate Inquiry (Day 5)

> 'I would like to make a little statement as regarding Mrs Mayer's [Meyer] statement in the newspapers about my drinking the whisky sir, and about the blankets. I was very cold, sir, and I was standing up in the boat. I had no hat on. A lady had a flask of whisky or brandy, or something of that description, given her by some gentleman on the ship before she left, and she pulled it out and gave me about a tablespoonful and I drank it. Another lady, who was lying in the bottom of the boat, in a rather weak condition gave me a half wet and half dry blanket to try [to] keep myself a little warm, as I was half-frozen. I think it was very unkind of her, sir, to make any statement criticising me. When we got to the ship I handled everyone as carefully as I could, and I was the last one to leave the boat, and I do not think I deserve anything like that to be put in the papers. That is what upset me and got on my nerves.'

Day 268

24 September | On this day in *Titanic* history

The following people were born on this day:
Frölicher-Stehli, Mr Maximilian Josef – 24 September 1851 – aged 60 when on *Titanic*; first-class passenger (ticket price: £79 4s); rescued in lifeboat 5

Deacon, Mr Percy William – 24 September 1894 – aged 17 when on *Titanic*; second-class passenger (ticket price: £73 10s); *Titanic* victim

Osén, Mr Olaf Elon – 24 September 1895 – aged 16 when on *Titanic*; third-class passenger (ticket price: £9 4s 4d); *Titanic* victim

The following people died on this day:
Bidois, Miss Rosalie – 24 September 1938 – aged 46 when on *Titanic*; personal maid to Mrs Madeleine Talmage Astor who was a first-class passenger (ticket price: £247 10s 6d); rescued in lifeboat 4

Murphy, Miss Catherine 'Kate' – 24 September 1968 – aged 18 when on *Titanic*; third-class passenger (ticket price: £15 10s); rescued in lifeboat 16

Titanic survivor quotes

Mr William Arthur Lucas

> 'I said to them, "Wait a minute, there is another collapsible boat being put down from the funnel. You had better get into that." I could not take them because my boat was full.'

Miss Rosalie Bidois, mentioned in *New York Times*, 22 April 1912.

> 'Lucas said that he and Mr Lightoller helped one elderly lady into the collapsible boat and then had to help her out again as she would not go without her husband. "There were several cases like that," he said. Mrs John Jacob Astor and her maid, Rosalie Bidois, who were saved in the disaster to the *Titanic*, have been able to give to members of the Astor household some account as to what took place in the lifeboat which carried them away from the wrecked steamship...
>
> 'In the work of rescue Mrs Astor's maid played a conspicuous part. Six men believed to be seamen were dragged by the women into the boat. One of the men was dead when they found him, and another died shortly after he was pulled into the boat...
>
> 'Mrs Astor, from the time she entered the lifeboat until the long interval which followed before she and her companions were picked up by the *Carpathia*, displayed the greatest courage and fortitude. After being taken aboard the rescue ship she experienced a nervous collapse...'

Day 269

25 September | On this day in *Titanic* history

The following people were born on this day:
Finoli, Mr Luigi – 25 September 1870 – aged 41 when on *Titanic*; third-class passenger (ticket price: £7 1s); rescued in lifeboat 15

Omont, Mr Alfred Fernand – 25 September 1882 – aged 29 when on *Titanic*; first-class passenger (ticket price: £25 14s 10d); rescued in lifeboat 7

Chaboisson, Mr Adrien Finnin – 25 September 1886 – aged 25 when on *Titanic*; roast cook; *Titanic* victim

McCarthy, Miss Catherine – 25 September 1887 – aged 24 when on *Titanic*; third-class passenger (ticket price: £7 15s); rescued in lifeboat 16

Sage, Master Anthony William – 25 September 1899 – aged 12 when on *Titanic*; third-class passenger (ticket price: £69 11s); *Titanic* victim

Titanic survivor quote

Mr Alfred Fernand Omont, quoted in *The Times*, Saturday, 20 April 1912

> 'In our little boat we were frozen with cold, having left the ship without overcoats or rugs. We shouted from time to time to attract the attention of the other boats, but obtained no reply. With the same object a German baron [Alfred Nourney] who was with us fired off all the cartridges in his revolver. This agonizing suspense lasted for many hours, until at last the *Carpathia* appeared. We shouted "Hurrah" and all the boats scattered on the sea made towards her. For us it was like coming back to life.
>
> 'A particularly painful episode occurred on board the *Titanic* after all the boats had left. Some of the passengers who had remained on the ship, realizing too late that she was lost, tried to launch a collapsible boat which they had great difficulty in getting into shape. Nevertheless they succeeded in lowering it. The frail boat was soon half full of water and the occupants one after the other either were drowned or perished with cold, the bodies of those who died being thrown out. Of the original 50 only 15 were picked up by the *Carpathia*, on board which we joined them.
>
> 'We cannot praise too highly the conduct of the officers and men of the *Carpathia*. All her passengers gave up their cabins to the rescued women and the sick, and we were received with every possible kindness. Similarly we bear sorrowful tribute to the brave dead of the *Titanic*. Colonel Astor and the others were admirable in their heroism and the crew fulfilled with sublime self-sacrifice all the dictates of humanity.'

Titanic fact

14,000 gallons of drinking water were consumed each day on *Titanic*. To put this into perspective, one gallon of water is equivalent to 16 cups. This means there could be 224,000 cups of water distributed among some 2,208 people every 24 hours.

Day 270

26 September | On this day in *Titanic* history

The following people were born on this day:
Butt, Major Archibald Willingham – 26 September 1865 – aged 46 when on *Titanic*; first-class passenger (ticket price: £26 11s); *Titanic* victim

Cumings, Mr John Bradley – 26 September 1872 – aged 39 when on *Titanic*; first-class passenger (ticket price: £71 5s 8d); *Titanic* victim

Major Archibald Willingham Butt, mentioned in *New York Times*, Sunday, 3 March 1912

> 'Major Archibald Butt, military aid to the President of the United States, sailed yesterday for Europe on the North German Lloyd liner *Berlin* for a rest in a suit of clothes that won the admiration of every passenger on the deck of the liner, including a deaf and dumb Greek sponge merchant from Patras. His cambric handkerchief was tucked up his left sleeve like Kipling's pukka Indian soldier man.
>
> 'He wore a bright copper-colored Norfolk jacket fastened by big ball-shaped buttons of red porcelain, a lavender tie, tall bay wing collar, trousers of the same material as the coat, a derby hat with broad, flat brim, and patent leather shoes with white tops. The Major had a bunch of lilies in his buttonhole, and appeared to be delighted at the prospect of going away. He said that he had lost twenty pounds in weight following the President in his strenuous tour through the West.
>
> 'When asked if it were true that he was engaged to Miss Dorothy Williams of Washington, Major Butt replied sadly: "I wish it were. This bachelorhood is a miserable existence. I have distress signals flying at the fore, and will refuse no reasonable offer to enter the matrimonial field. I'll do the best I can, and if this leap year gets away before I get back to my wife I shall feel very much discouraged."
>
> 'The gallant Major did not wear an overcoat, and he winced once or twice when he was posing the windswept deck for the photographers.
>
> 'Major Butt is bound for Rome, accompanied by his friend Francis D. Millet, the painter, who is to take charge of the American Academy of Arts in the Eternal City. Mr Millet said that the present buildings are to be reconstructed at a cost of $1,500,000, and new properties are to be taken over by the Academy. The central building of the Academy is the Villa Aurelia.'

Titanic fact

Six life jackets from the *Titanic* are thought to remain in existence in museums or private collections around the world.

Day 271

27 September | On this day in *Titanic* history

The following person was born on this day:
Reed, Mr Thomas Charles Prowse – 27 September 1857 – aged 54 when on *Titanic*; bedroom steward; *Titanic* victim

The following person died on this day:
Wells, Master Ralph Lester – 27 September 1972 – aged 2 when on *Titanic*; second-class passenger (ticket price: £23); rescued in lifeboat 14

Titanic survivor quote

Mr William Ward (saloon steward)

> 'I went to No. 9 boat and assisted to take the canvas cover off of her. Then we lowered her down to level with the boat deck, and a sailor came along with a bag and threw it in the boat. This man said he had been sent down to take charge of the boat by the captain. The boatswain's mate, Haynes [Haines], was there, and he ordered this man out of the boat, and the man got out again. He stayed there for three or four minutes, and I think the purser – I am not sure on that point – said "Are you all ready?" Haynes answered "Yes" – it was either the purser or Mr Murdoch – and with that he said: "Pass in the women and children that are here into that boat." There were several men standing around, and they fell back, and there was quite a quantity of women and children helped into the boat; I could not say how many. One old lady made a great fuss about it and absolutely refused to get into the boat. She went back to the companionway and forced her way in and would not get into the boat. We got away from the ship's side before No. 11 was in the water, and Haynes gave orders to pull away, and we had some difficulty in unlashing the oars on account of them being lashed up. No one had a knife, for some time. We pulled off about, I should say, a couple of hundred yards, and Haynes gave orders to lay on the oars, which we did.'

Titanic fact

The oldest female survivor in first class was Mrs Mary Compton, aged 64 years, 8 months and 8 days.

Day 272

28 September | On this day in *Titanic* history

The following people were born on this day:
Woolner, Mr Hugh – 28 September 1866 – aged 45 when on *Titanic*; first-class passenger (ticket price: £35 10s); rescued in collapsible lifeboat D

Harrison, Mr Aragõa Drummond – 28 September 1871 – aged 40 when on *Titanic*; saloon steward in first class; rescued in lifeboat 6

Chaudanson, Miss Victorine – 28 September 1875 – aged 36 when on *Titanic*; first-class passenger (ticket price: £262 7s 6d); rescued in lifeboat 4

Fay, Mr Thomas Joseph – 28 September 1881 – aged 30 when on *Titanic*; greaser; *Titanic* victim

Meyer, Mrs Leila – 28 September 1886 – aged 25 when on *Titanic*; first-class passenger (ticket price: £82 3s 5d); rescued in lifeboat 6

Ingram, Mr George – 28 September 1892 – aged 19 when on *Titanic*; trimmer; *Titanic* victim

The following people died on this day:
Scott, Mr Frederick William – 28 September 1915 – aged 28 when on *Titanic*; greaser; rescued in lifeboat 4

Collins, Mr John – 28 September 1950 – aged 38 when on *Titanic*; fireman; rescued in lifeboat 1

Titanic survivor quote

Mr Hugh Woolner, mentioned in *Toronto Daily Star*, Friday, 19 April 1912

> 'At the time of the collision a Mr Woolner and myself were seated in the cafe. There was only a slight jar, and we thought nothing of it until we heard the excitement on deck. There was an order issued for all women to come on deck. – Woolner and myself went two decks down and saw water rushing in there. A lifeboat was lowered with several women and children from the steerage, and as it passed where I was I jumped in. I grabbed an oar and assisted the two men in the boat to row away from the steamer. We were about 200 yards away when the *Titanic* went down. There was hardly any suction. We were picked up about five hours later suffering from the cold.'

Titanic fact

The oldest male survivor in first class was Mr Maximilian Josef Frölicher-Stehli, aged 60 years, 6 months and 21 days.

Day 273

29 September | On this day in *Titanic* history

The following people were born on this day:

Bochet, Mr Pierre Giuseppe – 29 September 1868 – aged 43 when on *Titanic*; waiter; *Titanic* victim

Carlsson, Mr Frank Olof – 29 September 1878 – aged 33 when on *Titanic*; first-class passenger (ticket price: £5); *Titanic* victim (it is possible his car broke down and as we say 'he missed the boat'. His name remained on the passenger list and when he didn't show on the *Carpathia*, he was listed as a casualty)

Clench, Mr Frederick Charles – 29 September 1878 – aged 33 when on *Titanic*; able seaman; rescued in lifeboat 12

Karlsson, Mr Julius Konrad Eugen – 29 September 1878 – aged 33 when on *Titanic*; engineer; *Titanic* victim

Abī-Al-Munà, Mr Nāsīf Qāsim – 29 September 1884 – aged 27 when on *Titanic*; third-class passenger (ticket price: £18 15s 9d); rescued in lifeboat 15

Hosgood, Mr Richard William – 29 September 1889 – aged 22 when on *Titanic*; fireman; *Titanic* victim

Tomlin, Mr Ernest Portage – 29 September 1889 – aged 21 when on *Titanic*; third-class passenger (ticket price: £8 1s); *Titanic* victim

Buckley, Mr Daniel – 29 September 1890 – aged 21 when on *Titanic*; third-class passenger (ticket price: £7 15s 17d); rescued in lifeboat 13

The following people died on this day:

Murphy, Miss Margaret Jane 'Mary' 'Maggie' – 29 September 1957 – aged 26 when on *Titanic*; first-class passenger (ticket price: £78 17s); rescued in lifeboat 6

Riordan, Miss Hannah – 29 September 1982 – aged 26 when on *Titanic*; first-class passenger (ticket price: £78 17s); rescued in lifeboat 6

Titanic survivor quote

Mr Harold S. Bride, witness at United States Senate Inquiry (Day 2)

> 'In the first place the Californian had called me, sir, with an ice report. I was rather busy just for the minute, and I did not take it. She did not call again. She transmitted the ice report to the *Baltic*, and as she was transmitting it to the *Baltic* I took it down. I took it to the captain; but it was not official because it was not intended for me afterwards.'

Day 274

30 September | On this day in *Titanic* history

The following people were born on this day:
Swift, Mrs Margaret Welles – 30 September 1865 – aged 46 when on *Titanic*; first-class passenger (ticket price: £25 18s 7d); rescued in lifeboat 8

O'Keefe, Mr Arthur – 30 September 1867 – aged 44 when on *Titanic*; third-class passenger (ticket price: £7 5s); *Titanic* victim

The following people died on this day:
Etches, Mr Henry Samuel – 30 September 1944 – aged 43 when on *Titanic*; bedroom steward in first class (to Thomas Andrews); rescued in lifeboat 5

Ilett, Miss Bertha – 30 September 1976 – aged 17 when on *Titanic*; second-class passenger (ticket price: £10 10s); rescued in lifeboat unknown

Mr Henry Samuel Etches, mentioned in *Chicago Record-Herald*, Sunday, 21 April 1912

> '"If anything should happen to me, tell my wife in New York that I've done my best in doing my duty."
>
> 'That was the last message of Benjamin Guggenheim of the famous family of capitalists, dictated to a steward only a short while before he sank to his death with the *Titanic.*
>
> 'It was reported today that the message was received by Mrs Guggenheim. It was delivered by James Etches, assistant steward in the first cabin of the *Titanic*, to whom Mr Guggenheim communicated it.
>
> 'Etches appeared at the St. Regis Hotel and inquired for Mrs Benjamin Guggenheim. He said that he had a message from Mr Guggenheim and that it had to be delivered in person.'

Titanic survivor quotes

Mr Harold Godfrey Lowe (fifth officer)

> 'I made the men on that collapsible turn those bodies over before I took them into my boat. I said, "Before you come on board here you turn those bodies over and make sure they are dead," and they did so.'

Mr John Poingdester (able seaman)

> 'There were hundreds gathered round waiting to get into the three boats. Nos. 12, 14, and 16.'

October

Day 275

1 October | On this day in *Titanic* history

The following people were born on this day:
Salomon, Mr Abraham Lincoln – 1 October 1868 – aged 43 when on *Titanic*; first-class passenger (ticket price: £26); rescued in lifeboat 1

Beattie, Mr Joseph John – 1 October 1871 – aged 40 when on *Titanic*; greaser; *Titanic* victim

Collyer, Mrs Charlotte Annie – 1 October 1880 – aged 30 when on *Titanic*; second-class passenger (ticket price: £26 5s); rescued in lifeboat 14

Allen, Miss Elisabeth Walton – 1 October 1882 – aged 29 when on *Titanic*; first-class passenger (ticket price: £211 60s 9d); rescued in lifeboat 2

Kreuchen, Miss Emilie – 1 October 1882 – aged 29 when on *Titanic*; first-class passenger (ticket price: £211 6s 9d); rescued in lifeboat 2

Jessop, Miss Violet Constance – 1 October 1887 – aged 24 when on *Titanic*; stewardess; rescued in lifeboat unknown

Johnson, Mr William Cahoone – 1 October 1892 – aged 19 when on *Titanic*; seaman; *Titanic* victim

Hippach, Miss Gertrude Isabelle – 1 October 1894 – aged 17 when on *Titanic*; first-class passenger (ticket price: £57 19s 7d); rescued in lifeboat 4

The following people died on this day:
Durán I Moné, Sra. Florentina – 1 October 1959 – aged 30 when on *Titanic*; second-class passenger (ticket price: £13 17s 2d); rescued in lifeboat 12

Ball, Mrs Ada E. – 1 October 1967 – aged 36 when on *Titanic*; second-class passenger (ticket price: £13); rescued in lifeboat 10

Titanic survivor quote

Mr Abraham Lincoln Salomon, quoted in *Newark Star*, Friday, 19 April 1912

> 'It was only after five men jumped into one of the boats that was to hold only women that an officer threatened to shoot the next man who showed himself a coward. To prove that he meant what he said he fired one shot into the air. This was sufficient to convince the men that the officer was in earnest.
>
> 'The lifeboats were absolutely unprepared for any emergencies. In fact some of the so-called collapsible boats could not be opened. This caused the death of a number of persons.'

Day 276

2 October | On this day in *Titanic* history

The following people were born on this day:
Holverson, Mr Alexander Oskar – 2 October 1869 – aged 42 when on *Titanic*; first-class passenger (ticket price: £52); *Titanic* victim

Evans, Mr George Richard – 2 October 1884 – aged 27 when on *Titanic*; saloon steward in first class; *Titanic* victim

Titanic survivor quote

Mrs Anna Louisa Hogeboom

'About dawn we saw a ship in the dim distance, seeming many miles away. This gave us our first hope, but at the same time the wind commenced to rise and the waves grew large. As our oarsmen and oarswomen were nearly exhausted, we did not know how long we could endure it. Had the wind increased, as it did a few hours afterward, we would never have escaped. Shortly after 8 o'clock the *Carpathia* reached near enough for us to row to it, we having rowed about nine miles, and being the last lifeboat to reach the rescue ship.

'With our frozen fingers and feet it was difficult to climb up the wet, slippery rope ladder, but a rope fastened around our waist protected us from slipping into the sea.'

Mr Henry Samuel Etches (bedroom steward to Thomas Andrews)

'She seemed to raise once as though she was going to take a violent dive, but sort of checked, as though she had scooped the water up and had levelled herself. She then seemed to settle very, very quiet, until the last, when she rose up, and she seemed to stand 20 seconds, stern in that position (indicating), and then she went down with an awful grating, like a small boat running off a shingly beach.'

Miss Daisy Minahan

'A stewardess who had been saved told me that after the *Titanic* left Southampton that there were a number of carpenters working to put the doors of the air-tight compartments in working order. They had great difficulty in making them respond, and one of them remarked that they would be of little use in case of accident, because it took so long to make them work.'

Day 277

3 October | On this day in *Titanic* history

The following people were born on this day:
Toms, Mr Fred – 3 October 1882 – aged 29 when on *Titanic*; saloon steward; rescued in lifeboat 15

Halford, Mr Richard – 3 October 1884 – aged 27 when on *Titanic*; steward in third class; rescued in lifeboat 15

Thaler, Mr Montague Donald – 3 October 1894 – aged 17 when on *Titanic*; steward in third class; *Titanic* victim

The following person died on this day:
Kenyon, Mrs Marion – 3 October 1958 – aged 40 when on *Titanic*; first-class passenger (ticket price: £51 17s 3d); rescued in lifeboat 8

Titanic survivor quote

Mrs Marion Estelle Kenyon, mentioned in *Oak Park Oak Leaves*, Saturday, 27 April 1912

> 'Mrs Kenyon afterwards told me that the shock of the impact with the iceberg was very loud and heavy, but immediate danger was not expected, and altho she and Mr Kenyon were then retiring, they had time to dress and she put on an overcoat and hood before they went to the deck.
>
> 'There was apparently no system, as the first few boats had been lowered without their full capacity, two of them having only sixteen people each. She was in the eighth boat and there were nearly sixty people in her boat, because by [that] time fear had commenced to take action.'
>
> 'The sailors handling the boats seemed to know very little about them. There was no assignment, evidently, to any particular station, and men grappled with the ropes that probably never handled a lifeboat before... She asked Mr Kenyon to accompany her, but he said he would not go aboard while there was a single woman or child on the *Titanic*. Mrs Kenyon wanted to remain, but he insisted that she go and said he would probably join her the next day: and kissed her good night. There were a sailor and three men on that boat and the rest of them were women and children. The three men had gone aboard with the understanding that they could row, but when they reached the water it was found that none of them had ever handled an oar. In consequence Mrs Kenyon and another lady helped row the boat miles through the ice drift...'

Titanic fact

The distance into the sea bed that the bow penetrated is thought to be about 60ft (18 metres).

Day 278

4 October | On this day in *Titanic* history

The following people were born on this day:
Kimball, Mrs Gertrude – 4 October 1866 – aged 45 when on *Titanic*; first-class passenger (ticket price: £52 11s 1d); rescued in lifeboat 5

Oxenham, Mr Percy Thomas – 4 October 1889 – aged 22 when on *Titanic*; second-class passenger (ticket price: £10 10s); rescued in lifeboat 13

Coe, Mr Harry – 4 October 1890 – aged 21 when on *Titanic*; trimmer; *Titanic* victim

Mr Percy Thomas Oxenham, mentioned in *Perth Amboy Evening News*, 26 April 1912

> 'He was one of the second cabin passengers and when the crash came he was in bed asleep. He said that he hastily dressed himself, putting on trousers, shirt and a coat, and came out on deck. The orders were given for all hands to put on lifebelts and that women and children should leave the ship in the lifeboats first. He said that he assisted in placing the lifebelts on some women and one of the officers asked him if he could handle an oar.
>
> 'When he replied in the affirmative he was placed in the next to the last boat that left the ship. After rowing six hours they were picked up by the *Carpathia*. After landing in New York Mr Oxenham went to the home of his brother Charles at New Dunham, from where he came to Pleasant Plains to visit his brother, Theodore Oxenham, who has resided there some time.'

Titanic survivor quotes

Mr Percy Thomas Oxenham, quoted in *Jersey Journal*, Saturday, 20 April 1912

> 'Oxenham was dressed only in his underclothing when he got into the lifeboat and when he was taken aboard the *Carpathia*. At the home of his brother last night he refused to discuss at length what happened. "My mind is in a sort of daze," he said. "It all seems like a nightmare to me – like some dream that I had, and to say it is only five days since the ship went down seems impossible. It feels more like an age has elapsed."'

Mr Lawrence Beesley

> 'Presently we heard the order, "All men stand back away from the boats. All ladies retire to the next deck below." The men all stood away and waited in absolute silence, some leaning against the end railings of the deck, others pacing slowly up and down.'

Day 279

5 October | On this day in *Titanic* history

The following person was born on this day:
Candee, Mrs Helen Churchill – 5 October 1859 – aged 52 when on *Titanic*; first-class passenger (ticket price: £27 8s 11d); rescued in lifeboat 6

We can also wish Kate Winslet a happy birthday!
Kate Elizabeth Winslet CBE – born 5 October 1975

Titanic survivor quote

Mrs Helen Churchill Candee, quoted in *Washington Times*, Friday, 19 April 1912

> 'Mrs Candee was in the last boat that left the sinking ship, and she waited until the last call before she took her place with those who braved the perils of the ocean in the lifeboat. Lulled into a false sense of security by the words of the officers, it was not until she saw that the last shred of hope had been blasted that Mrs Candee took her place.
>
> "As our boat was lowered into the water we were told that we would be [illegible] to go back on the *Titanic* again. As we pushed away from the ship the terrible scenes began. They had opened the steerage doors, and on the deck piled the foreigners in the steerage. Finns, Italians and others rushed on deck, and trampled down the women while on the air we heard the echoes of at least fifty shots. I saw four men fall, killed by the shots from the guns. But who fired the deadly bullets I do not know.
>
> "Men frantic in their desire to escape, leaped into the water. Some sank immediately, while others floated around until they were so chilled that they could no longer keep afloat, and gave up the fight.
>
> "Women, too, some of them with children in their arms, became panic-stricken, and, disregarding all orders, jumped into the water and drowned before our eyes. As we pushed away from the ship we ran across a raft which had seventeen aboard. They were taken into our lifeboat, and one of the passengers, a man, was dead when taken into our boat.
>
> "A few moments later we heard the crash of two explosions, and then the *Titanic* sank while the crew who manned our boat pulled away from the suction which the sinking ship made, as we were in danger every minute of being drawn into the whirlpools. We floated around for six hours, all of us clad only in light garments and some with nothing on but their night clothes.
>
> We were picked up by the *Carpathia* and saved."'

Titanic fact

Edward John Smith was often referred to as 'E.J.' by his crew-mates.

Day 280

6 October | On this day in *Titanic* history

The following people were born on this day:
Rice, Mrs Margaret – 6 October 1872 – aged 39 when on *Titanic*; third-class passenger (ticket price: £29 2s 6d); *Titanic* victim

Čačić, Mr Luka – 6 October 1873 – aged 38 when on *Titanic*; third-class passenger (ticket price: £8 13s 3d); *Titanic* victim

Vigott, Mr Philip Francis – 6 October 1879 – aged 32 when on *Titanic*; able seaman; rescued in lifeboat 13

Murphy, Miss Catherine – 6 October 1893 – aged 18 when on *Titanic*; third-class passenger (ticket price: £15 10s); rescued in lifeboat 16

Emanuel, Miss Virginia Ethel – 6 October 1905 – aged 6 when on *Titanic*; third-class passenger (ticket price: £12 9s 6d); rescued in lifeboat 13

The following people died on this day:
Hogeboom, Mrs Anna Louisa – 6 October 1947 – aged 51 when on *Titanic*; first-class passenger (ticket price: £77 19s 2d); rescued in lifeboat 10

Padron Manent, Mr Julian – 6 October 1968 – aged 41 when on *Titanic*; second-class passenger (ticket price: £13 17s 3d); rescued in lifeboat 9

Mr Philip Francis Vigott, mentioned in *Jersey Morning News*, Tuesday, 21 May 1912

> 'To meet and speak to one of the five hundred odd survivors of the *Titanic* disaster is something of an experience, and yesterday, when Mr Philip Vigott, who was an A.B. on the ill-fated leviathan, paid a visit to our offices, we could not help feeling that we were talking to someone who has gone through an experience which has riveted the attention of the whole civilised world.
>
> Mr Vigott, whose name was wrongly spelt in the list of survivors, and whose portrait we reproduce, informed us he is now living at Southampton, and is on a brief visit to a sister in Jersey. He, with two other seamen, was placed in charge of lifeboat No. 13 on the ill-fated night, and crossed over to England on the *Lapland*. He leaves the island on Wednesday morning, and will later appear as a witness at the inquiry now proceeding in London.
>
> 'Under the circumstances, Mr Vigott would be reticent concerning his share in helping to rescue the passengers, but we could not help noting that he gives a close observer unmistakable evidence of having gone through a remarkable experience.'

Day 281

7 October | On this day in *Titanic* history

The following people were born on this day:
Frauenthal, Mr Isaac Gerald – 7 October 1868 – aged 43 when on *Titanic*; first-class passenger (ticket price: £27 14s 5d); rescued in lifeboat 5

Bessette, Miss Nellie Mayo – 7 October 1872 – aged 39 when on *Titanic*; first-class passenger (ticket price: £63 7s 2d); rescued in lifeboat 8

Gillespie, Mr William Henry – 7 October 1880 – aged 31 when on *Titanic*; second-class passenger (ticket price: £13); *Titanic* victim

Parsons, Mr Frank Alfred – 7 October 1885 – aged 26 when on *Titanic*; senior fifth engineer; *Titanic* victim

Boughton, Mr Bernard John – 7 October 1887 – aged 24 when on *Titanic*; saloon steward in first class; *Titanic* victim

Grosclaude, Mr Gérald – 7 October 1887 – aged 24 when on *Titanic*; waiter; *Titanic* victim

Mäenpää, Mr Matti Alexanteri – 7 October 1889 – aged 22 when on *Titanic*; third-class passenger (ticket price: £7 2s 6d); *Titanic* victim

The following people died on this day:
Hardy, Mr John – 7 October 1953 – aged 40 when on *Titanic*; chief steward in second class; rescued in collapsible lifeboat D

Dick, Mrs Vera – 7 October 1973 – aged 17 when on *Titanic*; first-class passenger (ticket price: £57); rescued in lifeboat 3

Mr Isaac Gerald Frauenthal, mentioned in *New York Times*, Thursday, 30 May 1912

> 'The committee was headed by Frederick K. Seward, and included Mrs J. J. Brown of Denver, Lieut. H. B. Steffanson of the Swedish Army, F. O. Spedden, Karl Behr, the tennis player; I. G. Frauenthal, and George A. Harder. Chairman Seward then spoke to the crew, saying, in part: "The eyes of the world are upon you and were upon you when you came to us on the open ocean, when we saw the *Carpathia* coming to us out of the dawn, and to all of you we wish to give our heartfelt thanks. For your hospitality, for your devotion, for your unselfishness, and for all that was done for us we never can be adequately grateful, and as a slight token of that appreciation we wish you to accept the medals that we have had struck for every man and woman of this ship."

Day 282

8 October | On this day in *Titanic* history

The following people were born on this day:
Couch, Mr Joseph Henry – 8 October 1862 – aged 49 when on *Titanic*; greaser; *Titanic* victim

Crosby, Miss Harriette Rebecca – 8 October 1872 – aged 39 when on *Titanic*; first-class passenger (ticket price: £26 11s); rescued in lifeboat unknown

Carter, Mrs Lucile – 8 October 1875 – aged 36 when on *Titanic*; first-class passenger (ticket price: £120); rescued in lifeboat 4

Duffy, Mr William Luke – 8 October 1875 – aged 36 when on *Titanic*; writer / engineer's clerk; *Titanic* victim

Maynard, Mr Isaac Hiram – 8 October 1880 – aged 31 when on *Titanic*; entrée cook; rescued in collapsible lifeboat B

Sadlier, Mr Matthew – 8 October 1892 – aged 19 when on *Titanic*; third-class passenger (ticket price: £7 14s 7d); *Titanic* victim

Ryerson, Miss Emily Borie – 8 October 1893 – aged 18 when on *Titanic*; first-class passenger (ticket price: £262 7s 6d); rescued in lifeboat 4

The following people died on this day:
Chibnall, Mrs Edith Martha Bowerman – 8 October 1953 – aged 48 when on *Titanic*; first-class passenger (ticket price: £55); rescued in lifeboat 6

Del Carlo, Mrs Argene – 8 October 1970 – aged 24 when on *Titanic*; second-class passenger (ticket price: £27 14s 5d); rescued in lifeboat 11

Lenox-Conyngham, Miss Eileen – 8 October 1993 – aged 11 when on *Titanic*; first-class passenger (ticket price: £3); disembarked at Cherbourg

Mr Walter Lord – born 8 October 1917 – *Titanic* historian and the author of *A Night to Remember* (1955)

> 'The *Titanic* woke them up. Never again would they be quite so sure of themselves. In technology especially, the disaster was a terrible blow. Here was the "unsinkable ship" – perhaps man's greatest engineering achievement – going down the first time it sailed.
>
> 'But it went beyond that. If this supreme achievement was so terribly fragile, what about everything else? If wealth meant so little on this cold April night, did it mean so much the rest of the year?'

Titanic fact

There is a 1,970 ft gap between the wrecked bow and the stern (600 metres).

Day 283

9 October | On this day in *Titanic* history

The following people were born on this day:
Anderson, Mr Walter Yuill – 9 October 1861 – aged 50 when on *Titanic*; bedroom steward; *Titanic* victim

Hurst, Mr Charles John – 9 October 1871 – aged 40 when on *Titanic*; fireman; *Titanic* victim

Janin, Mr Claude Marie – 9 October 1882 – aged 29 when on *Titanic*; soup cook; *Titanic* victim

The following people died on this day:
Robinson, Mrs Annie – 9 October 1914 – aged 47 when on *Titanic*; stewardess; rescued in lifeboat 11

Pelham, Mr George – 9 October 1939 – aged 39 when on *Titanic*; trimmer; rescued in lifeboat 16

Goldenberg, Mrs Nella Carlynne – 9 October 1947 – aged 46 when on *Titanic*; first-class passenger (ticket price: £89 12s 1d); rescued in lifeboat 5

On 9 October 1912, the White Star Line withdrew *Olympic* from service. She was returned to her builders at Belfast to be refitted to improve safety and incorporate lessons learned from the *Titanic* disaster six months earlier.

Titanic survivor quote

Mr Harold Godfrey Lowe (fifth officer)

> 'I saw five boats go away without an Officer, and I told Mr Moody on my own that I had seen five boats go away, and an Officer ought to go in one of these boats. I asked him who it was to be – him or I – and he told me, "You go; I will get in another boat."'

Titanic fact

There were eight musicians on the *Titanic*, all employees of C.W. and F.N. Black, a Liverpool-based agency.

Day 284

10 October | On this day in *Titanic* history

The following people were born on this day:

Vandercruyssen, Mr Victor – 10 October 1865 – aged 46 when on *Titanic*; third-class passenger (ticket price: £9); *Titanic* victim

Asplund, Mrs Selma Augusta Emilia – 10 October 1873 – aged 38 when on *Titanic*; third-class passenger (ticket price: £31 7s 9d); rescued in lifeboat 15

Gibbons, Mr Jacob William – 10 October 1875 – aged 36 when on *Titanic*; steward in second class; rescued in lifeboat 11

Chambers, Mrs Bertha – 10 October 1879 – aged 32 when on *Titanic*; first-class passenger (ticket price: £53 2s); rescued in lifeboat 5

Newell, Miss Madeleine – 10 October 1880 – aged 31 when on *Titanic*; first-class passenger (ticket price: £113 5s 6d); rescued in lifeboat 6

Parr, Mr William Henry Marsh – 10 October 1882 – aged 29 when on *Titanic*; electrician; *Titanic* victim

Marks, Mr James – 10 October 1884 – aged 27 when on *Titanic*; assistant pantryman steward in first class; *Titanic* victim

Roth, Miss Sarah A. – 10 October 1885 – aged 26 when on *Titanic*; third-class passenger (ticket price: £8 1s); rescued in collapsible lifeboat C

Kiernan, Mr John Joseph – 10 October 1885 – aged 26 when on *Titanic*; third-class passenger (ticket price: £7 15s); *Titanic* victim

Myhrman, Mr Pehr Fabian Oliver Malkolm – 10 October 1893 – aged 18 when on *Titanic*; third-class passenger (ticket price: £7 15s); *Titanic* victim

Andersson, Miss Erna Alexandra – 10 October 1895 – aged 16 when on *Titanic*; third-class passenger (ticket price: £7 18s 6d); rescued in lifeboat unknown

The following person died on this day:

Lundström, Mr Thure Edvin – 10 October 1942 – aged 32 when on *Titanic*; third-class passenger (ticket price: £7 11s 7d); rescued in lifeboat 15

Mr Jacob William Gibbons, quoted in *Bournemouth Echo*, April 1912

> 'It has been denied by many that the band was playing but it was doing so and the strains of "Nearer my God to Thee" came clearly over the water with a solemnity so awful that words cannot express it.'

Day 285

11 October | On this day in *Titanic* history

The following people were born on this day:
Wheadon, Mr Edward Henry – 11 October 1846 – aged 65 when on *Titanic*; second-class passenger (ticket price: £10 10s); *Titanic* victim

Patchett, Mr George – 11 October 1892 – aged 18 when on *Titanic*; third-class passenger (ticket price: £14 10s); *Titanic* victim

Lee, Mr Herbert Henry – 11 October 1894 – aged 17 when on *Titanic*; trimmer; *Titanic* victim

The following people died on this day:
Goldenberg, Mr Samuel L. – 11 October 1936 – aged 47 when on *Titanic*; first-class passenger (ticket price: £89 2s 1d); rescued in lifeboat 5

Mayné, Mlle Berthe Antonine – 11 October 1962 – aged 24 when on *Titanic*; first-class passenger (ticket price: £49 10s 1d); rescued in lifeboat 6

Whiteley, Mr Thomas Arthur – 11 October 1944 – aged 18 when on *Titanic*; saloon steward; rescued in lifeboat unknown

Titanic survivor quotes

Mr Jacob William Gibbons

> 'I helped some of the passengers into boat No. 11, including two little children. Before doing this I had scanned the deck for others but could see nobody about... I saw one lady covered in furs complaining that she had several more left behind. She had a mascot in the shape of a little pig which played a tune and she would not leave the ship until she had secured her treasure... We drew away from the *Titanic* in charge of Mr Wheat, another steward, and when about half a mile away saw her sink.'

Mr Thomas Arthur Whiteley, quoted in *Stevens Point Journal*, 27 April 1912

> 'I floated on my life preserver for several hours… When the sun came up I saw the collapsible raft in the distance, just black with men. They were all standing up. Mr Lightoller, the second officer, was one of them. "It's 31 lives against yours," he said, "you can't come aboard. There's no room." I pleaded with him in vain, and then, I confess, I prayed that somebody might die so I could take his place. I was only human. And then someone did die and let me aboard.'

Day 286

12 October | On this day in *Titanic* history

The following people were born on this day:
Theissinger, Mr Alfred – 12 October 1865 – aged 46 when on *Titanic*; bedroom steward; rescued in lifeboat unknown

Wick, Mrs Mary – 12 October 1866 – aged 45 when on *Titanic*; first-class passenger (ticket price: £164 17s 4d); rescued in lifeboat 8

Etches, Mr Henry Samuel – 12 October 1868 – aged 43 when on *Titanic*; bedroom steward in first class; rescued in lifeboat 5

Barker, Mr Ernest Thomas – 12 October 1871 – aged 40 when on *Titanic*; saloon steward in first class; *Titanic* victim

Oliva Y Ocana, Doña Fermina – 12 October 1872 – aged 39 when on *Titanic*; first-class passenger (ticket price: £108 18s); rescued in lifeboat 8

Bradley, Mr Patrick Joseph – 12 October 1873 – aged 38 when on *Titanic*; fireman; *Titanic* victim

Laitinen, Miss Kristina Sofia – 12 October 1873 – aged 38 when on *Titanic*; third-class passenger (ticket price: £9 11s 9d); *Titanic* victim

Cavendish, Mr Tyrell William – 12 October 1875 – aged 36 when on *Titanic*; first-class passenger (ticket price: £78 17s); *Titanic* victim

Stokes, Mr Philip Joseph – 12 October 1887 – aged 24 when on *Titanic*; second-class passenger (ticket price: £10 10s); *Titanic* victim

Ilett, Miss Bertha – 12 October 1894 – aged 17 when on *Titanic*; second-class passenger (ticket price: £10 10s); rescued in lifeboat unknown

The following people died on this day:
Othen, Mr Charles Alfred – 12 October 1952 – aged 36 when on *Titanic*; fireman; rescued in lifeboat unknown

Wick, Miss Mary Natalie – 12 October 1944 – aged 31 when on *Titanic*; first-class passenger (ticket price: £164 17s 4d); rescued in lifeboat 8

Titanic survivor quote

Mrs Mary Peebles Wick, quoted in *Washington Times*, Friday, 19 April 1912

> 'After we had been on the top deck for a while, considerably more than an hour, I should say, the women were told to stand in a group by themselves and to be ready to get into the lifeboats. The men drew back and the women stood at the railing.'

Day 287

13 October | On this day in *Titanic* history

The following people were born on this day:
Hays, Mrs Clara Jennings – 13 October 1859 – aged 52 when on *Titanic*; first-class passenger (ticket price: £93 10s); rescued in lifeboat 3

Rousseau, Mr Pierre – 13 October 1863 – aged 48 when on *Titanic*; chef; *Titanic* victim

Hall, Mr Frank Alfred James – 13 October 1871 – aged 40 when on *Titanic*; scullion, *Titanic* victim

Burroughs, Mr Arthur Peckham – 13 October 1877 – aged 34 when on *Titanic*; fireman; *Titanic* victim

Fortune, Mr Charles Alexander – 13 October 1892 – aged 19 when on *Titanic*; first-class passenger (ticket price: £263); *Titanic* victim

Sage, Miss Dorothy – 13 October 1897 – aged 44 when on *Titanic*; third-class passenger (ticket price: £69 11s); *Titanic* victim

The following person died on this day:
Weisz, Mrs Mathilde Françoise – 13 October 1953 – aged 37 when on *Titanic*; second-class passenger (ticket price: £26); rescued in lifeboat 10

Sir Ernest Shackleton (explorer), witness at British Wreck Commissioner's Inquiry (Day 26)

> 'I should say that if ice was reported in any quantity, bergs and floes, when the vessel is anywhere near that latitude and longitude and had a late report of, say, even the day before, at night-time she should slow till she was past that latitude and longitude in which icebergs and floe ice were seen.'

Titanic survivor quote

Mr Masabumi Hosono

> 'After the ship sank there came back again frightful shrills and cries of those drowning in the water. Our lifeboat too was filled with sobbing, weeping children and women worried about the safety of their husbands and fathers. And I, too, was as much depressed and miserable as they were, not knowing what would become of myself in the long run.'

Titanic fact

In 1880, Edward Smith joined the White Star Line; he would stay with the company for the rest of his life.

Day 288

14 October | On this day in *Titanic* history

The following people were born on this day:
Bateman, Reverend Robert James – 14 October 1860 – aged 52 when on *Titanic*; second-class passenger (ticket price: £12 10s 6d); *Titanic* victim

Maxwell, Mr John – 14 October 1882 – aged 29 when on *Titanic*; carpenter / joiner; *Titanic* victim

Pulbaum, Mr Franz – 14 October 1884 – aged 27 when on *Titanic*; second-class passenger (ticket price: £15 8d); *Titanic* victim

Testoni, Sig. Ercole – 14 October 1888 – aged 23 when on *Titanic*; assistant glass man; *Titanic* victim

Johansson, Mr Erik – 14 October 1889 – aged 22 when on *Titanic*; third-class passenger (ticket price: £7 15s 11d); *Titanic* victim

Ilmakangas, Miss Ida Livija – 14 October 1894 – aged 17 when on *Titanic*; third-class passenger (ticket price: £7 18s 6d); *Titanic* victim

Lefebvre, Miss Jeannie – 14 October 1903 – aged 8 when on *Titanic*; third-class passenger (ticket price: £25 9s 4d); *Titanic* victim

The following person died on this day:
Hocking, Miss Ellen 'Nellie' – 14 October 1963 – aged 20 when on *Titanic*; second-class passenger (ticket price: £23); rescued in lifeboat 4

Miss Ellen Hocking, mentioned in *Primitive Methodist Leader*, April 1912

> 'Anxiously we have waited for tidings, and now they are come. George (Hocking) a member of the Young Mens Christian Association Choir, was accompanied on the voyage by his widowed mother and two sisters, one of whom was rejoining her husband with two infants. The three boys, Percy (Bailey), Harry (Cotterill) and George (Hocking) assisted the women and children into the boats. George's mother was last, and she begged him to come with them. No mother, he said, the men are good enough to stand back for you and I must do the same and let their wives and mothers go. He kissed her goodbye. So the three were left, with a number of Cornishmen, not one of whom was saved; indeed, Mrs J H Chapman of St. Neot, Cornwall, chose to stay beside her husband, refusing a place in the boat.'

Titanic fact

Captain Smith commanded eighteen White Star Line vessels during his career.

Day 289

15 October | On this day in *Titanic* history

The following people were born on this day:

Hosono, Mr Masabumi – 15 October 1870 – aged 41 when on *Titanic*; second-class passenger (ticket price: £13); rescued in lifeboat 10

Portaluppi, Mr Emilio Ilario Giuseppe – 15 October 1881 – aged 30 when on *Titanic*; second-class passenger (ticket price: £12 14s 9d); rescued in lifeboat 14

Marsden, Miss Evelyn – 15 October 1883 – aged 28 when on *Titanic*; stewardess; rescued in lifeboat 16

Fleet, Mr Frederick – 15 October 1887 – aged 24 when on *Titanic*; lookout; rescued in lifeboat 6

Harknett, Miss Alice Phoebe – 15 October 1890 – aged 21 when on *Titanic*; domestic kitchen servant; *Titanic* victim

The following people died on this day:

Buckley, Mr Daniel – 15 October 1918 – aged 21 when on *Titanic*; third-class passenger (ticket price: £7 15s 17d); rescued in lifeboat 13

Sparkman, Mr Henry William – 15 October 1947 – aged 36 when on *Titanic*; fireman; rescued in lifeboat unknown

Behr, Mr Karl Howell – 15 October 1949 – aged 26 when on *Titanic*; first-class passenger (ticket price: £30); rescued in lifeboat 5

Bright, Mr Arthur John – 15 October 1955 – aged 42 when on *Titanic*; quartermaster; rescued in collapsible lifeboat D

Mulvihill, Miss Bridget Elizabeth – 15 October 1959 – aged 25 when on *Titanic*; third-class passenger (ticket price: £7 15s); rescued in lifeboat 15

Titanic survivor quote

Mr Emilio Ilario Giuseppe Portaluppi, quoted in *The Times*, Saturday, 20 April 1912

> 'Many people seem to have slept through the shock of the collision, and the tale told by Emilio Portaluppi, a second cabin passenger, shows that he was first awakened by the explosion of one of the ship's boilers. He hurried up to deck one and strapped on a lifebelt. Following the example of others, he then leapt into the sea, and held on to an ice floe, with the help of which he managed to keep afloat until he was seen by those in the lifeboats and rescued.'

Titanic fact

Edward John Smith served with the White Star Line for 32 years.

Day 290

16 October | On this day in *Titanic* history

The following people were born on this day:

Goldschmidt, Mr George B. – 16 October 1840 – aged 71 when on *Titanic*; first-class passenger (ticket price: £34 13s 1d); *Titanic* victim

McGowan, Miss Catherine – 16 October 1869 – aged 42 when on *Titanic*; third-class passenger (ticket price: £7 15s); *Titanic* victim

Barry, Miss Julia – 16 October 1881 – aged 30 when on *Titanic*; third-class passenger (ticket price: £7 17s 7d); *Titanic* victim

Blake, Mr Percival Albert – 16 October 1889 – aged 22 when on *Titanic*; trimmer; rescued in lifeboat 15

Braf, Miss Elin Ester Maria – 16 October 1891 – aged 20 when on *Titanic*; third-class passenger (ticket price: £7 17s 1d); *Titanic* victim

Dornier, Mr Louis Auguste – 16 October 1891 – aged 20 when on *Titanic*; assistant fish cook; *Titanic* victim

Coutts, Master William Loch 'Willie' – 16 October 1902 – aged 9 when on *Titanic*; third-class passenger (ticket price: £15 18s); rescued in lifeboat 2

Tu'mah, Miss Mariyam – 16 October 1902 – aged 9 when on *Titanic*; third-class passenger (ticket price: £15 4s 11d); rescued in collapsible lifeboat C

The following people died on this day:

Kink-Heilmann, Mrs Luise – 16 October 1979 – aged 26 when on *Titanic*; third-class passenger (ticket price: £22 6d); rescued in lifeboat 2

West, Miss Barbara Joyce – 16 October 2007 – aged 10 months when on *Titanic*; second-class passenger (ticket price: £27 15s); rescued in lifeboat 11

Titanic survivor quote

Mr Samuel Hemming (lamp trimmer)

> 'I went on the boat deck. They were turning the boats out. As I went to the deck, I went where there were the least men, and helped to turn out the boats. Then I went to the boats on the port side, to do the same, until Mr Lightoller called me and said, "Come with me," and he said, "Get another good man." I says, "Foley is here somewhere." He says, "I have no time to stop for Foley." So he called a man himself, and he said, "Follow me." So we followed him, and he said: "Stand by to lower this boat." It was No. 4 boat.' We lowered the boat in line with the A deck, when I had an order come from the captain to see that the boats were properly provided with lights. I called Mr Lightoller and told him that I would have to leave the boat's fall; so he put another man in my place.'

Day 291

17 October | On this day in *Titanic* history

The following people were born on this day:
Van Der Brugge, Mr Wessel Adrianus – 17 October 1873 – aged 38 when on *Titanic*; fireman; *Titanic* victim

Silverthorne, Mr Spencer Victor – 17 October 1876 – aged 35 when on *Titanic*; first-class passenger (ticket price: £26 5s 9d); rescued in lifeboat 5

Hartnell, Mr Fred – 17 October 1890 – aged 21 when on *Titanic*; saloon steward; rescued in lifeboat unknown

The following people died on this day:
Archer, Mr Ernest Edward – 17 October 1917 – aged 35 when on *Titanic*; able seaman; rescued in lifeboat 16

O'Brien, Mrs Johanna 'Hannah' – 17 October 1918 – aged 27 when on *Titanic*; third-class passenger (ticket price: £15 10s); rescued in lifeboat unknown

Ismay, Mr Joseph Bruce – 17 October 1937 – aged 49 when on *Titanic*; first-class passenger (managing director of White Star Line); rescued in collapsible lifeboat C

Slade, Mr Bertram – 17 October 1943 – aged 28 when on *Titanic*; fireman; disembarked at Southampton and was listed as having deserted the ship

Persson, Mr Ernst Ulrik – 17 October 1951 – aged 25 when on *Titanic*; third-class passenger (ticket price: £7 15s 6d); rescued in lifeboat 15

Marvin, Mrs Mary Graham Carmichael – 17 October 1975 – aged 18 when on *Titanic*; third-class passenger (ticket price: £53 2s); rescued in lifeboat 10

Titanic survivor quote

Mrs Mahala Douglas

> 'Mr Boxhall had charge of the signal lights on the *Titanic*, [he] had put in the emergency boat a tin of green lights, like rockets. These he commenced to send off at intervals, and very quickly we saw the lights of the *Carpathia* ... we were the first boat to arrive at the *Carpathia*. When we pulled alongside Mr Boxhall called out, "shut down your engines and take us aboard. I have only one sailor." At this point I called out, "The *Titanic* has gone down with everyone on board," and Mr Boxhall told me to "shut up." This is not told in criticism; I think he was perfectly right. We climbed a rope ladder to the upper deck of the *Carpathia*. I at once asked the chief steward, who met us, to take the news to the captain. He said the officer was already with him.
>
> 'The history of our wonderful treatment on the *Carpathia* is known to the world. It has been underestimated. We reached the *Carpathia* at 4.10, and I believe by 10 o'clock all of the boats had been accounted for. We sailed away, leaving the Californian to cruise about the scene. We circled the point where the *Titanic* had gone down, and I saw nothing except quantities of cork, loose cork floating in the current, like a stream – nothing else.'

Day 292

18 October | On this day in *Titanic* history

The following people were born on this day:
Jones, Mr Harry Owen Glendower – 18 October 1875 – aged 36 when on *Titanic*; roast cook; *Titanic* victim

Guy, Mr Elgar John – 18 October 1882 – aged 29 when on *Titanic*; boots steward; rescued in lifeboat 5

Nilsson, Mr August Ferdinand – 18 October 1890 – aged 21 when on *Titanic*; third-class passenger (ticket price: £7 17s 1d); *Titanic* victim

The following people died on this day:
Chambers, Mrs Bertha – 18 October 1959 – aged 32 when on *Titanic*; first-class passenger (ticket price: £53 2s); rescued in lifeboat 5

Bowerman, Miss Elsie Edith – 18 October 1973 – aged 22 when on *Titanic*; first-class passenger (ticket price: £55); rescued in lifeboat 6

Mr Harry Owen Glendower Jones, mentioned in *Hampshire Observer County*, 20 April 1912

> 'Jones was one of the crew on the *Olympic* when she was struck by the *Hawke*. On every subsequent sailing he would tell his friends at home that his ship would "turn turtle".
>
> 'Jones was in the American Navy during the time of the war with the Spaniards and held the Santiago medal; he also held an award for another naval engagement with the Spaniards. He afterwards entered the service of the White Star Line having held berths on the *Adriatic* and *Olympic* as well as the St Paul and others.'

Titanic survivor quote

Mrs Henry Arthur Casebeer

> 'We saw the *Titanic* when it made its final plunge. The lights were burning until the very last moment and it was a spectacular as well as awesome sight. After the *Titanic* had sunk there were thousands of people struggling in the water crying piteously for help. Three times Officer Pitman ordered his men to turn about so that he could pick up some of them, but each time they were prevented from doing so by some of the passengers in the lifeboat who called upon the seamen frantically to go ahead and when they grasped the oars and interfered with the proper handling of the boat so that the seamen were finally forced to give up their efforts of turning back to rescue any of the unfortunates.'

Day 293

19 October | On this day in *Titanic* history

The following people were born on this day:
Threlfall, Mr Thomas – 19 October 1867 – aged 44 when on *Titanic*; leading fireman; rescued in lifeboat unknown

Roberts, Mrs Mary Kezia – 19 October 1870 – aged 41 when on *Titanic*; stewardess; rescued in lifeboat 16

Silvey, Mrs Alice – 19 October 1872 – aged 39 when on *Titanic*; first-class passenger (ticket price: £55 18s); rescued in lifeboat 11

Thompson, Mr Alexander Mellis – 19 October 1875 – aged 36 when on *Titanic*; third-class passenger (ticket price: £8 1s); *Titanic* victim

Niven, Mr John Brown – 19 October 1881 – aged 30 when on *Titanic*; fireman; *Titanic* victim

Long, Mr Milton Clyde – 19 October 1882 – aged 29 when on *Titanic*; first-class passenger (ticket price: £30); *Titanic* victim

Bainbrigge, Mr Charles Robert – 19 October 1889 – aged 22 when on *Titanic*; second-class passenger (ticket price: £10 10s); *Titanic* victim

The following person died on this day:
Carter, Miss Lucile Polk – 19 October 1962 – aged 14 when on *Titanic*; first-class passenger (ticket price: £120); rescued in lifeboat 4

Titanic survivor quote

Mr Thomas Threlfall, quoted in *Herts Advertiser & St. Albans Times*, Saturday, 4 May 1912

> 'Leading fireman T. Threlfall, one of the *Titanic* survivors, who was in boat No. 14, in the course of a narrative told to a pressman on landing at Plymouth, said: "From the wreckage we picked up four men. Then Mr Lowe called out, 'There's a boat sinking over there.' Although we were then towing a collapsible boat with about eighty people in her we reached the sinking boat just as the water was up to her gunwale and took twenty-six men and one woman, a Mrs Abbott off her. I held the woman in my arms till we reached the *Carpathia*."'

Mr Alfred Olliver, witness at United States Senate Inquiry (Day 7)

> 'She was well down at the head at first, when we got away from her at first, and to my idea she broke forward, and the afterpart righted itself and made another plunge and went right down. I fancied I saw her black form. It was dark, and I fancied I saw her black form going that way.'

Day 294

20 October | On this day in *Titanic* history

The following people were born on this day:

Sanderson, Mr Harold Arthur – 20 October 1859 – aged 52 when on *Titanic*; shipowner, embarked at Belfast and disembarked at Southampton

Anderson, Mr Harry – 20 October 1864 – aged 42 when on *Titanic*; first-class passenger (ticket price: £26 11s); rescued in lifeboat 3

De Messemaeker, Mrs Anna – 20 October 1875 – aged 36 when on *Titanic*; third-class passenger (ticket price: £17 8s); rescued in lifeboat 13

Penny, Mr William Farr – 20 October 1880 – aged 31 when on *Titanic*; assistant saloon steward in second class; *Titanic* victim

Laroche, Mme Juliette Marie Louise – 20 October 1889 – aged 22 when on *Titanic*; second-class passenger (ticket price: £41 11s 7d); rescued in lifeboat 14

Carter, Miss Lucile Polk – 20 October 1898 – aged 14 when on *Titanic*; first-class passenger (ticket price: £120); rescued in lifeboat 4

Titanic survivor quotes

Mrs Henry Arthur Casebeer

'I could not hear just what he said to me at the time on account of the din, but I saw him motion to me to get into the boat, which was about to be swung over the rail 90 feet above the water. I asked him why he did not get in also, and he said: "No, women and children first."

'Right here I wish to say that Bruce Ismay was there also, helping to load the women and children into the boat. He was dressed in pyjamas in slippers with a coat thrown over his shoulders and as the boat I was in [was] the sixth to leave the ship, you can see that reports that he was in one of the first boats are absolutely false.'

Miss Lucile Polk Carter, mentioned in *Washington Times*, Tuesday, 16 April 1912

'Mrs Stilson Hutchins, of this city, has received word that her cousin, Mrs Lucille Carter, of Philadelphia, has been rescued. Mr Carter and their two children also are among the saved. "I have just received a long distance phone from Baltimore," said Mrs Hutchins. "It tells me that news has come by wireless of the safety not alone of Mrs Carter, but of Mr Carter and the two children. They are on the *Carpathia*, according to the word I have received. I trust it is true." Mrs Carter was Miss Lucille Polk, of Baltimore, before her marriage to Mr Carter. They are both well known in Washington.'

Day 295

21 October | On this day in *Titanic* history

The following people were born on this day:
White, Mr William George – 21 October 1888 – aged 23 when on *Titanic*; trimmer; rescued in lifeboat 15

Asplund, Master Carl Edgar – 21 October 1906 – aged 5 when on *Titanic*; third-class passenger (ticket price: £31 7s 9d); *Titanic* victim

Asplund, Miss Lillian Gertrud – 21 October 1906 – aged 5 when on *Titanic*; third-class passenger (ticket price: £31 7s 9d); rescued in lifeboat 15

Mr William George White, mentioned in *Ballymena Observer*, Friday, 31 May 1912

> 'Florence Searle, a laundress, who appeared in court dressed in deep black, was sentenced three months' hard labour at Southampton on Tuesday for obtaining £2 from the mayor's *Titanic* Fund by false pretence.
>
> 'It was stated that apparently in great distress she represented herself as the widow of a trimmer named White, who survived the wreck, and told the court that he was a single man.'

Titanic survivor quote

Mr George Achilles Harder

> 'The boats were lowered very quickly, and from where we were located, there was apparent absolutely no panic, shooting or disorder. Men helped the women into the boats and I personally believe that there were only three Italians shot who insisted on getting into a boat. I do not know who shot them.
>
> 'We pulled away from the immediate vicinity of the *Titanic* as she was starting to list considerably and her bow was slowly tilting – but all the time the lights were burning brightly and the band was playing the "Star Spangled Banner". I did not hear any pistol shots, and although we were gradually pulling away from the boat I think we should have heard them.
>
> 'The boat sank within two hours after she struck. Five minutes before she dropped out of sight the lights went out. Then the bow dug deeper into the water until she was inundated right to midships, when she suddenly dove straight down, and thus made less suction than would have occurred had she sunk in the ordinary way. Just before she sank an awful screaming, ghastly and piercing, rent the air, and it was caused by a handful of steerage passengers, women, who had huddled into the aft end of the boat, just as an ostrich digs his head into the sand.'

Day 296

22 October | On this day in *Titanic* history

The following people were born on this day:
Harder, Mr George Achilles – 22 October 1886 – aged 25 when on *Titanic*; first-class passenger (ticket price: £5 8s 10d); rescued in lifeboat 5

Lievens, Mr René Aimé – 22 October 1887 – aged 24 when on *Titanic*; third-class passenger (ticket price: £9 10s); *Titanic* victim

Ödahl, Mr Nils Martin – 22 October 1888 – aged 23 when on *Titanic*; third-class passenger (ticket price: £9 4s 6d); *Titanic* victim

Pacey, Mr Reginald lvan – 22 October 1894 – aged 17 when on *Titanic*; lift steward; *Titanic* victim

The following person died on this day:
Mallet, Mrs Antonine Marie – 22 October 1974 – aged 41 when on *Titanic*; third-class passenger (ticket price: £7 1s); rescued in lifeboat 15

Titanic survivor quotes

Mr George Achilles Harder, quoted in *Brooklyn Daily Eagle*, Friday, 19 April 1912

> 'Mr Ismay helped us get into the boat and he acted splendidly. In our boat was a big box and a keg of water. Unlike the other boats, we had no lights at all, and we pulled away from the boat some distance, later taking turns at the oars. I was on the starboard side, the boat having listed that way. I am told the lowering of boats on the other side was a much more difficult proposition.
>
> 'I met Colonel Astor and his wife on the stairs and he was very calm, maintaining that there was no danger and that the boat could keep afloat for a long time. Later on deck, he started to help a little boy to get into one of the lifeboats, but the boy was pushed back with the statement that only women were being shipped then.
>
> 'Colonel Astor then went and picked up a woman's hat somewhere, put it on the boy's head and said: "Now you're a woman, get in there." And the boy is safe today as a result. Colonel Astor, his wife, and everyone else thoroughly believed that all would be saved. I found out afterwards that the ship's carpenter had been sent below to determine the extent of the damage and the depth of the water shipped but he never returned.'

Mr Lawrence Beesley

> 'I wonder where the lift-boy was that night. I would have been glad to find him in our boat, or on the *Carpathia* when we took count of the saved. He was quite young – not more than sixteen, I think – a bright-eyed, handsome boy, with a love for the sea and the games on deck and the view over the ocean – and he did not get any of them.'

Day 297

23 October | On this day in *Titanic* history

The following people were born on this day:
Simonius-Blumer, Mr Colonel (Oberst Alfons) – 23 October 1855 – aged 56 when on *Titanic*; first-class passenger (ticket price: £35 10s); rescued in lifeboat 3

Jacobsohn, Mr Sidney Samuel – 23 October 1869 – aged 42 when on *Titanic*; second-class passenger (ticket price: £27); *Titanic* victim

Johannesen, Mr Bernt Johannes – 23 October 1882 – aged 29 when on *Titanic*; third-class passenger (ticket price: £8 2s 3d); rescued in lifeboat 13

Coutts, Master Neville Leslie – 23 October 1908 – aged 3 when on *Titanic*; third-class passenger (ticket price: £15 18s); rescued in lifeboat 2

The following person died on this day:
Witter, Mr James William Cheetham – 23 October 1961 – aged 31 when on *Titanic*; smoke room steward in second class; rescued in lifeboat unknown

Titanic survivor quote

Master Neville Leslie Coutts, mentioned in *The Washington Post*, Saturday, 20 April 1912

> 'From the lips of the woman who was saved from the *Titanic* came today one of the most glowing tributes yet paid to the heroism and self-sacrifice of the brave men who gave their lives that women and children might be spared a watery grave.
>
> 'Mrs William Coutts, of this city, described in graphic manner how she and her two sons, Neville and William, 3 and 9 years old respectively, were rescued through the bravery of men aboard the doomed vessel: "My husband had sent me money to buy second-class passage for the children and myself… but I went in the steerage. I wanted to save the difference in passage money to help build up our home.
>
> "I was asleep when the ship struck. The crash was slight that I thought little of it. I lay awake for fully fifteen minutes before I got up. I dressed myself slowly, and then went out on deck to see what the trouble was. Every one was hurrying, but there was no disorder. I heard some talk about lifeboats, and then I hurried back to the children. I tied life preservers on the boys and then looked around for one for myself. There was none in sight.
>
> "I rushed out on deck with the children following me. Just when I had given up hope of finding my way a seaman came along and said Hurry now; all women and children to the lifeboats."'

Day 298

24 October | On this day in *Titanic* history

The following people were born on this day:
Warren, Mrs Anna Sophia – 24 October 1851 – aged 60 when on *Titanic*; first-class passenger (ticket price: £75 5s); rescued in lifeboat 5

Peñasco Y Castellana, Mr Victor – 24 October 1887 – aged 24 when on *Titanic*; first-class passenger (ticket price: £108 18s); *Titanic* victim

Garfirth, Mr John – 24 October 1889 – aged 21 when on *Titanic*; third-class passenger (ticket price: £14 10s); *Titanic* victim

Bulley, Mr Henry Ashburnham – 24 October 1890 – aged 41 when on *Titanic*; boots steward in second class; *Titanic* victim

Collins, Mr John – 24 October 1894 – aged 41 when on *Titanic*; scullion, rescued in collapsible lifeboat B

The following people died on this day:
Beane, Mr Edward – 24 October 1948 – aged 32 when on *Titanic*; second-class passenger (ticket price: £26); rescued in lifeboat 9

Mubārik, Master Jirjis – 24 October 1979 – aged 7 when on *Titanic*; third-class passenger (ticket price: £15 4s 11d); rescued in collapsible lifeboat C

Titanic survivor quote

Mrs Anna Sophia Warren, quoted in *Portland Oregonian*, Saturday, 27 April 1912

> 'With daylight the wind increased and the sea became choppy, and we saw icebergs in every direction; some lying low in the water and others tall, like ships and some of us thought they were. Our boat was picked up about 4.10 am by the *Carpathia* and too much cannot be said of the courtesy, kindness and unceasing care of the officers, crew and passengers of this vessel, who worked from morning until night and almost from night until morning in the relief of the survivors. I was in the second boat picked up. Others were adrift many hours longer and consequently suffered more. The captain of the *Carpathia* stayed until there were no more boats to pick up and he felt he must get out of the ice before sundown. We left the scene of the disaster about noon with the *Californian* still standing by, and as we turned back, as far as I could see in all directions, was a continuous floe of ice, marked by detached icebergs. It was well along towards night before we were clear of the field. The feeling among the rescued passengers was that the terrific speed of the *Titanic* and the lack of reasonable precautions were largely responsible for the accident.'

Day 299

25 October | On this day in *Titanic* history

The following people were born on this day:
Milling, Mr Jacob Christian – 25 October 1863 – aged 48 when on *Titanic*; second-class passenger (ticket price: £13); *Titanic* victim

Biétrix, Mr George Baptiste – 25 October 1883 – aged 28 when on *Titanic*; cook; *Titanic* victim

Al-Za'innī, Mr Fahīm Rūḥānā – 25 October 1886 – aged 25 when on *Titanic*; third-class passenger (ticket price: £7 4s 6d); rescued in lifeboat 6

Brailey, Mr William Theodore Ronald – 25 October 1887 – aged 24 when on *Titanic*; musician in *Titanic*'s orchestra; *Titanic* victim

The following person died on this day:
Serreplaà, Miss Augusta Jeanne – 25 October 1928 – aged 30 when on *Titanic*; first-class passenger (ticket price: £31); rescued in lifeboat 4

Titanic survivor quote

Mr Fahīm Rūḥānā al-Za'innī, mentioned in *Niles Daily News*, Thursday, 25 April 1912

> 'The sight of the sinking ship, after one plunge, bow downward, will never be forgotten by any of the people who were in the lifeboats. It was nearly five o'clock when they sighted the *Carpathia* coming toward them, and one great cry went up from the lifeboats, and lanterns were waved frantically.
>
> 'Mr Zanni tells of the kindness shown toward them by the captain and officers of the [illegible] boat and of the [illegible] trip to [illegible] New York. On their arrival they were shown every kindness by the waiting throngs. He was taken to the office of the Hebrew sheltering and immigration Aid Society where he was provided with clothing and rendered every assistance.
>
> 'An incident which tends to prove the utter selfishness of shallow-minded people is related by the young man. When the survivors were being raised into the *Carpathia* a woman who was in his lifeboat pleaded with him to save her dog, which she had clasped tightly in her arms since leaving the *Titanic*. Zanni informed her politely that human beings came first and she clung desperately to the little animal until someone lifted her to the deck of the boat. Mr Zanni was married four months ago and left his wife in France. He goes to Cincinnati soon where he will engage in business.'

Titanic fact

In July 1914, a memorial statue of Captain Smith was unveiled in Lichfield.

Day 300

26 October | On this day in *Titanic* history

The following people were born on this day:
Crosby, Mrs Catherine Elizabeth – 26 October 1847 – aged 64 when on *Titanic*; first-class passenger (ticket price: £71); rescued in lifeboat 7

Wright, Mr George – 26 October 1849 – aged 62 when on *Titanic*; first-class passenger (ticket price: £26 11s); *Titanic* victim

Guggenheim, Mr Benjamin – 26 October 1865 – aged 46 when on *Titanic*; first-class passenger (ticket price: £79 4s); *Titanic* victim

Orr, Mr James – 26 October 1872 – aged 39 when on *Titanic*; assistant vegetable cook; *Titanic* victim

Jaillet, Mr Henri Marie – 26 October 1873 – aged 38 when on *Titanic*; pastry cook; *Titanic* victim

Danbom, Mr Ernst Gilbert – 26 October 1877 – aged 34 when on *Titanic*; third-class passenger (ticket price: £14 8s); *Titanic* victim

The following people died on this day:
Spencer, Mrs Marie Eugenie – 26 October 1913 – aged 48 when on *Titanic*; first-class passenger (ticket price: £146 10s 5d); rescued in lifeboat 6

Brown, Mrs Margaret – 26 October 1932 – aged 44 when on *Titanic*; first-class passenger (ticket price: £27 14s 5d); rescued in lifeboat 6

Carter, Mrs Lucile – 26 October 1934 – aged 36 when on *Titanic*; first-class passenger (ticket price: £120); rescued in lifeboat 4

Titanic survivor quote

Mrs Catherine Elizabeth Crosby, quoted in *Chicago Record-Herald*, Tuesday, 23 April 1912

> 'Miss Crosby briefly discussed her experience in the shipwreck. She censured the conduct of the officers and crew of the *Titanic*, declaring they were too excited to give any attention to the women whom they thrust into the lifeboats. "Everybody was so excited," said Miss Crosby, "that they piled women and children into the lifeboats without knowing whether the plugs were in the bottom of the lifeboats."

Mr Frank Osman (able seaman)

> 'There was no suction whatever. When we were in the boat we shoved off from the ship, and I said to the officer, "See if you can get alongside to see if you can get any more hands, to see if you can squeeze any more hands in." So the women then started getting nervous after I said that, and the officer said "All right." The women disagreed to that. We pulled around to the starboard side of the ship and found we could not get to the starboard side because it was listing too far. We pulled astern that way again, and after we got astern we lay on our oars and saw the ship go down.'

Day 301

27 October | On this day in *Titanic* history

The following people were born on this day:
Walsh, Miss Catherine – 27 October 1869 – aged 42 when on *Titanic*; stewardess; *Titanic* victim

Donoghue, Mr Florence Thomas – 27 October 1880 – aged 31 when on *Titanic*; bedroom steward; *Titanic* victim

Tamlyn, Mr Frederick – 27 October 1888 – aged 23 when on *Titanic*; mess steward (deck department); *Titanic* victim

Brown, Miss Edith Eileen – 27 October 1896 – aged 15 when on *Titanic*; second-class passenger (ticket price: £39); rescued in lifeboat 14

The following people died on this day:
Williams, Mr Charles Eugene – 27 October 1935 – aged 23 when on *Titanic*; second-class passenger (ticket price: £13); rescued in lifeboat 14

Shutes, Miss Elizabeth Weed – 27 October 1949 – aged 40 when on *Titanic*; first-class passenger (ticket price: £153 9s 3d); rescued in lifeboat 3

Titanic survivor quote

Mrs Edith Graham, quoted in *New York Times*, Saturday, 20 April 1912

> 'With us, in an adjoining stateroom, was my daughter's companion, Miss E. W. Shutes, a teacher. She was the only other member in our party and was later saved with us...The trouble was that there was no one in command, and the two sailors couldn't do much. The men were silent, and that is why the women did most of the talking. There were sixteen oar locks in our boat, but we lost three oars right off because those who handled them didn't know anything about rowing. Then I took an oar myself. I don't think I helped very much. It was snappy cold and I was dressed very lightly. Everybody seemed rather dazed, but not so very excited. That came later.'

Mr Charles Eugene Williams, mentioned in *Chicago Daily Journal*, Friday, 19 April 1912

> 'Charles Williams, the racquet coach at Harrow, England, who is the professional champion of the world, was coming to New York to defend his title, said he was in the smoking room when the boat struck. He rushed out, saw the iceberg, which seemed to loom above the deck over 100 feet. It broke up amidships and floated away. He jumped from the boat deck on the starboard side as far away from the steamer as possible. He was nine hours in the small boat, standing in water to his knees. He said the sailors conducted themselves admirably.'

Titanic fact

After the rescue, the survivors on the *Carpathia* travelled directly on to New York, arriving at Pier 54 three days later.

Day 302

28 October | On this day in *Titanic* history

The following people were born on this day:
McElroy, Mr Hugh Walter – 28 October 1874 – aged 37 when on *Titanic*; purser; *Titanic* victim

Bourke, Mrs Catherine – 28 October 1879 – aged 32 when on *Titanic*; third-class passenger (ticket price: £15 10s); *Titanic* victim

Becker, Miss Ruth Elizabeth – 28 October 1899 – aged 12 when on *Titanic*; second-class passenger (ticket price: £39); rescued in lifeboat 13

The following person died on this day:
Birkhead, May R. – 28 October 1941 – aged 41 when on *Titanic*; third-class passenger (ticket price: £7 1s); rescued in lifeboat 15

Titanic survivor quote

Mr Benoît Picard

> 'I went and I found the door. There are always a few steps from this third class, with a movable door, and it is marked there [that] second-class passengers have no right to penetrate there. I found this door open so that I could go into the second class, where I did not find any people, only a few that climbed on the ladder and went into the first class, which I did. I found there only a few men and about two ladies. They had been putting them into lifeboats and as no women were there, we men sprang in the boat. We had only one woman and another young girl. There were two women. They just stood in front of me. We were lowered down, and when I was lowered down I saw the whole ship, as big as she was the right side a little bit sinking, and I was far from imagining that it was the beginning of the end. When I was going away from the ship, of course I was rather frightened; I was sorry at not being on the ship, and I said to the seaman, "I would rather be on the ship." He was laughing at me, and he said, "Do you not see we are sinking?" I was rather excited, and I said, "It is fortunate that the sea is nice, but perhaps in five minutes we will be turned over." So I was in the boat until 5 o'clock in the morning.'

Mme Léontine Pauline Aubart

> 'I had in my cabin jewels worth £4,000 as well as many trunks of dresses and hats. One does not come from Paris and buy one's clothes in America. That is understood, is it not? Nothing could I take with me; nothing at all.'

Titanic fact

There were two doctors in the crew: a surgeon and an assistant surgeon (neither survived).

Day 303

29 October | On this day in *Titanic* history

The following people were born on this day:
Ware, Mrs Florence Louise – 29 October 1880 – aged 32 when on *Titanic*; second-class passenger (ticket price: £21); rescued in lifeboat 10

Kemish, Mr George – 29 October 1889 – aged 22 when on *Titanic*; fireman; rescued in lifeboat unknown

Dibden, Mr William – 29 October 1893 – aged 18 when on *Titanic*; second-class passenger (ticket price: £73 10s); *Titanic* victim

The following people died on this day:
Rothschild, Mrs Elizabeth Jane Anne – 29 October 1943 – aged 54 when on *Titanic*; first-class passenger (ticket price: £59 8s); rescued in lifeboat 6

Futrelle, Mrs Lily May – 29 October 1967 – aged 35 when on *Titanic*; first-class passenger (ticket price: £53 2s); rescued in collapsible lifeboat D

Aubart, Mme Léontine Pauline – 29 October 1964 – aged 24 when on *Titanic*; first-class passenger (ticket price: £69 6s); rescued in lifeboat 9

Titanic survivor quote

Mrs Florence Louise Ware, quoted in *Bristol Times and Mirror*, Saturday, 11 May 1912

> 'In our boat there were a lot of women, and one steward and a fireman. None of the men knew anything about managing a small boat, so some of the women who were used to it took charge. I had nothing on but a nightgown and was very cold and I worked as hard as I could at an oar until we were picked up. There was nothing to eat or drink in our boat.
>
> 'With nothing on but a nightgown I helped row one of your boats for hours and hours,' said Mrs Ware, as she introduced herself to the office force of the White Star Line. 'Now how am I going to get back home?'

Titanic fact

There were 2,500 pounds of sausages; 25,000 pounds of poultry and 11,000 pounds of fresh fish on board.

Day 304

30 October | On this day in *Titanic* history

The following people were born on this day:

Faunthorpe, Mr Harry Bartram – 30 October 1880 – aged 31 when on *Titanic*; second-class passenger (ticket price: £26); *Titanic* victim

Salkjelsvik, Miss Anna Kristine – 30 October 1890 – aged 21 when on *Titanic*; third-class passenger (ticket price: £7 13s); rescued in collapsible lifeboat C

Tobin, Mr Roger – 30 October 1891 – aged 20 when on *Titanic*; third-class passenger (ticket price: £7 15s); *Titanic* victim

Doling, Miss Elsie – 30 October 1893 – aged 18 when on *Titanic*; second-class passenger (ticket price: £23); rescued in lifeboat unknown

Gilnagh, Miss Kate – 30 October 1894 – aged 17 when on *Titanic*; third-class passenger (ticket price: £7 14s 8d); rescued in lifeboat 16

Eklund, Mr Hans Linus – 30 October 1895 – aged 16 when on *Titanic*; third-class passenger (ticket price: £7 15s 6d); *Titanic* victim

The following people died on this day:

Christy, Miss Rachel Juli Cohen – 30 October 1931 – aged 25 when on *Titanic*; second-class passenger (ticket price: £30); rescued in lifeboat 12

Pugh, Mr Alfred – 30 October 1962 – aged 20 when on *Titanic*; steward in third class; rescued in lifeboat 14

Daly, Mr Eugene Patrick – 30 October 1965 – aged 29 when on *Titanic*; third-class passenger (ticket price: £7 15s); rescued in collapsible lifeboat B

Titanic survivor quote

Mr Eugene Patrick Daly (from the Frank Blackmarr Scrapbook)

> 'As I looked over my shoulder, as I was still hanging to this oar, I could see the enormous funnels of the *Titanic* being submerged in the water. These poor people that covered the water were sucked down in those funnels, each of which was twenty-five feet in diameter, like flies.
>
> 'I managed to get away and succeeded in reaching the same boat I had tried to set free from the deck of the *Titanic*. I climbed upon this, and with the other men balanced ourselves in water to our hips until we were rescued. People who came up beside us…begged to get on this upturned boat. As a matter of saving ourselves [we] were obliged to push them off. One man was alongside and asked if he could get upon it. We told him that if he did, we would all go down. His reply was "God bless you. Goodbye."'

Day 305

31 October | On this day in *Titanic* history

The following people were born on this day:
Icard, Miss Rose Amélie – 31 October 1872 – aged 39 when on *Titanic*; first-class passenger (ticket price: £80); rescued in lifeboat 6

Johnstone, Mr James – 31 October 1859 – aged 52 when on *Titanic*; saloon steward; rescued in lifeboat 2

Braund, Mr Owen Harris – 31 October 1889 – aged 22 when on *Titanic*; third-class passenger (ticket price: £7 5s); *Titanic* victim

The following people died on this day:
Lindqvist, Mr Eino William – 31 October 1958 – aged 20 when on *Titanic*; third-class passenger (ticket price: £7 18s 6d); rescued in lifeboat unknown

Judd, Mr Charles Edward – 31 October 1960 – aged 31 when on *Titanic*; fireman; rescued in unknown collapsible lifeboat

Titanic survivor quote

Mr Charles Edward Judd, quoted in *Bucks Free Press*, Friday, 3 May 1912

> 'I was alongside the ship when she went down like a stone. Her stern was 50ft or 60ft up in the air. We were the last to leave her; the water washed us all off her into the sea. Of course, we all thought a big ship like that could not sink, but the water was master, and always will be. We lost everything except what we had on and a change at home. All my poor mates are gone to the bottom – nearly all Shirley chaps on our watch. I do not know what pay we shall get, as we have only five days due to us, and we cannot claim any more. As soon as the ship strikes our pay stops – hard lines; but, never mind, I am only too thankful to be where I am.'

Mr James Johnstone, mentioned in *Liverpool Echo*, Monday, 22 April 1912

> 'It was reported last week that Mr James Johnston, one of the *Titanic*'s stewards, was missing. A cable, however, has been received by his wife that he has been saved, and is on board the *Lapland* returning to England. Mr Johnstone has his homer at 36, Church-road, Stanley, Liverpool. He is a native of Annan, Dumfriesshire, and was transferred from the *Olympic* to the *Titanic*. He was a steward on the *Olympic* at the time she collided with HMS *Hawke*. His wife resides at the above address with her five children, the eldest son, who also goes to sea, having yesterday celebrated his twentieth birthday. The last time Mr Johnstone was at home was five weeks ago last Monday, when he left for Belfast to join the *Titanic* prior to her going round to Southampton.'

November

Day 306

1 November | On this day in *Titanic* history

The following people were born on this day:
Trout, Mrs Jessie L. – 1 November 1885 – aged 27 when on *Titanic*; second-class passenger (ticket price: £12 13s); rescued in lifeboat 9

Blann, Mr Eustace Horatius – 1 November 1890 – aged 21 when on *Titanic*; fireman; *Titanic* victim

The following people died on this day:
Mullen, Miss Katherine 'Katie' – 1 November 1970 – aged 21 when on *Titanic*; third-class passenger (ticket price: £7 14s 8d); rescued in lifeboat 16

Cleaver, Miss Alice Catherine – 1 November 1984 – aged 22 when on *Titanic*; first-class passenger (ticket price: £151 16s); rescued in lifeboat 11

Titanic survivor quotes

Miss Alice Catherine Cleaver, mentioned in *New York Herald*, Saturday, 20 April 1912

> 'The nurse, Miss Alice Cleaver, was up at the time of the crash, while the others were in bed. Alarmed by the shock, and wishing to find the cause, but not daring to leave the baby alone, she wrapped him in his carriage rug, and went up to the deck. "This way. Get into the boat quick," was what the nurse heard when she reached the deck. Before she knew what was happening, attachés of the steamship had placed her and the baby in a lifeboat. The father and mother were asleep, as were the two maids. The maids, Mildred Brown and Sarah Daniels, were awakened and made their way to the deck. They likewise were bundled into boats and escaped. While they were being lowered away they saw Mr and Mrs Allison being placed in another lifeboat, but this, which was one of the last to leave the stricken steamship, foundered.'

Titanic fact

There were 15,000 bottles of ale, 1,000 bottles of wine and 850 bottles of liquor on board.

Day 307

2 November | On this day in *Titanic* history

The following people were born on this day:
Fortune, Mr Mark – 2 November 1847 – aged 64 when on *Titanic*; first-class passenger (ticket price: £263); *Titanic* victim

Byström, Mrs Karolina – 2 November 1872 – aged 39 when on *Titanic*; second-class passenger (ticket price: £13); rescued in lifeboat unknown

The following person died on this day:
Keen, Mr Percy Edward – 2 November 1954 – aged 30 when on *Titanic*; saloon steward in first class; rescued in lifeboat 15

Titanic survivor quotes

Mr Robert W. Daniel

> 'The bows of the ship were far beneath the surface and to me only the four monster funnels and the two masts were now visible. It was all over in an instant. The *Titanic*'s stern rose completely out of the water and went up 30, 40, 60 feet into the air. Then, with her body slanting at an angle of 45 degrees, slowly the *Titanic* slipped out of sight.'

Mr John Hardy (chief steward)

> 'I came in the last collapsible boat that left the ship and, greatly to my surprise, when I got on the *Carpathia* I saw Officer Lightoller coming in the following afternoon. When he stepped from this collapsible boat, I was sure he had gone down in the ship and I was greatly surprised when I saw him on the *Carpathia*.'

Mr Charles H. Lightoller (second officer)

> 'Although you cross the Atlantic for years and have ice reported and never see it, at other times it's not reported and you do see it.'

Titanic facts

For *Titanic's* maiden voyage, *Oceanic*, *Olympic* and *New York* provided some of their coal.

The name of the research ship that found the wreck of *Titanic* was called *Knorr*.

Day 308

3 November | On this day in *Titanic* history

The following people were born on this day:

Millet, Mr Francis Davis – 3 November 1846 – aged 65 when on *Titanic*; first-class passenger (ticket price: £26 11s); *Titanic* victim

Knight, Mr George – 3 November 1865 – aged 46 when on *Titanic*; saloon steward; rescued in lifeboat 13

Tucker, Mr Gilbert Milligan Jr – 3 November 1880 – aged 31 when on *Titanic*; first-class passenger (ticket price: £28 10s 9d); rescued in lifeboat 7

Cavendish, Mrs Julia Florence – 3 November 1886 – aged 25 when on *Titanic*; first-class passenger (ticket price: £78 17s); rescued in lifeboat 6

Fortune, Miss Mabel Helen – 3 November 1888 – aged 23 when on *Titanic*; first-class passenger (ticket price: £263); rescued in lifeboat 10

Lockyer, Mr Edward Thomas – 3 November 1892 – aged 19 when on *Titanic*; third-class passenger (ticket price: £7 17s 7d); *Titanic* victim

The following people died on this day:

Lindström, Mrs Sigrid – 3 November 1946 – aged 55 when on *Titanic*; first-class passenger (ticket price: £27 14s 5d); rescued in lifeboat 6

McDermott, Miss Bridget Delia – 3 November 1959 – aged 31 when on *Titanic*; third-class passenger (ticket price: £7 15s 8d); rescued in lifeboat 13

Sjöblom, Miss Anna Sofia – 3 November 1975 – aged 18 when on *Titanic*; third-class passenger (ticket price: £6 9s 11d); rescued in lifeboat 16

Titanic survivor quote

Mrs Julia Florence Cavendish, quoted in *Toronto Daily Star*, Friday, 19 April 1912

> 'I was in the second boat. My husband kissed me and bade me remain in the boat, declaring he was all right. There was no light, but the sky was clear. Bright skies illuminated the scene of the disaster. Just as the lifeboat was lowered I again kissed my husband. He assured me he would rather stay on the boat, thinking he would be safe. As the boat reached the water there were twenty-three women in the boat and two men to guide and row her. Many of us women implored men on the upper deck to come to our succor, but most of them said they could not row. One man there was about to get in the boat, but a sailor, after questioning him threw him aside.'

Day 309

4 November | On this day in *Titanic* history

The following person was born on this day:
Brown, Mrs Elizabeth Catherine – 4 November 1872 – aged 39 when on *Titanic*; second-class passenger (ticket price: £39); rescued in lifeboat 14

The following people died on this day:
Rostron, Captain Arthur Henry – 4 November 1940 – aged 42 when *Titanic* set sail; captain of *Carpathia*

Ross, Mr Horace Leopold – 4 November 1940 – aged 38 when on *Titanic*; scullion; rescued in lifeboat 13

McCormack, Mr Thomas Joseph – 4 November 1975 – aged 19 when on *Titanic*; third-class passenger (ticket price: £7 15s); rescued in lifeboat unknown

Captain Arthur H. Rostron, commander of *Carpathia*

> 'Icebergs loomed up and fell astern and we never slackened. It was an anxious time with the *Titanic*'s fateful experience very close in our minds. There were 700 souls on *Carpathia* and those lives as well as the survivors of the *Titanic* herself depended on the sudden turn of the wheel.
>
> 'When day broke, I saw the ice I had steamed through during the night. I shuddered, and could only think that some other hand than mine was on that helm during the night.
>
> 'At 8.30 all the people were on board. I wanted to hold a service, a short prayer of thankfulness for those rescued and a short burial service for those who were lost. While they were holding the service I manoeuvred around the scene of the wreckage. We saw nothing but one body.'

Mr Harold S. Bride, witness at United States Senate Inquiry (Day 2)

> 'C.Q.D. about half a dozen times; M.G.Y. half a dozen times.'

Titanic survivor quote

Mr Thomas Joseph McCormack, mentioned in *Newark Evening News*, Saturday, 20 April 1912

> 'The story that he was beaten by sailors when he swam to a lifeboat is told by Thomas McCormack, one of the *Titanic* survivors, who is now at St. Vincents Hospital, New York.
>
> 'McCormack, who is nineteen years old and whose home is at 36 West Twentieth street, this city, says that he had jumped into the sea and was barely able to crawl into a lifeboat, despite the assault by the sailors. His face, arms and body were bruised.'

Day 310

5 November | On this day in *Titanic* history

The following people were born on this day:
Burke, Mr William – 5 November 1872 – aged 39 when on *Titanic*; saloon steward; rescued in lifeboat unknown

Jenkin, Mr Stephen Curnow – 5 November 1879 – aged 32 when on *Titanic*; second-class passenger (ticket price: £10 10s); *Titanic* victim

Barrows, Mr Edward William Peter – 5 November 1879 – aged 32 when on *Titanic*; saloon steward; *Titanic* victim

Dodds, Mr Henry Watson – 5 November 1885 – aged 26 when on *Titanic*; junior assistant fourth engineer; *Titanic* victim

Colbert, Mr Patrick – 5 November 1887 – aged 26 when on *Titanic*; junior assistant fourth engineer; *Titanic* victim

Linehan, Mr Michael – 5 November 1890 – aged 21 when on *Titanic*; third-class passenger (ticket price: £7 17s 7d); *Titanic* victim

Hocking, Miss Ellen 'Nellie' – 5 November 1891 – aged 20 when on *Titanic*; second-class passenger (ticket price: £23); rescued in lifeboat 4

The following people died on this day:
Greenfield, Mrs Blanche – 5 November 1936 – aged 45 when on *Titanic*; first-class passenger (ticket price: £63 7s 2d); rescued in lifeboat 7

Ryan, Mr Edward – 5 November 1974 – aged 24 when on *Titanic*; third-class passenger (ticket price: £7 15s); rescued in lifeboat 14

Titanic survivor quote

Mr Alfred White

> 'I heard the chief engineer tell one of his subordinates that number six bulkhead had given way. At that time things began to look bad… I was told to go up and see how things were, and made my way up a dummy funnel to the bridge deck. By that time all the boats had left the ship, yet everyone in the engine room was at his post. I was near the captain and heard him say, "Well boys, it's every man for himself now."'

Titanic fact

Perfume salesman Mr Adolphe Saalfeld lost his sample case of bottles in the sinking. He survived. Decades later, in 2000, the bottles and the scents inside were recovered from the *Titanic* wreck and exhibited.

Day 311

6 November | On this day in *Titanic* history

The following people were born on this day:

Revell, Mr William James Francis – 6 November 1880 – aged 31 when on *Titanic*; saloon steward in first class; *Titanic* victim

Mannion, Miss Margaret – 6 November 1883 – aged 28 when on *Titanic*; third-class passenger (ticket price: £7 14s 9d); rescued in lifeboat 16

Sobey, Mr Samuel James Hayden – 6 November 1886 – aged 25 when on *Titanic*; second-class passenger (ticket price: £13); *Titanic* victim

Del Carlo, Mrs Argene – 6 November 1887 – aged 24 when on *Titanic*; second-class passenger (ticket price: £27 14s 5d); rescued in lifeboat 11

Kennedy, Mr John – 6 November 1887 – aged 24 when on *Titanic*; third-class passenger (ticket price: £7 15s); rescued in lifeboat unknown

Titanic survivor quotes

Miss Margaret Mannion, quoted in *Chicago American*, Tuesday, 23 April 1912

'I should not have been saved except for Mrs Burke's refusal to leave her husband and the Misses Burke saying they would not go if their uncle and aunt could not go with them... I went in the very last boat and I was the very last passenger. The officer said there was room for just one more.

'I was aroused by the call of the stewardess, who told us all to dress as quickly as we could, though she did not explain what was the trouble. I dressed and went upon the second deck. Annie McGowan was with me when I was going up the stairs, but she became separated from me at the head of the stairway, and was carried by the throng over to the other side of the ship. I did not see her again until I was on the *Carpathia*.

'On the side where I was carried, some wild-looking men were trying to rush into the boats, and the officers and crew fired at them. Some of the men fell. Others were beaten back by the officers who used their pistols on them.'

Mr Augustus Henry Weikman

'The water was too cold for me to swim and I was hardly more than one hundred feet away when the ship went down. The suction was not what one would expect and only rocked the water around me. I was picked up after two hours. I have done with the sea.'

Day 312

7 November | On this day in *Titanic* history

The following people were born on this day:
Foreman, Mr Benjamin Laventall – 7 November 1881 – aged 30 when on *Titanic*; first-class passenger (ticket price: £27 15s); *Titanic* victim

Wyeth, Mr James Robert – 7 November 1886 – aged 25 when on *Titanic*; fireman; *Titanic* victim

The following person died on this day:
Weikman, Mr Augustus Henry – 7 November 1924 – aged 52 when on *Titanic*; barber; rescued in lifeboat unknown

Titanic survivor quotes

Mr George Symons (able seaman)

> 'I came on deck and I went into the mess room in the course of ordinary events to see if there was any coffee. From there I heard the water coming in to No. 1 hold. I looked down No. 1 hold, and hardly had I looked down there when the order came for "All hands on the boat deck."'

Mr Augustus Henry Weikman, quoted in *Camden Post-Telegram*, Friday, 19 April 1912

> 'The explosions,' Weikman said, 'were caused by the rushing of the icy water on the boilers. A bundle of deck chairs, roped together, was blown off the deck with me, and struck my back, injuring my spine, but it served as a temporary raft.
>
> 'The crew and passengers had faith in the bulkhead system to save the ship and we were lowering a collapsible boat, all confident the ship would get through, when she took a terrific dip forward and the water rushed up and swept over the deck and into the engine rooms.
>
> 'The bow went clean down and I caught the pile of chairs as I was washed up against the rail. Then came the explosions and [they] blew me fifteen feet.'

Titanic facts

On 24 May 1912, 473 musicians, drawn from seven London orchestras, played at the *Titanic* Band Memorial Concert at the Royal Albert Hall.

Titanic's departure was delayed by one hour because of a near collision.

Day 313

8 November | On this day in *Titanic* history

The following people were born on this day:
Leitch, Miss Jessie Wills – 8 November 1880 – aged 31 when on *Titanic*; second-class passenger (ticket price: £33); rescued in lifeboat 11

Carrau, Mr Francisco Mauro Severiano – 8 November 1884 – aged 27 when on *Titanic*; first-class passenger (ticket price: £47 2s); *Titanic* victim

Sandström, Mrs Agnes Charlotta – 8 November 1887 – aged 24 when on *Titanic*; third-class passenger (ticket price: £16 14s); rescued in lifeboat 13

Parkes, Mr Francis – 8 November 1890 – aged 21 when on *Titanic*; plumber; *Titanic* victim

Thomas/Tannous, Master Assad Alexander – 8 November 1911 – aged 5 months when on *Titanic*; third-class passenger (ticket price: £8 10s 4d); rescued in lifeboat 16

The following person died on this day:
Lenox-Conyngham, Master Dennis Hugh – 8 November 1928 – aged 10 when on *Titanic*; first-class passenger (ticket price: £3); disembarked at Cherbourg

Titanic survivor quote

Miss Jessie Wills Leitch, quoted in *Chicago Tribune*, Sunday, 21 April 1912

'The following is Miss Leitch's account in brief:

"About midnight Mr Harper came to our stateroom and told us that the vessel had struck an iceberg. While I was dressing he went to learn further particulars and returned to say that the order had been given to put on the lifebelts. We did so, and, picking up Nana in his arms, he took her up to the deck. There the women were ordered to the upper deck. I had to climb a vertical iron ladder and Mr Harper brought Nana after me up the ladder and the men at the top lifted her up to me again.

"There was no opportunity for farewell, and, in fact, even then we did not realise the danger, as we were assured again and again that the vessel could not sink, that the *Olympic* would be alongside at any minute, and that the women and children were to be put into the boats first and the men to follow, and that there were boats sufficient for all. Our boat was well manned – it was the eleventh to leave the vessel. After about half an hour the *Titanic* went down. We were about a mile away."'

Titanic fact

The weight of each anchor was 15 tons.

Day 314

9 November | On this day in *Titanic* history

The following people were born on this day:

Thayer, Mrs Marian Longstreth – 9 November 1872 – aged 39 when on *Titanic*; first-class passenger (ticket price: £110 17s 8d); rescued in lifeboat 4

Beckwith, Mr Richard Leonard – 9 November 1874 – aged 37 when on *Titanic*; first-class passenger (ticket price: £52 11s 1d); rescued in lifeboat 5

Lindsay, Mr William Charles – 9 November 1881 – aged 30 when on *Titanic*; fireman; rescued in collapsible lifeboat B

Björnström-Steffansson, Mr Mauritz Hokan – 9 November 1883 – aged 28 when on *Titanic*; first-class passenger (ticket price: £26 11s); rescued in collapsible lifeboat D

Hebb, Mr William Albert – 9 November 1889 – aged 22 when on *Titanic*; trimmer; rescued in collapsible lifeboat B

Binstead, Mr Walter William – 9 November 1891 – aged 20 when on *Titanic*; trimmer; rescued in lifeboat 3

Sage, Miss Elizabeth Ada – 9 November 1901 – aged 10 when on *Titanic*; third-class passenger (ticket price: £69 11s); *Titanic* victim

The following person died on this day:

Noel, Major Gerard Thomas – 9 November 1941 – aged 55 when on *Titanic*; first-class passenger (ticket price: £3); disembarked at Cherbourg

Titanic survivor quote

Mr William Lindsay, Monday, 22 April 1912 (unidentified newspaper)

> 'I expect you have heard before now about the *Titanic* going down, but thank God we got picked up all right. I don't know how mother and wife feel about it, but I do hope that they have heard by now that we are safe, what few there are of us. There were about 60 firemen and trimmers saved out of 400. Stewart Palmer and Charles Newman were drowned. Fred Palmer was saved. He picked us up in his boat. I was on the ship till the water came up to the funnel and got away on a raft. There were 39 of us on it. Oh, but the sight was awful. I shall never forget it, for she broke in three pieces. I only hope that the wife and mother know by now, for the company sent a cable for us. They ought to get that Saturday. Well, Jack, I feel too upset to write more, so will close. Write home as soon as you can. William Lindsay has been on the seas probably eight years and was on the *Olympic* when his local brother heard from him before the disaster. His brother did not know he was on the *Titanic*.'

Day 315

10 November | On this day in *Titanic* history

The following person was born on this day:
Daniels, Miss Sarah Rebecca – 10 November 1875 – aged 36 when on *Titanic*; first-class passenger (ticket price: £151 16s); rescued in lifeboat 8

The following person died on this day:
Richards, Mrs Emily – 10 November 1972 – aged 24 when on *Titanic*; second-class passenger (ticket price: £18 15s); rescued in lifeboat 4

Titanic survivor quote

Mr Ernest Archer, witness at United States Senate Inquiry (Day 7)

> 'I was in my bunk, asleep. I heard a kind of a crush, something similar to when you let go the anchor; it wounded like the cable running through the hawse pipe.'

Miss Sarah Rebecca Daniels, quoted in *Chicago Daily Tribune*, Saturday, 20 April 1912

> 'Miss Daniels, although stunned by the terrible sights she had witnessed, was able to tell, even graphically, of the story of Mrs Allison to her husband.
>
> "My sister [*sic*] died rather than leave her husband when the officers refused to let both into the lifeboat. She said life was not worth living alone and she went down, even smiling, with her arm around Herbert.
>
> "When the boat struck few realised there was any danger. Herbert and Bessie laughed and went to dress. When we stood together at the rail, as the boats were being sent away, I was standing with Wilbur and Loraine was with Herbert and Bessie. Then came the order. 'Women only' and an officer tried to put Bessie in the boats. 'Not without my husband!' she cried.
>
> 'You must,' cried the officer, but Bessie threw her arms around Herbert's neck and refused to leave him. Then after the officer stopped trying to force her into the boat, she ran to me, pushed me into the boat, and threw little Wilbur in after me.
>
> "The boat was full and she grasped Loraine in one arm, her husband with the other, and stood waving her hand, and it seemed to me, smiling as she saw us rowing away.
>
> "The last I saw of her, just as the boat started to plunge to the bottom, was Bessie turning to her husband for a farewell kiss. As the water washed to their knees, Loraine was holding [on] to her mother's skirts."

Day 316

11 November | On this day in *Titanic* history

The following people were born on this day:

Beauchamp, Mr Henry James – 11 November 1868 – aged 43 when on *Titanic*; second-class passenger (ticket price: £26); *Titanic* victim

Blair, Mr David – 11 November 1874 – aged 37 when on *Titanic*; second officer; disembarked at Southampton

Parsons, Mr Edward – 11 November 1874 – aged 37 when on *Titanic*; chief storekeeper in first class; *Titanic* victim

Stewart, Mr John – 11 November 1883 – aged 28 when on *Titanic*; veranda steward; rescued in lifeboat 15

Russell, Mr Boysie Richard – 11 November 1894 – aged 17 when on *Titanic*; saloon steward in second class; *Titanic* victim

We can also wish Leonardo DiCaprio a happy birthday!
Leonardo Wilhelm DiCaprio – born 11 November 1974

Titanic survivor quote

Miss Elizabeth Dowdell

'On the *Carpathia* we were looked over by the officers, and those who apparently had nothing were all ordered down into the steerage. With my charge I was put in with the rest. We were fed on hard-tack. Many of us refused to eat, and when the *Carpathia*'s officers saw this, they had us come into the second cabin. Before this we had been down in the steerage for a full day, rubbing arms with Chinese immigrants. We were desperately hungry by the time they had decided to take us out of this place.

'All of the survivors who were recognized to have been of prominence or means were well taken care of and given choice treatment, while we were accorded anything but that.

'Those who had money could send messages, and those without funds had to go without doing so. I had $5 in my pocketbook, and I was charged a dollar a word to send word to Mrs Emanuel, in England, telling of our rescue and the safety of her daughter. I sent the message on Tuesday.

'One man, a barber, had but $1.25 with him, and he handed over one dollar of this to send the word "safe" to his mother.'

Day 317

12 November | On this day in *Titanic* history

The following people were born on this day:
Brereton, Mr George Andrew (George Arthur Brayton / Bradley) – 12 November 1874 – aged 37 when on *Titanic*; first-class passenger (ticket price: £26 11s); rescued in lifeboat 9

Middleton, Mr Mark Victor – 12 November 1887 – aged 24 when on *Titanic*; saloon steward in second class; *Titanic* victim

The following people died on this day:
McCarthy, Miss Catherine 'Katie' – 12 November 1948 – aged 24 when on *Titanic*; third-class passenger (ticket price: £7 15s); rescued in lifeboat 16

Greenfield, Mr William Bertram – 12 November 1949 – aged 23 when on *Titanic*; first-class passenger (ticket price: £63 7s 2d); rescued in lifeboat 7

Titanic survivor quote

Miss Catherine McCarthy, quoted in *Cork Examiner*, Saturday, 11 May 1912

'I was in the last boat to leave the ship, and was the second last person put into it. This was a short time before the ship went down. We were only just out of the way when the ship split in two and sank. We remained in the boat all night until near eight o'clock next morning, when we were rescued by the *Carpathia*. Our boat was so full I thought it would go down every moment, and one of the boats capsized when we were leaving the sinking ship. I did not, however, feel at all frightened, and did not fully realize the danger and the full nature of the awful tragedy until I was safe on board the *Carpathia*. When we were put on board the *Carpathia* we were immediately given restoratives and put to bed. I slept for an hour and then got up, feeling all right.'

Mr James Widgery (bath steward)

'The purser sent me along to No. 9. They had taken the canvas off of No. 9 and lowered it, and just then some biscuits came up from the storekeeper. I helped him put one of the boxes into the bottom of the boat, and the purser took hold of my arm and said, "Get in the boat." He said, "Get in the boat and help the boatswain's mate pass the ladies in." So I got in the boat, and stepped on the side, and we passed the ladies in. We thought we had them all in, and the purser called out, "Are there any more women?" Just then some one said, "Yes." This woman came along, rather an oldish lady, and she was frightened, and she gave me her hand. I took one hand, and gave it to the boatswain's mate, and he caught hold of the other hand, and she pulled her hand away, and went back to the door and would not get in. One of them went after her, but she had gone down the stairs.'

Day 318

13 November | On this day in *Titanic* history

The following people were born on this day:
Skoog, Mrs Anna Bernhardina – 13 November 1868 – aged 43 when on *Titanic*; third-class passenger (ticket price: £27 18s); *Titanic* victim

Wynn, Mr Walter – 13 November 1870 – aged 41 when on *Titanic*; quartermaster; rescued in lifeboat 9

Akerman, Mr Albert Edward – 13 November 1880 – aged 31 when on *Titanic*; steward in third class; *Titanic* victim

Smither, Mr Harry James – 13 November 1889 – aged 22 when on *Titanic*; fireman; *Titanic* victim

The following person died on this day:
Hawksford, Mr Walter James – 13 November 1922 – aged 45 when on *Titanic*; first-class passenger (ticket price: £30); rescued in lifeboat 3

Titanic survivor quote

Mr Charles Joughin (chief baker)

> 'Just as it was breaking daylight I saw what I thought was some wreckage, and I started to swim towards it slowly. When I got near enough, I found it was a collapsible, not properly upturned but on its side, with an Officer and I should say about twenty or twenty-five men standing on the top of it. I tried to get on it, but I was pushed off it, and I what you call hung around it. I eventually got round to the opposite side, and a cook that was on the collapsible recognised me, and held out his hand and held me – a chap named Maynard. We were hanging on to this collapsible, and eventually a lifeboat came in sight. They got within about 50 yards and they sung out that they could only take 10. So I said to this Maynard, "Let go my hand," and I swam to meet it, so that I would be one of the 10.'

Titanic facts

On the night of the sinking, second-class passenger Selena Cook and her roommates were kept awake by the people in the cabin next door having a pillow fight.

57,000 photographs were taken during the first manned submersible mission to the *Titanic* wreck site.

Day 319

14 November | On this day in *Titanic* history

The following people were born on this day:
Young, Mr Francis James – 14 November 1879 – aged 32 when on *Titanic*; fireman; *Titanic* victim

Allison, Mrs Bessie Waldo – 14 November 1886 – aged 25 when on *Titanic*; first-class passenger (ticket price: £151 16s); *Titanic* victim

Najib Kiamie, Miss Adele 'Jane' – 14 November 1896 – aged 15 when on *Titanic*; third-class passenger (ticket price: £7 4s 6d); rescued in collapsible lifeboat C

Andersson, Miss Ebba Iris Alfrida – 14 November 1905 – aged 6 when on *Titanic*; third-class passenger (ticket price: £31 5s 6d); *Titanic* victim

The following people died on this day:
Harper, Mrs Myna – 14 November 1923 – aged 49 when on *Titanic*; first-class passenger (ticket price: £76 14s 7d); rescued in lifeboat 3

Gieger, Miss Amalie – 14 November 1933 – aged 35 when on *Titanic*; first-class passenger (ticket price: £211 10s); rescued in lifeboat 4

Hippach, Miss Jean Gertrude – 14 November 1974 – aged 17 when on *Titanic*; first-class passenger (ticket price: £57 19s 7d); rescued in lifeboat 4

Titanic survivor quote

Miss Gertrude Isabelle Hippach, mentioned in *New York Times*, Monday, 22 April 1912

> 'Mrs Ida S. Hippach and her daughter, Jean, survivors of the *Titanic*, who arrived home today, said that they were saved by Col. John Jacob Astor, who forced the crew of the last lifeboat to wait for them.
>
> "We saw Col. Astor place Mrs Astor in a boat and heard him assure her that he would follow later," said Mrs Hippach. "He turned to us with a smile and said 'Ladies, you are next.' The officer in charge of the boat protested that the craft was full and the seamen started to lower it. Col. Astor exclaimed 'Hold that boat,' in the voice of a man to be obeyed, and the men did as he ordered. The boat had been ordered past the upper deck, and the Colonel took us to the next deck below and put us in the boat, one after the other, through a porthole."
>
> 'Mrs Hippach said that after the impact the *Titanic* lay broadside on an iceberg that seemed, she said, to be a hundred feet high and extended farther than she could see.'

Titanic fact

6,611 tons of coal were stored in the ship's bunkers.

Day 320

15 November | On this day in *Titanic* history

The following people were born on this day:
Elliott, Mr Everett Edward – 15 November 1888 – aged 23 when on *Titanic*; trimmer; *Titanic* victim

Beane, Mrs Ethel – 15 November 1892 – aged 22 when on *Titanic*; second-class passenger (ticket price: £26); rescued in lifeboat 9

Jones, Mr Thomas William – 15 November 1877 – aged 34 when on *Titanic*; able seaman; rescued in lifeboat 8

Moor, Master Meier – 15 November 1904 – aged 7 when on *Titanic*; third-class passenger (ticket price: £12 9s 6d); rescued in lifeboat 14

The following people died on this day:
Fryer, Mr Albert Ernest – 15 November 1944 – aged 29 when on *Titanic*; trimmer; rescued in lifeboat unknown

Doel, Mr Frederick – 15 November 1940 – aged 22 when on *Titanic*; fireman; rescued in lifeboat unknown

Nourney, Mr Alfred – 15 November 1972 – aged 20 when on *Titanic*; first-class passenger (ticket price: £13 17s 3d); rescued in lifeboat 7

Titanic survivor quote

Mr Thomas William Jones, quoted in *New York Times*, 20 April 1912

> '"There was a woman in my boat as was a woman," he said yesterday, straightening up in her honor. "She was the Countess of Rothes, and let me tell you about her. I was one of those who was ordered to man the boats, and my place was in No. 8. There were thirty-five of us in that boat, mostly women, but there were some men along.
>
> "I was in command, but I had to row, and I wanted someone at the tiller. And I saw the way she was carrying herself, and I heard the quiet, determined way she spoke to the others, and I knew she was more of a man than any we had on board. And I put her in command. I put her at the tiller, and she was at the tiller when the *Carpathia* came along five hours later. And there was another woman on board who was strong in the work we had to do. She was at the oar with me, and, though I never learned her name, she was helping every minute. It was she who suggested that we should sing. Sing, you say? I should think we did. To keep up our spirits, I guess. We sang as we rowed, all of us, starting out with 'Pull for the Shore', and we were still singing when we saw the lights of the *Carpathia*. Then we stopped singing and prayed."'

Day 321

16 November | On this day in *Titanic* history

The following people were born on this day:
Lurette, Miss Eugénie Elise – 16 November 1852 – aged 59 when on *Titanic*; first-class passenger (ticket price: £146 10s 5d); rescued in lifeboat 6

Poingdestre, Mr John Thomas – 16 November 1884 – aged 27 when on *Titanic*; able seaman; rescued in lifeboat 12

Danbom, Master Gilbert Sigvard Emanuel – 16 November 1911 – aged 4 months when on *Titanic*; third-class passenger (ticket price: £14 8s); *Titanic* victim

The following people died on this day:
Frauenthal, Mr Isaac Gerald – 16 November 1932 – aged 43 when on *Titanic*; first-class passenger (ticket price: £27 14s 5d); rescued in lifeboat 5
Dowdell, Miss Elizabeth – 16 November 1962 – aged 31 when on *Titanic*; third-class passenger (ticket price: £12 9s 6d); rescued in lifeboat 13

Mr Isaac Gerald Frauenthal, mentioned in *New York Times*, Wednesday, 17 April 1912

> 'One of the first to appear at the office of the (White Star) Company was Edward Frauenthal, of 786 Lexington Ave., who had two brothers on the *Titanic*, Dr Hyman and J.C. Frauenthal. Both are reported saved. When he was told that his brothers' names appeared on the *Carpathia*'s list of survivors, Mr Frauenthal was so overcome that he could hardly walk to the telephone to send the good news to his home. In a voice that was scarcely audible to those near him he called his home number. When Mrs Frauenthal answered, he broke down and sobbed.
>
> "I tell you they are saved!" he cried in a hoarse voice. "Yes! Yes! They are safe! Do you hear what I say?" The telephone receiver fell from his hand and Mr Frauenthal sank to the floor, completely overcome by the good news. Helped to his feet, he turned to the reporters.'
>
> "Young men," he said in a broken voice. "Pardon me, but I could not help it. My nerves are all unstrung."'

Titanic survivor quote

Miss Elizabeth Dowdell, quoted in *Jersey Journal*, Saturday, 20 April 1912

> 'Prior to the wreck there was open gambling aboard the *Titanic* every day. No effort was made to conceal it. Even on Sunday the tables were crowded with men of the first and second cabins, and the games were open to anyone who wished to enter. I felt then that something dreadful would happen.'

Day 322

17 November | On this day in *Titanic* history

The following people were born on this day:
Martin, Mrs Annie Martha – 17 November 1872 – aged 39 when on *Titanic*; stewardess; rescued in lifeboat 11

Long, Mr Frank – 17 November 1880 – aged 31 when on *Titanic*; trimmer; *Titanic* victim

The following people died on this day:
Thresher, Mr George Terrill – 17 November 1939 – aged 25 when on *Titanic*; fireman; rescued in lifeboat 9

Hopkins, Mr Robert John – 17 November 1943 – aged 43 when on *Titanic*; able seaman; rescued in lifeboat 13

Mr Robert John Hopkins, mentioned in *Newark Star*, Wednesday, 24 April 1912

> 'Reports that bribes were offered the sailors who manned the boat in which C. E. Henry Stengel of Newark; Sir Cosmo Duff-Gordon and Lady Duff-Gordon and several others left the *Titanic* are branded as absolutely untrue by Mr Stengel. The report was circulated by an able seaman, Robert Hopkins, when he asked for assistance at the City Hall, New York. Hopkins declared that had he been in what he described as the "money boat" he would not have to apply for assistance. Questioned, he said that all of the sailors in the dinghy in which Mr Stengel and Sir and Lady Duff-Gordon escaped, were paid for putting off as quickly as possible. Others in the boat were Miss Francatelli, of London, and A. L. Solomon, a wholesale stationer, of New York.'

Titanic survivor quote

Mrs Margaretta Corning Spedden

> 'It was so dark that we couldn't find our light, and the bow of our boat was so packed we could not move. Altogether there were thirty-two of us, ten of them being stokers and sailors, but we had no officer with us. I shall never forget the sight of THAT WONDERFUL SHIP GOING SLOWLY DOWN nor the cries of the poor people who couldn't get off. It was too dreadful, and yet almost impossible to realise.'

Titanic fact

The breadth of *Titanic* was 92 feet (28 metres).

Day 323

18 November | On this day in *Titanic* history

The following people were born on this day:
Wiseman, Mr Philippe – 18 November 1858 – aged 53 when on *Titanic*; third-class passenger (ticket price: £7 5s); *Titanic* victim

Klaber, Mr Herman – 18 November 1870 – aged 41 when on *Titanic*; first-class passenger (ticket price: £26 11s); *Titanic* victim

Davidson, Mrs Orian – 18 November 1884 – aged 27 when on *Titanic*; first-class passenger (ticket price: £52); rescued in lifeboat 3

Nakid, Mrs Waika 'Mary' – 18 November 1892 – aged 19 when on *Titanic*; third-class passenger (ticket price: £15 14s 10d); rescued in lifeboat unknown

The following people died on this day:
Yearsley, Mr Harry – 18 November 1939 – aged 40 when on *Titanic*; saloon steward in first class; rescued in lifeboat 9

Perreault, Miss Mary Anne – 18 November 1968 – aged 33 when on *Titanic*; first-class passenger (ticket price: £93 10s); rescued in lifeboat 3

Titanic survivor quote

Mr Charles E. Stengel

> 'I saw two, a certain physician in New York and his brother, jump into the same boat my wife was in. Then the officer or the man that was loading the boat, said "I will stop that. I will go down and get my gun." He left the deck momentarily and came right back again. Afterwards I heard about five shots; that is while we were afloat. Four of them I can account for in this way, that when the green lights were lit on the boat they were lashed to my wife's boat – the man shot off a revolver four times, thinking it was a vessel. The man in charge said, "You had better save all your revolver shots, you had better save all your matches, and save everything. It may be the means of saving your life." After that I heard another shot that seemed to be aboard the *Titanic*. It was explained to me afterwards that that was the time that one of the men shot off his revolver – that is, the mate or whoever had charge of the boat shot off his revolver – to show the men that his revolver was loaded and he would do what he said; that any man who would step into the lifeboat he would shoot.'

Day 324

19 November | On this day in *Titanic* history

The following people were born on this day:
Stengel, Mr Charles Emil Henry – 19 November 1857 – aged 54 when on *Titanic*; first-class passenger (ticket price: £55 8s 10d); rescued in lifeboat 1

Spedden, Mrs Margaretta Corning – 19 November 1872 – aged 39 when on *Titanic*; first-class passenger (ticket price: £134 10s); rescued in lifeboat 3

Beane, Mr Edward – 19 November 1879 – aged 32 when on *Titanic*; second-class passenger (ticket price: £26); rescued in lifeboat 9

Righini, Mr Sante – 19 November 1883 – aged 28 when on *Titanic*; first-class passenger (ticket price: £135 12s 8d); *Titanic* victim

Pettersson, Mr Johan Emil – 19 November 1886 – aged 25 when on *Titanic*; third-class passenger (ticket price: £7 15s 6d); *Titanic* victim

Daniels, Mr Sidney Edward – 19 November 1893 – aged 18 when on *Titanic*; steward in third class; rescued in lifeboat unknown

Spedden, Master Robert Douglas – 19 November 1905 – aged 6 when on *Titanic*; first-class passenger (ticket price: £134 10s); rescued in lifeboat 3

The following people died on this day:
Thorne, Miss Gertrude Maybelle – 19 November 1947 – aged 37 when on *Titanic*; first-class passenger (ticket price: £80 4s); rescued in collapsible lifeboat D

Lindsay, Mr William Charles – 19 November 1960 – aged 30 when on *Titanic*; fireman; rescued in collapsible lifeboat B

Öhman, Miss Velin – 19 November 1966 – aged 22 when on *Titanic*; third-class passenger (ticket price: £7 15s 6d); rescued in collapsible lifeboat C

Odell, Master Jack – 19 November 1995 – aged 11 when on *Titanic*; first-class passenger (ticket price: £24 1s); disembarked at Queenstown, Ireland

Mrs Margaretta Corning Spedden, quoted in *Devon and Exeter Gazette*, Tuesday, 14 May 1912

> 'Luckily the sea was as smooth as glass, (otherwise not a soul would have been saved), and we had the stars to guide us. We tied up to another boat once, but the rope got broke, and we also lost two oars, and got pretty well separated from the others. Then some of them lighted their lanterns after a while and we kept track of them.'

Day 325

20 November | On this day in *Titanic* history

The following people were born on this day:

Chaffee, Mr Herbert Fuller – 20 November 1865 – aged 46 when on *Titanic*; first-class passenger (ticket price: £61 3s 6d); *Titanic* victim

Clifford, Mr George Quincy – 20 November 1871 – aged 40 when on *Titanic*; first-class passenger (ticket price: £52); *Titanic* victim

West, Mr Edwy Arthur – 20 November 1875 – aged 36 when on *Titanic*; second-class passenger (ticket price: £27 15s); *Titanic* victim

Pitman, Mr Herbert John – 20 November 1877 – aged 34 when on *Titanic*; third officer; rescued in lifeboat 5

Pearce, Mr John – 20 November 1878 – aged 33 when on *Titanic*; fireman; rescued in lifeboat unknown

Weller, Mr William Clifford – 20 November 1881 – aged 30 when on *Titanic*; able seaman; rescued in lifeboat 7

Ward, Mr Arthur – 20 November 1888 – aged 41 when on *Titanic*; junior assistant fourth engineer; *Titanic* victim

Titanic survivor quote

Mr Herbert John Pitman, mentioned in *Brighton Argus*, Wednesday, 24 April 1912

> 'Mr Pitman, the third officer, who confirmed the statement that only two boats were lowered at the Board of Trade inspection. He did not see any ice before the disaster, but knew a wireless warning had been received. After the receipt of the warning a special lookout was ordered. After the collision he met Mr Ismay, who told him to get the women and children into the boats and assisted him in doing so. Witness left the *Titanic* in a boat containing 40 people although it would have held 60.
>
> 'When the ship disappeared, he heard cries of distress and ordered the men to row in that direction. Passengers demurred, however, and he yielded to their importunities. The cries continued for an hour.
>
> 'The witness, who appeared distressed, gave no further explanation as to not going to the rescue of the drowning people. He bore out Mr Boxhall's statement regarding a strange vessel which declined to answer the *Titanic* signals.'

Titanic fact

Colonel John Jacob Astor IV was the wealthiest passenger on board, and was believed to be one of the richest people in the world at the time.

Day 326

21 November | On this day in *Titanic* history

The following person was born on this day:
Lowe, Mr Harold Godfrey – 21 November 1882 – aged 29 when on *Titanic*; fifth officer; rescued in lifeboat 14

The following people died on this day:
Martin, Mrs Annie Martha – 21 November 1936 – aged 39 when on *Titanic*; stewardess; rescued in lifeboat 11

Finoli, Mr Luigi – 21 November 1938 – aged 41 when on *Titanic*; third-class passenger (ticket price: £7 1s); rescued in lifeboat 15

Titanic survivor quotes

Mr George Symons (able seaman)

> 'After we left Southampton and got clear of the Nab Lightship I went up to the Officers' mess -room and asked for glasses. I asked Mr Lightoller, and he went into another Officers' room, which I presume was Mr Murdoch's, and he came out and said, "Symons, there are none." With that I went back and told my mates.'

Mr Thomas Ranger (greaser)

> 'We only just got away in time before the ship went down. The forward end of the ship went underneath and seemed to break off, and the afterpart came back on a level keel. It turned up and went down steadily.'

Mr John Hart (steward in third class)

> 'I met several on the deck directing them the way to the boat deck. There was one man at the foot of the companion leading from the sleeping accommodation to the after-well deck; there was one man at the end of the companion leading from the well deck to the E deck, and there were others along the saloon and second cabin deck showing them the way to the boat deck. So that there was no difficulty for anybody who wanted to get to the boats to find their way there.'

Titanic fact

The internal usable volume of the *Titanic* (referred to as the 'gross register tons' or GRT) was 46,328 tons.

Day 327

22 November | On this day in *Titanic* history

The following people were born on this day:
White, Mr Frank Leonard – 22 November 1884 – aged 27 when on *Titanic*; trimmer; *Titanic* victim

Strandberg, Miss Ida Sofia – 22 November 1889 – aged 22 when on *Titanic*; third-class passenger (ticket price: £9 16s 9d); *Titanic* victim

The following people died on this day:
Frölicher-Stehli, Mr Maximilian Josef – 22 November 1913 – aged 60 when on *Titanic*; first-class passenger (ticket price: £79 4s); rescued in lifeboat 5

Nye, Mrs Elizabeth – 22 November 1963 – aged 29 when on *Titanic*; second-class passenger (ticket price: £10 10s); rescued in lifeboat 11

Wennerström, Mr August – 22 November 1950 – aged 27 when on *Titanic*; third-class passenger (ticket price: £7 15s 11d); rescued in collapsible lifeboat A

Titanic survivor quote

Mrs Elizabeth Nye, quoted in *Folkestone Herald*, 4 May 1912

> 'We were all in bed on Sunday night at about 11.30, when we felt an awful jerk, and the boat grazed something along its side, and the sea seemed to splash right over the deck. The men in the next cabin slipped on their coats and ran up to see what it was, and came and told us the ship had run into an iceberg nearly as large as herself. Most of the people went back to bed again, but then came an order "get up and put something warm on, put on a lifebelt and come on deck." So I got one underskirt on and a skirt, and stockings, and shoes and coat, and ran up to find a lifebelt, because there were only three in our berth for four of us. A boy from the next cabin stole one from ours, but he went down with it, poor boy. We did not have time to go back to our cabins again to get anything, and we did not dream it was serious. I thought I should get back to get more clothes on and get a few other things, but we were put into the lifeboats, and pushed off at once.
>
> 'When we got away from the ship we could understand the hurry and the order to get half a mile away as soon as possible. For the *Titanic* was half in the water. We watched the port holes go under until half the ship, only the back half, stuck up. Then the lights went out, and the boilers burst and blew up. There was a sickening roar like hundreds of lions, and we heard no more but THE MOANING AND SHOUTING for help from the hundreds of men and a few women who went down with her. There were not enough boats for so many people.'

Day 328

23 November | On this day in *Titanic* history

The following people were born on this day:

Taylor, Mr John Henry – 23 November 1862 – aged 49 when on *Titanic*; fireman; *Titanic* victim

Davies, Mrs Elizabeth Agnes Mary – 23 November 1863 – aged 48 when on *Titanic*; second-class passenger (ticket price: £36 15s); rescued in lifeboat 14

Fry, Mr Richard Thomas – 23 November 1872 – aged 39 when on *Titanic*; personal valet to Mr Joseph Bruce Ismay, *Titanic* victim

Botsford, Mr William Hull – 23 November 1886 – aged 25 when on *Titanic*; second-class passenger (ticket price: £13); *Titanic* victim

Sage, Mr Douglas Bullen – 23 November 1893 – aged 18 when on *Titanic*; third-class passenger (ticket price: £69 11s); *Titanic* victim

The following people died on this day:

Perry, Mr Edgar Lionel – 23 November 1940 – aged 19 when on *Titanic*; trimmer; rescued in lifeboat unknown

Flynn, Mr John Irwin – 23 November 1951 – aged 36 when on *Titanic*; first-class passenger (ticket price: £26 5s 9d); rescued in lifeboat 5

Anderson, Mr Harry – 23 November 1951 – aged 42 when on *Titanic*; first-class passenger (ticket price: £26 11s); rescued in lifeboat 3

Lines, Miss Mary Conover – 23 November 1975 – aged 16 when on *Titanic*; first-class passenger (ticket price: £39 8s); rescued in lifeboat 9

Titanic survivor quote

Mr Joseph T. Wheat (assistant second steward)

> 'First I told the men off to make sure that the plug in No. 11 was in tight, and then I told five or six men, I cannot tell which, to get into the boat to hand the women and children in. Then the order was passed to pass the women and children along. After the women and children were all passed in we filled her up with as many as the boat would possibly hold, and Mr Murdoch, looking over the top, said, "You have got enough there."'

Titanic fact

The height of *Titanic* was 175 feet (53.3 metres) from the top of the funnels to the keel.

Day 329

24 November | On this day in *Titanic* history

The following people were born on this day:
Phillips, Mr Escott Robert – 24 November 1868 – aged 43 when on *Titanic*; second-class passenger (ticket price: £21); *Titanic* victim

Andersson, Miss Ida Augusta Margareta – 24 November 1873 – aged 38 when on *Titanic*; third-class passenger (ticket price: £7 15s 6d); *Titanic* victim

McCarty, Mr Frederick James – 24 November 1873 – aged 38 when on *Titanic*; bedroom steward in first class; *Titanic* victim

Ross, Mr John Hugo – 24 November 1875 – aged 36 when on *Titanic*; first-class passenger (ticket price: £40 2s 6d); *Titanic* victim

Webb, Mr Samuel Francis – 24 November 1883 – aged 28 when on *Titanic*; trimmer; *Titanic* victim

Forward, Mr James – 24 November 1884 – aged 27 when on *Titanic*; able seaman; rescued in lifeboat 16

McCoy, Mr Bernard – 24 November 1887 – aged 24 when on *Titanic*; third-class passenger (ticket price: £23 5s); rescued in lifeboat 16

Hurst, Mr Walter – 24 November 1888 – aged 23 when on *Titanic*; fireman; rescued in collapsible lifeboat B

Dyker, Mrs Anna Elisabeth Judith – 24 November 1889 – aged 22 when on *Titanic*; third-class passenger (ticket price: £13 18s); rescued in lifeboat 16

Duquemin, Mr Joseph Pierre – 24 November 1887 – aged 24 when on *Titanic*; third-class passenger (ticket price: £7 11s); rescued in collapsible lifeboat D

Mr Walter Hurst, mentioned in *Western Morning News*, Monday, 29 April 1912

> 'A fireman named Walter Hurst, of Southampton, said he was off watch at the time, and made his way to the boat deck. He was, of course, told to stand back and about five minutes before the liner sank he jumped the 75 feet to the sea, having a lifebelt on. He was in the water no less than three hours, and was not picked up until daylight. He could not have lasted much longer. He was numbed to the hips, and still felt the effects of it. On duty in the engine room and stokehold were ten engineers, five leading firemen, ten greasers, 25 trimmers, and 53 firemen. Of these all but six were drowned. My mates, he said, received orders to remain below and rake out the fires to prevent explosions. They obeyed them at the sacrifice of their lives.'

Titanic survivor quote

Mrs Charlotte Caroline Collyer
'"Women and children first!" Someone was shouting these last few words over and over again. They meant my own safety but they also meant the greatest loss I've ever suffered, the life of my husband.'

Day 330

25 November | On this day in *Titanic* history

The following people were born on this day:
May, Mr Arthur William – 25 November 1852 – aged 59 when on *Titanic*; fireman; *Titanic* victim

McCarthy, Mr Timothy John – 25 November 1857 – aged 54 when on *Titanic*; first-class passenger (ticket price: £51 17s 3d); *Titanic* victim

Hippach, Mrs Ida Sophia – 25 November 1867 – aged 44 when on *Titanic*; first-class passenger (ticket price: £57 19s 7d); rescued in lifeboat 4

Beattie, Mr Thomson – 25 November 1875 – aged 36 when on *Titanic*; first-class passenger (ticket price: £75 4s 10d); *Titanic* victim

Cunningham, Mr Alfred Fleming – 25 November 1890 – aged 21 when on *Titanic*; fitter; *Titanic* victim

Taussig, Miss Ruth – 25 November 1893 – aged 18 when on *Titanic*; first-class passenger (ticket price: £79 13s); rescued in lifeboat 8

The following people died on this day:
Lenox-Conyngham, Mrs Barbara Josephine – 25 November 1937 – aged 44 when on *Titanic*; first-class passenger (ticket price: £3); disembarked at Cherbourg

Sinkkonen, Miss Anna – 25 November 1963 – aged 30 when on *Titanic*; second-class passenger (ticket price: £13); rescued in lifeboat 10

Titanic survivor quotes

Mr A. H. Weikman (barber)

> 'I was proceeding to launch the next boat when the ship suddenly sank at the bow and there was a rush of water that washed me overboard… I started to swim.. for some dark object in the water. It was dark. This was about 1.50 am toward the stern… my watch was stopped at that time by the water. I was about 15 feet away from the ship when I heard a second explosion. I think the boilers blew up about in the middle of the ship. The explosion blew me along with a wall of water toward the dark object I was swimming to, which proved to be a bundle of deck chairs, which I managed to climb on.'

Miss Annie Robinson (stewardess in first class)

> 'Mr Andrews told me to put my lifebelt on after I had been on E deck. Mr Andrews said to me, "Put your lifebelt on and walk about and let the passengers see you." I said to him, "It looks rather mean," and he said, "No, put it on," and then after that he said to me, "Well, if you value your life put your belt on."'

Day 331

26 November | On this day in *Titanic* history

The following person was born on this day:
Collyer, Mr Harvey – 26 November 1880 – aged 31 when on *Titanic*; second-class passenger (ticket price: £26 5s); *Titanic* victim

Titanic survivor quotes

Mrs Henry Arthur Casebeer

> 'It was not long after this, however, that the pursers started to go among the passengers ordering them to go below and put on warm clothing and be prepared to embark in the lifeboats. I hurried below and dressed and when I came on deck again I found that the deck had started to list in a very alarming manner. I had already donned a life preserver which I found with some difficulty, and when I reached the deck I met Mr Andrews again and he took me by the arm and led me to the lifeboat.
>
> 'I could not hear just what he said to me at the time on account of the din, but I saw him motion to me to get into the boat, which was about to be swung over the rail 90 feet above the water. I asked him why he did not get in also, and he said: "No, women and children first."'

Mr Herbert John Pitman

> 'I went to the bridge and saw Captain Smith and told him that I thought it was Mr Ismay that wanted me to get the boat away with women and children in it and he said "Go ahead; carry on." I came along and brought in my boat. I stood in it and said: "Come along, ladies." There was a big crowd. Mr Ismay helped get them along. We got the boat nearly full and I shouted out for any more ladies. None were to be seen so I allowed a few men to get into it. Then I jumped on the ship again. Mr Murdoch said: "You go in charge of this boat and hang around the after gangway."
>
> 'As soon as the ship disappeared I said: "Now, men, we will pull toward the wreck." Everyone in my boat said it was a mad idea because we had far better save what few I had in my boat than go back to the scene of the wreck and be swamped by the crowds that were there. My boat would have accommodated a few more – about sixty in all. I turned No. 5 boat around to go in the direction from which these cries came but was dissuaded from my purpose by the passengers.

Titanic fact

Second-class passenger Harvey Collyer, who died in the disaster, sewed all his valuables into his clothes.

Day 332

27 November | On this day in *Titanic* history

The following people were born on this day:
Picard, Mr Benoît – 27 November 1878 – aged 33 when on *Titanic*; third-class passenger (ticket price: £8 1s); rescued in lifeboat 9

Bochatay, Mr Alexis Joseph – 27 November 1881 – aged 30 when on *Titanic*; chef; *Titanic* victim

Toshack, Mr James Adamson – 27 November 1881 – aged 30 when on *Titanic*; saloon steward in first class; *Titanic* victim

Wrapson, Mr Frederick Bernard – 27 November 1893 – aged 18 when on *Titanic*; assistant pantryman steward; *Titanic* victim

Mubārik, Master Jirjis – 27 November 1904 – aged 7 when on *Titanic*; third-class passenger (ticket price: £15 4s 11d); rescued in collapsible lifeboat C

The following person died on this day:
Meyer, Mrs Leila – 27 November 1957 – aged 25 when on *Titanic*; first-class passenger (ticket price: £82 3s 5d); rescued in lifeboat 6

Titanic survivor quote

Mrs Leila Meyer, mentioned in *Le Journal*, Saturday, 20 April 1912

> 'Mrs Meyer told reporters that her husband and herself hurried to the lifeboats as soon as the first jolt was felt. She wanted to stay with him but he strongly pushed her inside a boat and told her about their nine-year-old son, for whom she had to survive.

Mrs Leila Meyer, mentioned in *New York Tribune*, Sunday, 1 December 1912

> 'Mrs John Jacob Astor, who became a widow through the *Titanic* disaster and whose baby was born only a short time after her rescue at sea, sent her congratulations to Mrs Eloise Smith today.
>
> 'Mrs Smith is only eighteen years old, and her husband was twenty-four. She and Mrs Edgar J. Meyer escaped from the *Titanic* in a boat containing only one man. She declared that the boat pushed off with room for twenty more passengers, yet neither Mr Smith nor Mr Meyer were allowed aboard. Mr and Mrs Smith were awakened by the crash. To calm her fears he ate an apple while she dressed. Then, assuring her that he would follow by another boat after the women were all safely off, he induced her to leave him.'

Day 333

28 November | On this day in *Titanic* history

The following people were born on this day:

Allsop, Mr Frank Richard – 28 November 1868 – aged 43 when on *Titanic*; saloon steward; *Titanic* victim

Hogg, Mr Charles William – 28 November 1868 – aged 43 when on *Titanic*; bedroom steward in first class; *Titanic* victim

Homer, Mr Harry – 28 November 1871 – aged 40 when on *Titanic*; first-class passenger (ticket price: £26 11s); rescued in lifeboat 15

Williamson, Mr James Bertram – 28 November 1876 – aged 35 when on *Titanic*; postal clerk / postman; *Titanic* victim

Crispin, Mr William – 28 November 1878 – aged 33 when on *Titanic*; glory hole steward; *Titanic* victim

Lewy, Mr Ervin G. – 28 November 1881 – aged 30 when on *Titanic*; first-class passenger (ticket price: £27 14s 5d); *Titanic* victim

The following people died on this day:

Collyer, Mrs Charlotte Caroline – 28 November 1916 – aged 30 when on *Titanic*; second-class passenger (ticket price: £26 5s); rescued in lifeboat 14

Howard, Miss May Elizabeth – 28 November 1958 – aged 26 when on *Titanic*; third-class passenger (ticket price: £8 1s); rescued in collapsible lifeboat C

Titanic survivor quote

Mr Harry Haven Homer, mentioned in *The Witney Gazette*, Saturday, 11 May 1912

> 'The two men, says *The Daily Chronicle*'s New York correspondent, are known as "Doc Owen" and "Kid Homer", and they were playing with a third man when the crash came. Learning that there was no hope for the *Titanic*, they decided to try to get away in one of the boats. Those in authority, however, were allowing only women and children to go. "Doc Owen" therefore got hold of a steward who, it is alleged, had been paid to keep the identity of the gamblers secret during the voyage, and, giving him a roll of bank notes, got him to furnish women's clothing and hats. Dressed in these clothes, the three men hurried to the deck and leaped into a lifeboat filled with women just as it was being lowered.
>
> 'Afterwards they stripped themselves of the women's clothes, which they threw overboard. The boat they were in was filled with immigrant women and children, and did not have enough men to work the oars. Accordingly, their assistance was welcomed.'

Day 334

29 November | On this day in *Titanic* history

The following people were born on this day:
Cassebeer, Mrs Eleanor Genevieve – 29 November 1875 – aged 36 when on *Titanic*; first-class passenger (ticket price: £27 14s 5d); rescued in lifeboat 5

Givard, Mr Hans Kristensen – 29 November 1884 – aged 27 when on *Titanic*; second-class passenger (ticket price: £13); *Titanic* victim

Mrs Eleanor Genevieve Cassebeer, quoted in *Binghamton Press*, Monday, 29 April 1912

> 'Another thing that is not generally known is that the *Titanic* was not ready to sail at the time she did. Mr Andrews told me himself and said that the only reason they allowed her to go when they did was that the sailing date had already been fixed and they just simply had to start. While the ship was fitted up most sumptuously one could not help but notice that she was not prepared to sail. There were none of the usual printed notices in the cabins. The frames for them were on the walls, but the notices themselves were not there and when I tried to find a life preserver I did not know where to look for it and was compelled to inquire of some stewards who showed me where to find it.'

Titanic survivor quotes

Mr Thomas Jones (seaman)

> 'I got the collapsible boat on the port side ready. I got my own boat No. 8, ready. An officer sent me for a lamp, and as I was going forward there was a man coming with two or three lamps in his hand. I went back again, and this No. 8 boat was there, all swung out, and there were about 35 ladies in it. I jumped in the boat. The captain asked me was the plug in the boat, and I answered, "Yes, sir." "All right," he said, "Any more ladies?" There was one lady came there and left her husband. She wanted her husband to go with her, but he backed away, and the captain shouted again – in fact, twice again – "Any more ladies?" There were no more there, and he lowered away.'

Mr Joseph G. Boxhall, witness at United States Senate Inquiry (Day 3)

> 'On such a night as that, even if there is no moon, you can very, very often see an iceberg by the water on the sides of it; that is, if there is a little breeze. But when the water is in one of those oily calms...'

Day 335

30 November | On this day in *Titanic* history

The following people were born on this day:
Taylor, Mrs Juliet Cummins – 30 November 1862 – aged 49 when on *Titanic*; first-class passenger (ticket price: £52); rescued in lifeboat 5

Partner, Mr Austin – 30 November 1871 – aged 40 when on *Titanic*; first-class passenger (ticket price: £28 10s); *Titanic* victim

Hosking, Mr George Fox – 30 November 1875 – aged 36 when on *Titanic*; senior third engineer; *Titanic* victim

Maybery, Mr Frank Hubert – 30 November 1875 – aged 36 when on *Titanic*; second-class passenger (ticket price: £16); *Titanic* victim

Morris, Mr Frank Herbert – 30 November 1883 – aged 28 when on *Titanic*; bathroom steward in first class; rescued in lifeboat 14

Smillie, Mr John Downing – 30 November 1883 – aged 28 when on *Titanic*; saloon steward; *Titanic* victim

Østby, Miss Helen Ragnhild – 30 November 1889 – aged 22 when on *Titanic*; first-class passenger (ticket price: £61 19s 7d); rescued in lifeboat 5

Ratti, Sig. Enrico – 30 November 1890 – aged 21 when on *Titanic*; waiter; *Titanic* victim

Rice, Master George Hugh – 30 November 1903 – aged 8 when on *Titanic*; third-class passenger (ticket price: £29 2s 6d); *Titanic* victim

Mrs Juliet Cummins Taylor, mentioned in *Atlantic City Daily Press*, Friday, 19 April 1912

> 'Mr Taylor is the inventor of the paper cups manufactured by the American Mono-Service Company, which is furnishing the drinking vessels to local hotels and business places and hotels and railroads all over the Eastern states.
>
> Mr Taylor described the night of Sunday as one of great horror and could scarcely give many coherent details. He and Mrs Taylor were among the first who were placed in the lifeboats that were sent off from the ill-fated *Titanic*. When their boat, which contained about thirty people, started from the side of the ship they found the sea was filled with cakes of floating ice. A short time after their small boat had pulled from the side of the ship the lights of the vessel went out and then the only illumination from the sinking ship were from torches and rockets, which were sent up until a short time before the ship took her final plunge into the icy ocean.'

December

Day 336

1 December | On this day in *Titanic* history

The following people were born on this day:

Harris, Mr Henry Birkhardt – 1 December 1866 – aged 45 when on *Titanic*; first-class passenger (ticket price: £83 9s 6d); *Titanic* victim

Hopkins, Mr Robert John – 1 December 1868 – aged 43 when on *Titanic*; able seaman; rescued in lifeboat 13

Frauenthal, Mrs Clara – 1 December 1869 – aged 42 when on *Titanic*; first-class passenger (ticket price: £133 13s); rescued in lifeboat 5

Panula, Mrs Maija Emelia Abrahamintytar – 1 December 1870 – aged 41 when on *Titanic*; third-class passenger (ticket price: £39 13s 9d); *Titanic* victim

Reeves, Mr David – 1 December 1875 – aged 36 when on *Titanic*; second-class passenger (ticket price: £10 10s); *Titanic* victim

Taylor, Mr William Henry – 1 December 1883 – aged 28 when on *Titanic*; fireman; rescued in lifeboat 15

Pitfield, Mr William James – 1 December 1886 – aged 25 when on *Titanic*; greaser; *Titanic* victim

Leader, Mr Archibald – 1 December 1889 – aged 22 when on *Titanic*; confectioner; *Titanic* victim

The following people died on this day:

Wilson, Miss Helen Alice – 1 December 1939 – aged 33 when on *Titanic*; first-class passenger (ticket price: £134 10s); rescued in lifeboat 3

Johannesen, Mr Bernt Johannes – 1 December 1962 – aged 29 when on *Titanic*; third-class passenger (ticket price: £8 2s 3d); rescued in lifeboat 13

Sandström, Mrs Agnes Charlotta – 1 December 1985 – aged 24 when on *Titanic*; third-class passenger (ticket price: £16 14s); rescued in lifeboat 13

Harder, Mrs Dorothy – 1 December 1926 – aged 21 when on *Titanic*; first-class passenger (ticket price: £55 8s 10d); rescued in lifeboat 5

Titanic survivor quote

Mr Alfred Crawford (bedroom steward in first class)

> 'He pointed in the direction of the two lights, and said: "Pull for that vessel; land your people and return to the ship." Those were Captain Smith's words.'

Titanic fact

First-class passenger Mrs Eleanor Widener wore a famous multi-strand pearl necklace, then valued at $250,000.

Day 337

2 December | On this day in *Titanic* history

The following people were born on this day:

Karun, Mr Franz – 2 December 1872 – aged 39 when on *Titanic*; third-class passenger (ticket price: £13 8s 4d); rescued in lifeboat 15

Rondberg, Mr Edvard Larsson – 2 December 1889 – aged 22 when on *Titanic*; third-class passenger (ticket price: £7 15s 6d); *Titanic* victim

Kalvik, Mr Johannes Halvorsen – 2 December 1890 – aged 21 when on *Titanic*; third-class passenger (ticket price: £8 8s 8d); *Titanic* victim

Shillabeer, Mr Charles Frederick – 2 December 1892 – aged 19 when on *Titanic*; trimmer; *Titanic* victim

The following person died on this day:

Brice, Mr Walter Thomas – 2 December 1928 – aged 42 when on *Titanic*; able seaman; rescued in lifeboat 11

Titanic survivor quote

Mr Franz Karun, quoted in *Galesburg Evening Mail*, Tuesday, 23 April 1912

'Suddenly he saw a lifeboat being lowered over the side. He sprang forward and with his daughter in his arms, dived head first over the side of the deck, lighting ten feet below in the boat, which was being rapidly let over the side. There were already fifty passengers in the boat, but room was made for him, although this was done under protest. The report yesterday that Kurun was among those on the raft with Col. Gracie was unfounded but the circumstances of his escape as above outlined were taken from the survivor's own lips this morning.

'When asked whether any shots were fired at men trying to get on the lifeboats, he replied that he had not heard or seen anything of the sort, but had seen the officers draw revolvers. The lifeboat in which he was seated was not far off from the liner when it sank. Kurun did not see the boat break in two, but he saw the big hole in its side. He said that before he left the boat, the water was coming up through the steerage hatchways and caused the people to be greatly frightened. When the *Titanic* sank, he heard the cries of the people and saw men struggling in the water after the boat went down. There was much ice floating about. Of the fifty-two persons on board the lifeboat, most of them were women and children, there being over a dozen children. It was four hours and a half before the *Carpathia* was sighted and almost five hours before the half-frozen survivors were taken aboard.'

Day 338

3 December | On this day in *Titanic* history

The following people were born on this day:
Ford, Mrs Margaret Ann Watson – 3 December 1857 – aged 54 when on *Titanic*; third-class passenger (ticket price: £34 7s 6d); *Titanic* victim

Brady, Mr John Bertram – 3 December 1870 – aged 41 when on *Titanic*; first-class passenger (ticket price: £30 10s); *Titanic* victim

Yearsley, Mr Harry – 3 December 1871 – aged 40 when on *Titanic*; saloon steward in first class; rescued in lifeboat 9

Mackie, Mr George William – 3 December 1877 – aged 34 when on *Titanic*; bedroom steward in second class; *Titanic* victim

Cornaire, Mr Marcel Raymond André – 3 December 1892 – aged 19 when on *Titanic*; assistant roast cook; *Titanic* victim

King, Mr Alfred John Moffett – 3 December 1893 – aged 18 when on *Titanic*; lift steward; *Titanic* victim

The following people died on this day:
Symons, Mr George Thomas Macdonald – 3 December 1950 – aged 24 when on *Titanic*; lookout; rescued in lifeboat 1

Prangnell, Mr George Alexander – 3 December 1953 – aged 30 when on *Titanic*; greaser; rescued in collapsible lifeboat B

Burgess, Mr Charles – 3 December 1960 – aged 18 when on *Titanic*; extra third baker; rescued in lifeboat unknown

Ayyūb-Dāhir, Miss Bannūrah – 3 December 1970 – aged 15 when on *Titanic*; third-class passenger (ticket price: £7 4s 7d); rescued in collapsible lifeboat C

Dodge, Master Washington – 3 December 1974 – aged 4 when on *Titanic*; first-class passenger (ticket price: £81 17s 2d); rescued in lifeboat 5

Troutt, Miss Edwina Celia – 3 December 1984 – aged 27 when on *Titanic*; second-class passenger (ticket price: £10 10s); rescued in lifeboat 16

Mr Charles Burgess, mentioned in *The Daily Banner*, Thursday, 16 May 1912

> 'Just upon half-past seven the second local survivor from the terrible disaster arrived from Southampton in the person of Mr Chas. Burgess, second class baker of the *Titanic*. A large crowd had assembled at the station to give him a welcome home.'

Day 339

4 December | On this day in *Titanic* history

The following people were born on this day:
Deeble, Mr Alfred Arnold – 4 December 1877 – aged 34 when on *Titanic*; saloon steward in first class; *Titanic* victim

Jewell, Mr Archie – 4 December 1888 – aged 23 when on *Titanic*; lookout; rescued in lifeboat 7

Cavell, Mr George Henry – 4 December 1889 – aged 22 when on *Titanic*; trimmer; rescued in lifeboat 15

Doel, Mr Frederick – 4 December 1889 – aged 22 when on *Titanic*; fireman; rescued in lifeboat unknown

The following people died on this day:
Compton, Mrs Mary Eliza – 4 December 1930 – aged 64 when on *Titanic*; first-class passenger (ticket price: £83 3s 2d); rescued in lifeboat 14

Andrews, Miss Kornelia Theodosia – 4 December 1913 – aged 62 when on *Titanic*; first-class passenger (ticket price: £77 19s 2d); rescued in lifeboat 10

Gracie, Colonel Archibald – 4 December 1912 – aged 53 when on *Titanic*; first-class passenger (ticket price: £28 10s); rescued in collapsible lifeboat B

Abelseth, Mr Olaus Jørgensen – 4 December 1980 – aged 25 when on *Titanic*; third-class passenger (ticket price: £7 13s); rescued in collapsible lifeboat A

Al-Za'innī, Mr Fahīm Rūḥānā – 4 December 1927 – aged 25 when on *Titanic*; third-class passenger (ticket price: £7 4s 6d); rescued in lifeboat 6

Richards, Master Sibley George – 4 December 1987 – aged 9 months when on *Titanic*; second-class passenger (ticket price: £18 15s); rescued in lifeboat 4

Titanic survivor quotes

Mr Joseph Scarrott (able seaman)

> 'I did not feel any direct impact, but it seemed as if the ship shook in the same manner as if the engines had been suddenly reversed to full speed astern, just the same sort of vibration, enough to wake anybody up if they were asleep.'

Mr George Symons (able seaman)

> 'What awakened me was a grinding sound on her bottom. I thought at first she had lost her anchor and chain, and it was running along her bottom. I would not get up. The others got up, but I would not get up as I thought there was nothing the matter.'

Day 340

5 December | On this day in *Titanic* history

The following people were born on this day:
Larsson, Mr Bengt Edvin – 5 December 1882 – aged 29 when on *Titanic*; third-class passenger (ticket price: £7 15s); *Titanic* victim

Reynaldo, Sra. Encarnación – 5 December 1882 – aged 29 when on *Titanic*; second-class passenger (ticket price: £13); rescued in lifeboat 9

The following people died on this day:
Mills, Mr Christopher – 5 December 1930 – aged 51 when on *Titanic*; butcher; rescued in collapsible lifeboat C

Lehmann, Miss Bertha – 5 December 1967 – aged 17 when on *Titanic*; second-class passenger (ticket price: £12); rescued in lifeboat 12

Titanic survivor quote

Mr William T. Sloper

> 'Two hours after our lifeboat was launched, the sailors estimated that we had drifted more than two miles from where the *Titanic* was sinking. The ship remained until 2 or 3 minutes before she sank as brilliantly lighted as she was directly after the accident occurred and all the lights had been turned on. Then suddenly (like the house lights in a brilliantly lighted theatre just before the curtain goes up) all the lights dipped simultaneously to a pale glow. A moment or two later everyone watching in the lifeboats saw silhouetted against the starlit sky the stern of the ship rise perpendicularly into the air from about midship. Then with a prolonged rush and a roar like the ten thousand tons of coal sliding down a metal chute several hundred feet long, the great ship went down out of sight and disappeared beneath the surface of the ocean. Then a great cry arose on the air from the surface of the calm sea where the ship had been. One of the sailors cut the ropes and divided the rugs among the women, some of whom were not too warmly dressed. The night air was very cold and Dorothy dressed in a summer dress with only a sweater and a polo coat pulled on over it soon felt the cold very much. I used Sunday night as an excuse for not changing at dinner time into my evening clothes. I had been wearing a brand new suit of heavy woollen material which had just been finished for me just before sailing by a London tailor. When I went down to get my life preserver I had pulled on a heavy Shetland wool V neck sweater and my winter weight overcoat. With my life preserver I was cumbersomely dressed that after a few minutes of pulling an oar in the lifeboat threw me into a dripping perspiration.'

Day 341

6 December | On this day in *Titanic* history

The following people were born on this day:
Reuchlin, Mr Jonkheer Johan George – 6 December 1874 – aged 37 when on *Titanic*; first-class passenger (complimentary ticket); *Titanic* victim

Herman, Miss Alice – 6 December 1887 – aged 24 when on *Titanic* (twin of Kate Herman); second-class passenger (ticket price: £65); rescued in lifeboat 9

Herman, Miss Kate – 6 December 1887 – aged 24 when on *Titanic* (twin of Alice Herman); second-class passenger (ticket price: £65); rescued in lifeboat 9

Hays, Miss Margaret Bechstein – 6 December 1887 – aged 24 when on *Titanic*; first-class passenger (ticket price: £83 3s 2d); rescued in lifeboat 7

Gradidge, Mr Ernest Edward – 6 December 1889 – aged 22 when on *Titanic*; fireman; *Titanic* victim

The following people died on this day:
Harder, Mr William – 6 December 1947 – aged 39 when on *Titanic*; window cleaner; rescued in lifeboat 14

Burrage, Mr Arthur Victor Edwards – 6 December 1951 – aged 20 when on *Titanic*; plate steward; rescued in lifeboat 13

Mr Jonkh Johan George Reuchlin, mentioned in *Evening Telegraph*, Tuesday, 23 April 1912

> 'It was learned today that Jonkheer J. G. Reuchlin, managing director of the Holland-American Steamship Line, had gone down with the *Titanic*. Mr Reuchlin was coming to this country to arrange business relative to the opening of the Panama Canal.'

Miss Alice Herman and **Miss Kate Herman**, mentioned in *Newark Evening News*, Saturday, 20 April 1912

> 'Mrs Herman and her daughters were in great distress of mind over Mr Herman's fate and their own experience. They were at a loss to understand how the disaster occurred on such clear night. They said they were near enough to the big ship when she sank to feel the suction. Members of the crew who manned their boat thought that the boat was full…Mr Herman was fifty years old, the same age as his wife. The daughters are each twenty-four years old.'

Titanic survivor quote

Mr Charles H. Lightoller (second officer)
'Many died from the cold during the night. And a mighty long time it seemed before daylight broke, standing wet through and up to our knees in icy water on the upturned boat. Frankly I didn't think many of us expected to see daylight.'

Day 342

7 December | On this day in *Titanic* history

The following people were born on this day:
Meanwell, Mrs Marian – 7 December 1849 – aged 62 when on *Titanic*; third-class passenger (ticket price: £8 1s); *Titanic* victim

Burns, Miss Elizabeth Margaret – 7 December 1870 – aged 41 when on *Titanic*; first-class passenger (ticket price: £134 10s); rescued in lifeboat 3

Ryerson, Mr William Edwy – 7 December 1878 – aged 33 when on *Titanic*; saloon steward in second class; rescued in lifeboat unknown

Hellström, Miss Hilda Maria – 7 December 1889 – aged 22 when on *Titanic*; third-class passenger (ticket price: £8 19s 3d); rescued in collapsible lifeboat C

The following people died on this day:
Peuchen, Major Arthur Godfrey – 7 December 1929 – aged 52 when on *Titanic*; first-class passenger (ticket price: £30 10s); rescued in lifeboat 6

Seward, Mr Frederic Kimber – 7 December 1943 – aged 34 when on *Titanic*; first-class passenger (ticket price: £26 11s); rescued in lifeboat 7

Pitman, Mr Herbert John – 7 December 1961 – aged 34 when on *Titanic*; third officer; rescued in lifeboat 5

Titanic survivor quote

Major Arthur Godfrey Peuchen, mentioned in *The Toronto World*, Saturday, 20 April 1912

> 'Second Officer Lightoller, who had charge of the lifeboats:
>
> "In the fourth lifeboat I was running short of seamen. I put two seamen in and one of them jumped out. That was the first boat I had to put a man passenger in. He was standing near by and said he would go if I needed him.
>
> "I said: 'Are you a sailor?' and he replied that he was a yachtsman. Then I told him that if he was a sailor to get over the bulwarks to the lifeboat – to go ahead. He did, and proved himself afterwards to be a very brave man."
>
> "Who was he? Did you know him?"
>
> "I didn't know him then, but afterward I looked him up. He was Major Peuchen of Toronto."
>
> "Had you ever seen him before?"
>
> "Never."'

Titanic fact

Each day, approximately 100 tons of ash was dumped into the sea from the *Titanic*.

Day 343

8 December | On this day in *Titanic* history

The following person died on this day:
Lightoller, Mr Charles Herbert – 8 December 1952 – aged 38 when on *Titanic*; second officer; rescued in collapsible lifeboat B

Mr Charles Herbert Lightoller, mentioned in *Southampton Times*, Saturday, 20 April 1912

> 'Mr C. H. Lightoller, the second officer on the ill-fated *Titanic*, who is reported to be among the survivors, lived at Netley Abbey, and on Wednesday one of our representatives called on his wife at their residence at Hound to convey congratulations at Mr Lightoller's providential escape. Mrs Lightoller, who has two little boys, had just returned from Southampton, where she had been visiting the wives of other officers less fortunate than herself, and said how terribly anxious she had been, but how thankful she felt now that she was assured of her husband's safety. Mr Lightoller, who comes from Liverpool, has seen service with the White Star Line on the liners *Teutonic* and *Oceanic*, but recently has been engaged on the *Majestic*. Mrs Lightoller is a native of Australia, and, with her husband, have been living at Netley since the Company came to Southampton.
>
> 'Mrs Lightoller said that she had endeavoured to see the Mayor of Southampton, to ask what she might do to help in collecting at Netley, as she thought she ought to do everything in her power now she herself was relieved from anxiety. In conclusion, Mrs Lightoller said she had been besieged with telegrams of inquiry and congratulations, and wished to thank those who had sent, for she could not possibly reply to so many.'

Mr Charles Herbert Lightoller, mentioned in *Daily Sketch*, Monday, 13 May 1912

> 'The surviving officers of the *Titanic* – Messrs. Lightoller, Lowe, Boxhall and Pitman – reached Liverpool by the same ship as Mr Ismay – the *Adriatic*.

The fifth officer, Mr Lowe, who told the American Court of Inquiry that he asked Mr Ismay, while the boats were being lowered, not to let his anxiety to help hinder the men, attracted much attention from the crowd. It was Mr Lowe to whom Senator Smith, the chairman of the inquiry, apologised for having made – through a misunderstanding – an unfounded suggestion. Mr Lowe was frank in his criticisms of some of the newspapers on the other side. Also among the passengers on the *Adriatic* were Mrs Dean and her two children, survivors of the wreck. Mrs Dean's husband was lost, and over £60 was collected for her on the *Adriatic*.'

Day 344

9 December | On this day in *Titanic* history

The following people were born on this day:

Whitford, Mr Alfred Henry – 9 December 1872 – aged 39 when on *Titanic*; saloon steward in second class; *Titanic* victim

Wallcroft, Miss Ellen 'Nellie' – 9 December 1875 – aged 36 when on *Titanic*; second-class passenger (ticket price: £21); rescued in lifeboat 14

Allison, Mr Hudson Joshua Creighton – 9 December 1881 – aged 30 when on *Titanic*; first-class passenger (ticket price: £151 16s); *Titanic* victim

Banfi, Sig. Ugo – 9 December 1887 – aged 41 when on *Titanic*; waiter; *Titanic* victim

Lennon, Mr Denis – 9 December 1891 – aged 20 when on *Titanic*; third-class passenger (ticket price: £15 10s); *Titanic* victim

Corr, Miss Ellen (Helen) – 9 December 1895 – aged 16 when on *Titanic*; third-class passenger (ticket price: £7 15s); rescued in lifeboat 14 or 16

The following people died on this day:

Ryerson, Mr William Edwy – 9 December 1949 – aged 33 when on *Titanic*; saloon steward in second class; rescued in lifeboat unknown

Joughin, Mr Charles John – 9 December 1956 – aged 33 when on *Titanic*; chief baker; rescued in collapsible lifeboat B

Snyder, Mrs Nelle – 9 December 1983 – aged 23 when on *Titanic*; first-class passenger (ticket price: £82 5s 4d); rescued in lifeboat 7

Tu'mah, Master Jirjis Yūsuf – 9 December 1991 – aged 8 when on *Titanic*; third-class passenger (ticket price: £15 4s 11d); rescued in collapsible lifeboat C

Titanic survivor quote

Miss Edith Russell

> 'Well, I got out on deck. I noticed a gentleman standing by the rail and several people and again, this greyish mass. I said "What on earth is that?" "Well, madam, that's an iceberg." "Iceberg? Gracious! I have always wanted to see one of those things since I was a child." "Well you're seeing a corker now, there's one eight above the water and seven-eighths below and believe me that's some iceberg." So, after that we picked up bits of ice, played snowballs for a little while and was very very cold. I asked one of the officers if there was any danger. He said no so I went back to bed.'

Day 345

10 December | On this day in *Titanic* history

The following people were born on this day:
Cornell, Mrs Malvina Helen – 10 December 1856 – aged 55 when on *Titanic*; first-class passenger (ticket price: £25 14s 10d); rescued in lifeboat 15

Chapman, Mr Joseph Charles – 10 December 1879 – aged 32 when on *Titanic*; boots steward in second class; rescued in lifeboat 9

Vanderplancke, Mrs Emelie Maria – 10 December 1880 – aged 31 when on *Titanic*; third-class passenger (ticket price: £18); *Titanic* victim

Ranger, Mr Thomas – 10 December 1882 – aged 29 when on *Titanic*; greaser; rescued in lifeboat 4

Titanic survivor quotes

Mr James Johnstone

'The *Titanic* was going 23 knots an hour. She ripped herself clear apart. There was no panic. It was very quiet. When the boats were lowered there were many who refused to go. One of the most pathetic scenes was the refusal of Mrs Isidor Straus to leave her husband. She remained with him to the last.

'When the first signal was given to lower the boats, some of the crew pressed forward. It was then that the rallying cry came through the megaphone from the bridge, "Be British, my men." It was Capt. Smith's voice. Every man obeyed the command and faced death calmly. They knew that there was no hope and as the big, strong English seamen assisted the women and children into the boats they gave no sign that they realized that Capt. Smith's words "Be British," had sealed their fate. They remained at their posts and died like men.'

Mr George Symons (able seaman)

'Being the master of the situation, I used my own discretion. I said nothing to anybody about the ship being doomed, in my opinion. I pulled a little further away to escape, if there was any suction. A little while after that we pulled a little way and lay on the oars again. The other boats were around us by that time, and some were pulling further away from us. I stood and watched it till I heard two sharp explosions in the ship. What they were I could not say. Then she suddenly took a top cant, her stern came well out of the water then.'

Titanic fact

The nickname of the underwater robot used to explore the wreck of *Titanic* was JJ.

Day 346

11 December | On this day in *Titanic* history

The following people were born on this day:
Bowker, Miss Ruth Harwood – 11 December 1880 – aged 31 when on *Titanic*; cashier; rescued in lifeboat 6

Middleton, Mr Alfred Pirrie – 11 December 1884 – aged 27 when on *Titanic*; assistant electrician; *Titanic* victim

Andersson, Mr Johan Samuel – 11 December 1885 – aged 26 when on *Titanic*; third-class passenger (ticket price: £7 15s 6d); *Titanic* victim

Klasén, Miss Gertrud Emilia – 11 December 1910 – aged 1 when on *Titanic*; third-class passenger (ticket price: £12 3s 8d); *Titanic* victim

The following person died on this day:
Ibrāhīm, Mrs Sāfiyah – 11 December 1976 – aged 18 when on *Titanic*; third-class passenger (ticket price: £7 4s 7d); rescued in collapsible lifeboat C

Titanic survivor quote

Miss Ruth Harwood Bowker, quoted in *The Observer* (Cheshire), Saturday, 18 May 1912

> 'It seemed an insult to human intelligence and human handicraft that such a gigantic craft should ever sink. At length it was obvious that her bow was dropping slowly. Tier by tier the lights went out, and then suddenly there was what seemed a series of explosions. The ship broke in two, and there was an astounding rattling noise, as though the machinery had broken loose. Finally she sank quietly. The ensuing cries, Miss Bowker declared, still rang in her ears, and never would be forgotten.
>
> 'Her boat was one of the last to be picked up, and before the longed-for *Carpathia* arrived the sea and wind had got up so that the lifeboat threatened to capsize. Their lifebelts probably saved them from pneumonia. On the *Carpathia* they were treated with every possible kindness, though the discomfort on such a small ship was naturally considerable. It was indeed a ship of mourning. Terrible stories were told by some of the survivors who had remained on the *Titanic* to the last.
>
> 'On entering New York harbour, frantic efforts to obtain news were made by the American journalists, many of whom had chartered tugs, from which questions were bawled at the *Carpathia* through megaphones. 'Did John Jacob Astor die a hero?' is a fair specimen of these.
>
> 'At New York most of the crew and staff were transferred direct to the *Lapland*. A delightful passage to England greatly benefitted shattered nerves and exhausted bodies.'

Day 347

12 December | On this day in *Titanic* history

The following people were born on this day:

Appleton, Mrs Charlotte – 12 December 1858 – aged 53 when on *Titanic*; first-class passenger (ticket price: £51 9s 7d); rescued in lifeboat 2

Ismay, Mr Joseph Bruce – 12 December 1862 – aged 49 when on *Titanic*; first-class passenger; rescued in collapsible lifeboat C

Rothschild, Mr Martin – 12 December 1865 – aged 46 when on *Titanic*; first-class passenger (ticket price: £59 8s); *Titanic* victim

Rowe, Mr Edgar Maurice – 12 December 1880 – aged 31 when on *Titanic*; saloon steward in first class; *Titanic* victim

Prangnell, Mr George Alexander – 12 December 1881 – aged 30 when on *Titanic*; greaser; rescued in collapsible lifeboat B

Swane, Mr George – 12 December 1892 – aged 19 when on *Titanic*; second-class passenger (ticket price: £13); *Titanic* victim

McCormack, Mr Thomas Joseph – 12 December 1892 – aged 19 when on *Titanic*; third-class passenger (ticket price: £7 15s); rescued in lifeboat unknown

Asplund, Master Filip Oscar – 12 December 1898 – aged 13 when on *Titanic*; third-class passenger (ticket price: £31 7s 9d); *Titanic* victim

The following people died on this day:

Buley, Mr Edward John – 12 December 1917 – aged 26 when on *Titanic*; able seaman; rescued in lifeboat 10

Seward, Mr Wilfred Deable – 12 December 1963 – aged 25 when on *Titanic*; chief pantry steward in second class; rescued in lifeboat 3

Mr Joseph Bruce Ismay, mentioned in *Washington Times*, Friday, 19 April 1912

> 'Not only did J. Bruce Ismay, managing director of the White Star Line, get into one of the first lifeboats to be launched, but he was escorted and assisted by several seamen, while the women had to tumble in and take care of themselves, according to Mrs Lucien P. Smith, daughter of Congressman Hughes of West Virginia, and widowed bride of Lucien P. Smith.'

Titanic fact

Titanic had four funnels, however, only three of them were functional. The fourth funnel was for ventilation only.

Day 348

13 December | On this day in *Titanic* history

The following person was born on this day:
Sloper, Mr William Thompson – 13 December 1883 – aged 28 when on *Titanic*; first-class passenger (ticket price: £35 10s); rescued in lifeboat 7

The following person died on this day:
Smith, Miss Marion Elsie – 13 December 1940 – aged 49 when on *Titanic*; second-class passenger (ticket price: £13); rescued in lifeboat unknown

Titanic survivor quotes

Mr William Thompson Sloper

> 'Luckily for both Seward and me, Dorothy held onto my hand and demanded that we get into the boat with them. "We won't go unless you do," she said. "What do you say?" I asked Seward. "What's the difference we may as well go along with them." Finding seats for ourselves we sat in the lifeboat designed for 65 persons for about ten minutes, looking up into the rim of the faces of the passengers looking down at us trying to make up their minds to get in with us. After 19 people had finally made up their minds and had been lowered into the boat, the first officer asked for the last time through his megaphone, "Are there anymore who would like to get into this boat before we lower away?" When no one else made the move toward him, he gave the signal to lower us away. Then began a jerky descent to the surface of the ocean 60 feet below. Fortunately for us the three sailors knew their business, for in a few minutes they skilfully launched our boat on the surface of the ocean without accident.
>
> 'The sea was perfectly calm – not even a ripple on the surface. For the next hour and a half we just sat there and drifted farther and farther away.'

Mr Frank Herbert Morris (steward for first-class bathroom)

> 'Oh, we saw hundreds in the water, but they were not crying for help; they might have been unconscious, they might have been dead, we could not say to that.'

Titanic facts

Titanic was scheduled to leave New York for her return voyage on 20 April.

The last lifeboat arrived at the rescue ship *Carpathia* at 8.30 am.

Day 349

14 December | On this day in *Titanic* history

The following people were born on this day:
Witt, Mr Henry Dennis – 14 December 1874 – aged 37 when on *Titanic*; fireman; *Titanic* victim

Casswill, Mr Charles – 14 December 1877 – aged 34 when on *Titanic*; saloon steward; *Titanic* victim

Moss, Mr Albert Johan – 14 December 1882 – aged 29 when on *Titanic*; third-class passenger (ticket price: £7 15s 6d); rescued in lifeboat unknown

Clarke, Mrs Ada Maria – 14 December 1883 – aged 28 when on *Titanic*; second-class passenger (ticket price: £27); rescued in lifeboat 14

Fioravante, Sig. Giuseppe Bertoldo – 14 December 1888 – aged 23 when on *Titanic*; assistant scullion, *Titanic* victim

The following person died on this day:
Madigan, Miss Margaret – 14 December 1968 – aged 21 when on *Titanic*; third-class passenger (ticket price: £7 15s); rescued in lifeboat 15

Titanic survivor quotes

Mr James R. McGough

> 'Our boat was launched with 28 people; we, however, transferred 5 from one of the other boats after we were out in the ocean, which was some time after the ship went down.
>
> … we were directed to row away a short distance… feeling it would be but a short time until we would be taken back on the *Titanic*. We then rested on our oars; but after realising that the *Titanic* was really sinking, we rowed away for about half a mile, being afraid that the suction would draw us down.'

Mr Frederick Scott (greaser)

> 'I saw two boats then, and one of the boats was where the Officer pulled a revolver out and shot it between the ship and the boat and said, "If any man jumps into the boat I will shoot him like a dog."'

Titanic facts

The U.S. Senate Inquiry led to a recommendation that all U.S. passenger vessels carry 'sufficient lifeboats to accommodate every passenger and every member of the crew.'

The smallest number of people in *Titanic*'s lifeboats was twelve.

Day 350

15 December | On this day in *Titanic* history

The following people were born on this day:
Nieminen, Miss Manta Josefina – 15 December 1882 – aged 29 when on *Titanic*; third-class passenger (ticket price: £7 18s 6d); *Titanic* victim

White, Mr Edward Joseph – 15 December 1884 – aged 27 when on *Titanic*; glory hole steward in third class; *Titanic* victim

Thorn, Mr Harry – 15 December 1886 – aged 25 when on *Titanic*; assistant ship's cook; *Titanic* victim

Taylor, Mr George – 15 December 1888 – aged 23 when on *Titanic*; fireman; rescued in lifeboat 1

Welch, Mr William Harold – 15 December 1890 – aged 21 when on *Titanic*; cook; *Titanic* victim

Ström, Miss Telma Matilda – 15 December 1909 – aged 2 when on *Titanic*; third-class passenger (ticket price: £10 9s 3d); *Titanic* victim

Wells, Master Ralph Lester – 15 December 1909 – aged 2 when on *Titanic*; second-class passenger (ticket price: £23); rescued in lifeboat 14

The following people died on this day:
Hansen, Mrs Jennie Louise – 15 December 1952 – aged 42 when on *Titanic*; third-class passenger (ticket price: £14 2s 2d); rescued in lifeboat 11

Allen, Miss Elisabeth Walton – 15 December 1967 – aged 29 when on *Titanic*; first-class passenger (ticket price: £211 60s 9d); rescued in lifeboat 2

Sap, Mr Julius ('Jules') – 15 December 1966 – aged 21 when on *Titanic*; third-class passenger (ticket price: £9 10s); rescued in lifeboat 11

Bentham, Miss Lillian W. – 15 December 1977 – aged 19 when on *Titanic*; second-class passenger (ticket price: £13); rescued in lifeboat 12

Titanic survivor quote

Mrs Jennie Louise Hansen, quoted in *Racine Journal News*, Wednesday, 24 April 1912

> 'I stood there with Peter and Henry and when an officer told me to get into a lifeboat, I was willing but when they wouldn't let Peter go with me I just hung onto him and begged and begged them to let me stay too. An officer grabbed hold of me as I kissed Peter, and threw me into a lifeboat and I got an awful bump on the head that stunned me for a long while. There were 40 women in the little craft and it was crowded. As we were lowered to the water someone threw a baby over the deck, but instead of it going into the lifeboat it hit the water, and was drowned. Peter threw me my pony coat and I caught it all right.'

Day 351

16 December | On this day in *Titanic* history

The following people were born on this day:
Slemen, Mr Richard James – 16 December 1876 – aged 35 when on *Titanic*; second-class passenger (ticket price: £10 10s); *Titanic* victim

Curtis, Mr Arthur – 16 December 1885 – aged 26 when on *Titanic*; fireman; *Titanic* victim

Mallet, Mrs Antonine Marie – 16 December 1887 – aged 24 when on *Titanic*; second-class passenger (ticket price: £37 1d); rescued in lifeboat 10

Smith, Mr James William – 16 December 1887 – aged 24 when on *Titanic*; assistant baker; *Titanic* victim

Dyker, Mr Adolf Fredrik – 16 December 1888 – aged 23 when on *Titanic*; third-class passenger (ticket price: £13 18s); *Titanic* victim

McNamee, Mrs Eileen – 16 December 1892 – aged 19 when on *Titanic*; third-class passenger (ticket price: £16 2s); *Titanic* victim

Ryerson, Master John Borie – 16 December 1898 – aged 13 when on *Titanic*; first-class passenger (ticket price: £262 7s 6d); rescued in lifeboat 4

The following people died on this day:
O'Keeffe, Mr Patrick – 16 December 1939 – aged 21 when on *Titanic*; third-class passenger (ticket price: £7 15s); rescued in collapsible lifeboat B

Johnstone, Mr James – 16 December 1943 – aged 52 when on *Titanic*; saloon steward; rescued in lifeboat 2

Davies, Master John Morgan Jr – 16 December 1951 – aged 8 when on *Titanic*; second-class passenger (ticket price: £36 15s); rescued in lifeboat 14

Mr Patrick O'Keeffe, mentioned in *Cork Examiner*, Thursday, 16 May 1912

> 'An act of heroism was performed by Mr Patrick O'Keefe who, plunging into the sea from the steerage deck, managed to capture a collapsible raft on which he first pulled an Englishman from Southampton then a Guernsey Islander, and after that with the assistance of those he had already rescued, some 20 other men and women, who were finally landed safely on board the *Carpathia*. O'Keeffe's success in rescuing lives on the raft was one of the many providential avenues of escape provided for the steerage passengers.'

Day 352

17 December | On this day in *Titanic* history

The following people were born on this day:
Wittevrongel, Mr Camilius Aloysius – 17 December 1875 – aged 36 when on *Titanic*; third-class passenger (ticket price: £9 10s); *Titanic* victim

Edvardsson, Mr Gustaf Hjalmar – 17 December 1893 – aged 18 when on *Titanic*; third-class passenger (ticket price: £7 15s 6d); *Titanic* victim

The following people died on this day:
Lines, Mrs Elizabeth Lindsey – 17 December 1942 – aged 50 when on *Titanic*; first-class passenger (ticket price: £39 8s); rescued in lifeboat 9

Hurst, Mr Walter – 17 December 1964 – aged 23 when on *Titanic*; fireman; rescued in collapsible lifeboat B

Avery, Mr James Albert – 17 December 1915 – aged 20 when on *Titanic*; trimmer; rescued in lifeboat 15

Wednesday, 17 December 1958 Opens at Criterion

A NIGHT TO REMEMBER, screen play by Eric Ambler, from the book of the same name by Walter Lord; directed by Roy Baker; produced by William MacQuitty for the J. Arthur Rank Organization. At the Criterion Theatre, Seventh Avenue and Forty-fourth Street.

Running time: 122 minutes.

Cast

Lightoller .... Kenneth More
Mr Clarke .... Ronald Allen
Peuchen................ Robert Ayres
Mrs Lucas .... Honor Blackman
Captain Rostron ... Anthony Bushell
Murphy................ John Cairney
Colonel Gracie...... James Dyrenforth
Chairman........... Frank Lawton
Murdoch .... Richard Leech
Lucas.................. John Merivale
Captain Smith..... Laurence Naismith
Sir Richard* .... Patrick Waddington
Guggenheim .... Harold Goldblatt
Mrs Straus.......... Helen Misener

* a fictional character

Titanic fact

The two bronze outer propellers weighed 38 tons while the central propeller, made of a moulded construction, was 22 tons.

Day 353

18 December | On this day in *Titanic* history

The following people were born on this day:
Østby, Mr Engelhart Cornelius – 18 December 1847 – aged 64 when on *Titanic*; first-class passenger (ticket price: £61 19s 7d); *Titanic* victim

Douton, Mr William Joseph – 18 December 1856 – aged 55 when on *Titanic*; second-class passenger (ticket price: £26); *Titanic* victim

Lindström, Mrs Sigrid – 18 December 1856 – aged 55 when on *Titanic*; first-class passenger (ticket price: £27 14s 5d); rescued in lifeboat 6

White, Mrs Ella – 18 December 1856 – aged 55 when on *Titanic*; first-class passenger (ticket price: £135 12s 8d); rescued in lifeboat 8

Brewster, Mr George Henry – 18 December 1859 – aged 52 when on *Titanic*; bedroom steward in first class; *Titanic* victim

Taussig, Mrs Tillie – 18 December 1872 – aged 39 when on *Titanic*; first-class passenger (ticket price: £79 13s); rescued in lifeboat 8

Boyes, Mr John Henry – 18 December 1875 – aged 36 when on *Titanic*; saloon steward in first class; *Titanic* victim

Sharp, Mr Percival – 18 December 1882 – aged 29 when on *Titanic*; second-class passenger (ticket price: £26); *Titanic* victim

Peduzzi, Mr Giuseppe – 18 December 1887 – aged 24 when on *Titanic*; waiter; *Titanic* victim

Bowerman, Miss Elsie Edith – 18 December 1889 – aged 22 when on *Titanic*; first-class passenger (ticket price: £55); rescued in lifeboat 6

Pekoniemi, Mr Edvard – 18 December 1890 – aged 21 when on *Titanic*; third-class passenger (ticket price: £7 18s 6d); *Titanic* victim

Kallio, Mr Nikolai Erland – 18 December 1894 – aged 17 when on *Titanic*; third-class passenger (ticket price: £7 2s 6d); *Titanic* victim

The following person died on this day:
Caldwell, Master Alden Gates – 18 December 1992 – aged 10 months when on *Titanic*; second-class passenger (ticket price: £29); rescued in lifeboat 13

Titanic fact

The weight of each of *Titanic*'s funnels was 60 tons.

Day 354

19 December | On this day in *Titanic* history

The following people were born on this day:

Bannon, Mr John Joseph – 19 December 1877 – aged 34 when on *Titanic*; greaser; *Titanic* victim

Lindahl, Miss Agda Thorilda Viktoria – 19 December 1886 – aged 25 when on *Titanic*; third-class passenger (ticket price: £7 15s 6d); *Titanic* victim

Birkeland, Mr Hans Martin Monsen – 19 December 1890 – aged 21 when on *Titanic*; third-class passenger (ticket price: £7 15s 6d); *Titanic* victim

Rugg, Miss Emily – 19 December 1890 – aged 22 when on *Titanic*; second-class passenger (ticket price: £10 10s); rescued in lifeboat 12

Goldsmith, Master Frank John William – 19 December 1902 – aged 9 when on *Titanic*; third-class passenger (ticket price: £20 10s 6d); rescued in collapsible lifeboat C

Pålsson, Miss Torborg Danira – 19 December 1903 – aged 8 when on *Titanic*; third-class passenger (ticket price: £21 1s 6d); *Titanic* victim

The following person died on this day:

Johnson, Mrs Elisabeth Vilhelmina – 19 December 1968 – aged 27 when on *Titanic*; third-class passenger (ticket price: £11 2s 8d); rescued in lifeboat 15

Titanic survivor quote

Miss Emily Rugg, mentioned in *North American*, Saturday, 20 April 1912

> 'She was asleep when the ship struck the berg, and the jar aroused her. Looking out she saw a mass of ice. Throwing a coat about her, she went on deck and saw lifeboats being lowered. Returning to the cabin, she dressed, and then went to an adjoining cabin and aroused two women friends. Following this Miss Rugg ran up on deck and was taken in charge by some of the crew, who dragged her toward a lifeboat. She was lifted into the third boat from the last to leave the ship.
>
> 'She said that there seemed to be nearly seventy-five persons in the boat and that it was very much crowded. In the meantime a panic had started among those who remained on the *Titanic*. An Italian jumped from the steerage deck and fell into a lifeboat, landing upon a woman who had a baby in her arms. Miss Rugg saw the *Titanic* go down and declares but for the horror of it all, it might have been termed one of the grandest sights she ever saw. The boat seemed to have broken in half, and with all the lights burning brightly, the stern arose into the air, the lights being extinguished as it did so. A moment later the ship plunged beneath the surface.'

Day 355

20 December | On this day in *Titanic* history

The following people were born on this day:
McKane, Mr Peter David – 20 December 1865 – aged 46 when on *Titanic*; second-class passenger (ticket price: £26); *Titanic* victim

Hansen, Mrs Jennie Louise – 20 December 1866 – aged 45 when on *Titanic*; third-class passenger (ticket price: £14 2s 2d); rescued in lifeboat 11

Chisholm, Mr Roderick Robert Crispin – 20 December 1868 – aged 43 when on *Titanic*; draughtsman; *Titanic* victim

Gumery, Mr George – 20 December 1887 – aged 24 when on *Titanic*; mess steward; *Titanic* victim

Dean, Mr George Fox Hopkins – 20 December 1892 – aged 19 when on *Titanic*; assistant saloon steward; *Titanic* victim

The following people died on this day:
Daniel, Mr Robert Williams – 20 December 1940 – aged 27 when on *Titanic*; first-class passenger (ticket price: £30 10s); rescued in lifeboat 3

Turja, Miss Anna Sofia – 20 December 1982 – aged 18 when on *Titanic*; third-class passenger (ticket price: £9 16s 10d); rescued in lifeboat 15

Keane, Miss Nora Agnes – 20 December 1944 – aged 46 when on *Titanic*; second-class passenger (ticket price: £12 7s); rescued in lifeboat 10

Titanic survivor quote

Robert Williams Daniel, mentioned in *Worcester Evening Gazette*, Friday, 19 April 1912

> 'Among the passengers I talked with was Robert W. Daniel, formerly of Virginia, but known as a Philadelphia banker. Mr Daniel said: "I was partially undressed when the jar came. It was about 12 o'clock. It was hardly noticeable to shake you badly, but still enough to disturb you. I went up on deck, and a lot more came running up. Some of them went back to bed." Mr Daniel was smooth shaven and looked in fair shape. He wore a raincoat, a soft white collar, but was without a tie.'

Titanic fact

It took *Carpathia* three and a half hours to arrive to pick up survivors.

Day 356

21 December | On this day in *Titanic* history

The following people were born on this day:
Jacobsen, Mrs Gunhilde Oline 'Lena' – 21 December 1848 – aged 63 when on *Titanic*; third-class passenger (ticket price: £8 2s 3d); *Titanic* victim

Rosenshine, Mr George – 21 December 1866 – aged 45 when on *Titanic*; first-class passenger (ticket price: £79 4d); *Titanic* victim

Dulles, Mr William Crothers – 21 December 1872 – aged 39 when on *Titanic*; first-class passenger (ticket price: £29 14s); *Titanic* victim

Henriksson, Miss Jenny Lovisa – 21 December 1883 – aged 28 when on *Titanic*; third-class passenger (ticket price: £7 15s 6d); *Titanic* victim

Malachard, Mr Jean-Noël – 21 December 1886 – aged 25 when on *Titanic*; second-class passenger (ticket price: £15 11d); *Titanic* victim

Dyer, Mr Henry Ryland – 21 December 1887 – aged 24 when on *Titanic*; senior assistant fourth engineer, *Titanic* victim

Maioni, Miss Roberta Elizabeth Mary – 21 December 1891 – aged 20 when on *Titanic*; first-class passenger (ticket price: £86 10s); rescued in lifeboat 8

The following person died on this day:
Clark, Mrs Virginia Estelle – 21 December 1958 – aged 26 when on *Titanic*; first-class passenger (ticket price: £136 15s 7d); rescued in lifeboat 4

Titanic survivor quote

Miss Roberta Elizabeth Mary Maioni, quoted in *Daily Express*, 1926

> 'I was carried by a swarm of other passengers to the boat deck, and shall never forget the strange sight that met my eyes. There were pieces of ice all over the deck and groups of men and women, looking gaunt and fearful in their night attire or in odd garments hastily donned. Some of them were talking calmly, firmly believing that her watertight compartment would save the *Titanic* from sinking. Others were frantic with excitement or dumb with terror, huddled closely together in silence as though they knew they were about to be parted by death.
>
> 'There were men swearing horribly and women quietly sobbing and I knew that many of them were praying as they never prayed before. But there was no panic, and I, with the fortitude of youth, looked on in wonder.'

Titanic fact

The name of the submersible used to discover *Titanic* in 1985 was Alvin.

Day 357

22 December | On this day in *Titanic* history

The following people were born on this day:
O'Connor, Mr John – 22 December 1882 – aged 29 when on *Titanic*; trimmer; rescued in collapsible lifeboat B

Jensen, Mr Svend Lauritz – 22 December 1895 – aged 16 when on *Titanic*; third-class passenger (ticket price: £7 17s 1d); *Titanic* victim

Titanic survivor quotes

Mr Norman C. Chambers

At the time of the collision I was in bed, and I noticed no very great shock, the loudest noise by far being that of jangling chains whipping along the side of the ship. This passed so quickly that I assumed something had gone wrong with the engines on the starboard side.'

Captain Arthur H. Rostron (captain of the *Carpathia*)

'The wireless operator had taken the message and run with it up to the bridge, and gave it to the first officer who was in charge, with a junior officer with him, and both ran down the ladder to my door and called me. I had only just turned in. It was an urgent distress signal from the *Titanic*, requiring immediate assistance and giving me his position.

'Immediately on getting the message, I gave the order to turn the ship around, and immediately I had given that order I asked the operator if he was absolutely sure it was a distress signal from the *Titanic*. I asked him twice.

'I did not ask him. He simply told me that he had received a distress signal from the *Titanic*, requiring immediate assistance, and gave me his position; and he assured me he was absolutely certain of the message.

'In the meantime I was dressing, and I picked up our position on my chart, and set a course to pick up the *Titanic*. The course was north 52 degrees west true 58 miles from my position. I then sent for the chief engineer. In the meantime I was dressing and seeing the ship put on her course. The chief engineer came up. I told him to call another watch of stokers and make all possible speed to the *Titanic*, as she was in trouble. He ran down immediately and told me my orders would be carried out at once.

'After that I gave the first officer, who was in charge of the bridge, orders to knock off all work which the men were doing on deck, the watch on deck, and prepare all our lifeboats, take out the spare gear, and have them all ready for turning outboard.'

Day 358

23 December | On this day in *Titanic* history

The following people were born on this day:
Coleman, Mr Albert Edward – 23 December 1883 – aged 28 when on *Titanic*; saloon steward in first class; *Titanic* victim

Sivier, Mr William – 23 December 1888 – aged 23 when on *Titanic*; steward in third class; *Titanic* victim

The following person died on this day:
Toomey, Miss Ellen Mary – 23 December 1933 – aged 50 when on *Titanic*; second-class passenger (ticket price: £10 10s); rescued in lifeboat 9

Titanic survivor quote

Miss Ellen Mary Toomey, quoted in *Indianapolis Star*, Tuesday, 23 April 1912

> 'A lady by the name of Watts and her daughter were in the cabin I occupied... Before we knew it we found ourselves in a lifeboat and being lowered. There were three sailors in the boat and two men... There were thirty-five in our boat, which was the second or third boat to put off... The *Titanic* was gradually sinking and when we were about two miles away, we could hear the awful shrieks of the drowning people...we were in the small boats about seven hours...'

Arthur H. Rostron (captain of the *Carpathia*)

> 'The silence after the first frantic appeal for relief was ominous to me. Our Marconi sent out rays that scraped the sky in vain, but there was no response whatever to any of our inquiries. I swung the *Carpathia* around straight to the position the poor *Titanic*'s first aerogram said she occupied. Our engines were put at top speed.
>
> 'The silence of the air so far as the *Titanic* was concerned made me shudder as we sped on our way to the rescue. I realised what it meant. On and on we sped. Our stokers never worked harder. When the first faint daylight came and I knew we were still miles from the spot of the tragedy, I felt we should arrive too late to be of any service.'

Mrs Helen Bishop, witness at United States Senate Inquiry (Day 11)

> 'The captain told him something in an undertone. He came back and told six of us, who were standing with his wife, that we had better put on our lifebelts. I had gotten down two flights of stairs to tell my husband, who had returned to the stateroom for a moment, before I heard the captain announce that the lifebelts should be put on. That was about three or four minutes later that the captain announced the lifebelts should be put on. We came back upstairs and found very few people up.'

Day 359

24 December | On this day in *Titanic* history

The following people were born on this day:
Hemming, Mr Samuel Ernest – 24 December 1868 – aged 43 when on *Titanic*; lamp trimmer; rescued in lifeboat 4

Äijö-Nirva, Mr Isakki Antonpoika – 24 December 1870 – aged 41 when on *Titanic*; third-class passenger (ticket price: £7 2s 6d); *Titanic* victim

Scanlan, Mr James – 24 December 1891 – aged 20 when on *Titanic*; third-class passenger (ticket price: £7 14s 6d); *Titanic* victim

Thayer, Mr John Borland Jr – 24 December 1894 – aged 17 when on *Titanic*; first-class passenger (ticket price: £110 17s 8d); rescued in collapsible lifeboat B

The following people died on this day:
Stephenson, Mrs Martha – 24 December 1934 – aged 52 when on *Titanic*; first-class passenger (ticket price: £78 5s 4d); rescued in lifeboat 4

Daly, Mr Peter Dennis – 24 December 1932 – aged 51 when on *Titanic*; first-class passenger (ticket price: £26 11s); rescued in lifeboat unknown

Joseph Bruce Ismay, witness at United States Senate Inquiry (Day 1)

> 'The ship was specially constructed so that she would float with any two compartments full of water. I think I am right in saying that there are very few ships – perhaps I had better not say that, but I will continue, now that I have begun it – I believe there are very few ships today of which the same can be said. When we built the *Titanic* we had that especially in mind. If this ship had hit the iceberg stern on, in all human probability she would have been here today.'

Titanic survivor quote

Mr Archibald Gracie

> 'The crew seemed to resent my working with them, but they were very glad when I worked with them later on. Every opportunity I got to help, I helped. When I arrived on the other side, as I have said, there were these women, and of a sudden I heard the cry that there was room for more women on the port side; so I grabbed by the arm these two ladies, Miss Evans and Mrs Browne, and conducted them to the port side. But I did not get but half way – that is, directly at the bow – when the crew made what you might call a dead line, and said, "No men are allowed beyond this line." So I let the ladies go beyond, and then about six ladies followed after the two that I had particular charge of.'

Day 360

25 December | On this day in *Titanic* history

The following people were born on this day:
White, Mr Percival Wayland – 25 December 1857 – aged 54 when on *Titanic*; first-class passenger (ticket price: £77 5s 9d); *Titanic* victim

Hart, Mr Benjamin – 25 December 1864 – aged 47 when on *Titanic*; second-class passenger (ticket price: £26 5s); *Titanic* victim

Head, Mr Christopher – 25 December 1869 – aged 42 when on *Titanic*; first-class passenger (ticket price: £42 10s); *Titanic* victim

Andersson, Mrs Alfrida Konstantia Brogren – 25 December 1872 – aged 39 when on *Titanic*; third-class passenger (ticket price: £31 5s 6d); *Titanic* victim

Rothes, Lucy Noël Martha, Countess of – 25 December 1878 – aged 33 when on *Titanic*; first-class passenger (ticket price: £86 10s); rescued in lifeboat 8

Copperthwaite, Mr Albert Harry – 25 December 1883 – aged 28 when on *Titanic*; fireman; *Titanic* victim

Foley, Mr William – 25 December 1891 – aged 20 when on *Titanic*; third-class passenger (ticket price: £7 15s); *Titanic* victim

Light, Mr Christopher William – 25 December 1891 – aged 20 when on *Titanic*; fireman; *Titanic* victim

Tannūs, Mrs Thamīn – 25 December 1895 – aged 16 when on *Titanic*; third-class passenger (ticket price: £8 10s 4d); rescued in lifeboat 14

The following people died on this day:
Ward, Miss Annie Moore – 25 December 1955 – aged 37 when on *Titanic*; first-class passenger (ticket price: £512 6s 7d); rescued in lifeboat 3

Coutts, Master William Loch 'Willie' – 25 December 1957 – aged 9 when on *Titanic*; third-class passenger (ticket price: £15 18s); rescued in lifeboat 2

Titanic survivor quote

Mrs Lucy Noël Martha, Countess of Rothes

> 'While on board the *Carpathia*, her Ladyship earned the title of "the plucky little Countess", by the crew, for she was instantly ready to help the sick in steerage and helped make clothes for the babies. A stewardess said, "You have made yourself famous by rowing the boats," and her Ladyship replied, "I hope not; I have done nothing."

Day 361

26 December | On this day in *Titanic* history

The following people were born on this day:
Graham, Mr Thomas Gibson – 26 December 1883 – aged 28 when on *Titanic*; fireman; rescued in lifeboat unknown

Carr, Mr Richard Stephen – 26 December 1884 – aged 27 when on *Titanic*; trimmer; *Titanic* victim

Bolhuis, Mr Hendrik – 26 December 1884 – aged 27 when on *Titanic*; larder cook; *Titanic* victim

Peltomäki, Mr Nikolai Johannes – 26 December 1886 – aged 25 when on *Titanic*; third-class passenger (ticket price: £7 18s 6d); *Titanic* victim

White, Mr Richard Frasar – 26 December 1890 – aged 21 when on *Titanic*; first-class passenger (ticket price: £77 5s 9d); *Titanic* victim

Enander, Mr Ingvar – 26 December 1890 – aged 21 when on *Titanic*; second-class passenger (ticket price: £13); *Titanic* victim

Talbot, Mr George Frederick Charles – 26 December 1891 – aged 41 when on *Titanic*; steward; *Titanic* victim

Lefebvre, Miss Ida – 26 December 1908 – aged 3 when on *Titanic*; third-class passenger (ticket price: £25 9s 4d); *Titanic* victim

Titanic survivor quote

Mr William Burke, witness at United States Senate Inquiry (Day 9)

> 'When there were no more women to be had around the deck the chief officer gave the order for the boat to be lowered. I might say that about the last woman that was about to be passed in slipped, and was about to fall between the ship and the boat when I caught her. I just saved her from falling. Her head passed toward the next deck below. A passenger caught her by the shoulders and forced me to leave go. It was my intention to pull her back in the boat. He would not let go of the woman, but pulled her right on the ship.
>
> 'The boat was lowered then into the water. One of the sailors took an oar, and I took an oar, and the only other member of the crew, a fireman, got an oar. The sailor steered the boat, and we rowed away from the ship. We got probably about a quarter of a mile away, and remained there. We saw pretty well the last of the ship – the *Titanic*.
>
> 'We remained drifting about practically all night. At one time we were tied up with three boats together, until I gave the order myself in that boat to cut us adrift that we might go to a collapsible boat that was in distress.'

Day 362

27 December | On this day in *Titanic* history

The following people were born on this day:
Franklin, Mr Alan Vincent – 27 December 1883 – aged 28 when on *Titanic*; saloon steward in second class; *Titanic* victim

Jensen, Mr Charles Valdemar – 27 December 1887 – aged 24 when on *Titanic*; saloon steward; *Titanic* victim

Reed, Mr James George – 27 December 1893 – aged 18 when on *Titanic*; third-class passenger (ticket price: £7 5s); *Titanic* victim

The following people died on this day:
Allen, Mr Ernest Frederick – 27 December 1968 – aged 24 when on *Titanic*; trimmer; rescued in collapsible lifeboat B

Kelly, Miss Mary – 27 December 1950 – aged 22 when on *Titanic*; third-class passenger (ticket price: £7 15s); rescued in collapsible lifeboat D

Nilsson, Miss Berta Olivia – 27 December 1976 – aged 18 when on *Titanic*; third-class passenger (ticket price: £7 15s 6d); rescued in collapsible lifeboat D

Titanic survivor quote

Mr Norman C. Chambers, witness at United States Senate Inquiry (Day 13)

> 'All this time I considered that the lifeboats were merely a precaution and, upon my wife's suggestion, we moved up forward of the entry from the deck house.
>
> 'There were still quite a number of passengers coming out, the stewards standing there directing them to the boats aft. Instead of going aft, we stepped behind the projection of this entry, which was of the vestibule type and waited until people had apparently ceased coming and the steward was no longer there. Then we started forward again, and, as nearly as I can remember, stopped at the last one of the forward starboard group of lifeboats. This was already swung out level with the deck, and to my eyes, appeared sufficiently loaded.
>
> 'However, my wife said that she was going in that boat, and proceeded to jump in, calling me to come. As I knew she would get out again had I not come, I finally jumped into the boat, although I did not consider it, from the looks of things, safe to put very many more people in that boat. As I remember it, there were two more men, both called by their wives, who jumped in after I did. One of them – a German, I believe – told me, as I recollect it, later on the *Carpathia* that he had looked around and had seen no one else and no one to ask whether he should go in or not, and he jumped in.'

Day 363

28 December | On this day in *Titanic* history

The following people were born on this day:
Buss, Miss Kate – 28 December 1875 – aged 36 when on *Titanic*; second-class passenger (ticket price: £13); rescued in lifeboat 9

Paintin, Mr James Arthur – 28 December 1882 – aged 29 when on *Titanic*; captain's steward (Tiger); *Titanic* victim

Wright, Mr William – 28 December 1864 – aged 47 when on *Titanic*; glory hole steward; rescued in lifeboat 13

Becker, Miss Marion Louise – 28 December 1907 – aged 4 when on *Titanic*; second-class passenger (ticket price: £39); rescued in lifeboat 11

The following people died on this day:
Taylor, Mr George – 28 December 1931 – aged 23 when on *Titanic*; fireman; rescued in lifeboat 1

Ryerson, Mrs Emily Maria – 28 December 1939 – aged 48 when on *Titanic*; first-class passenger (ticket price: £262 7s 6d); rescued in lifeboat 4

Hämäläinen, Mrs Anna Lydia Amanda – 28 December 1945 – aged 23 when on *Titanic*; second-class passenger (ticket price: £14 10s); rescued in lifeboat unknown

Driscoll, Miss Bridget – 28 December 1976 – aged 27 when on *Titanic*; third-class passenger (ticket price: £7 15s); rescued in collapsible lifeboat D

McCoy, Miss Alicia – 28 December 1959 – aged 26 when on *Titanic*; third-class passenger (ticket price: £23 5s); rescued in lifeboat 16

Kelly, Miss Anna Katherine – 28 December 1969 – aged 20 when on *Titanic*; third-class passenger (ticket price: £7 15s); rescued in lifeboat 16

Titanic survivor quote

Mr Archie Jewell (lookout)

> 'We put all the women in that was there, and children. Up to that time there was not many people; we could not get them up; they were rather afraid to go into the boat; they did not think there was anything wrong.
>
> 'We stopped there and watched her gradually sink away. We could see the people about on the deck before the lights went out. As she went away by the head so the lights went out, and we heard some explosions as she was going down. But all the lights went out and we could only see a black object in front of us.'

Day 364

29 December | On this day in *Titanic* history

The following people were born on this day:

Brice, Mr Walter Thomas – 29 December 1869 – aged 42 when on *Titanic*; able seaman; rescued in lifeboat 11

Noss, Mr Henry – 29 December 1881 – aged 30 when on *Titanic*; fireman; rescued in lifeboat unknown

Windeløv, Mr Einar – 29 December 1890 – aged 21 when on *Titanic*; third-class passenger (ticket price: £7 5s); *Titanic* victim

The following person died on this day:

Graham, Mrs Edith – 29 December 1924 – aged 58 when on *Titanic*; first-class passenger (ticket price: £153 9s 3d); rescued in lifeboat 3

Titanic survivor quote

Mrs Edith Graham, quoted in *New York Times*, Saturday, 20 April 1912

> 'Some seemed to think that we ought to stay very near, because, they said, the ship wouldn't sink anyway. Others were in favor of going away out.
>
> 'We went out about three-quarters of a mile, I think, following another boat which carried some green lanterns. That was the only thing we had to go by. Behind us the lights on the *Titanic* went out, and in an hour and a half the big ship went down. It was in that hour and a half that the passengers got their fright. We couldn't tell what was going on on the ship – but those shrieks and cries! I'll never forget them. And there were many shots, I don't know how many.
>
> 'I saw many dead. That was frightful. I saw Mrs Harris on the *Carpathia*. She appeared dazed and didn't say anything. I saw Mrs Astor, too. She didn't appear ill when I saw her. Throughout the journey to New York I didn't see Mr Ismay. You see, he remained in his cabin.'

Mr Graham, at this point, asked his wife whether she ever wanted to go abroad again.

> 'Never,' she replied, without hesitating. 'And how about you, Margaret?' he asked.
>
> 'Well, replied the daughter, hesitating a moment, 'at least not immediately.'

Day 365

30 December | On this day in *Titanic* history

The following people were born on this day:

De Messemaeker, Mr Guillaume Joseph – 30 December 1875 – aged 36 when on *Titanic*; third-class passenger (ticket price: £17 8s); rescued in lifeboat 15

Mernagh, Mr Robert – 30 December 1883 – aged 28 when on *Titanic*; third-class passenger (ticket price: £7 15s); *Titanic* victim

Shine, Miss Ellen Natalia – 30 December 1891 – aged 20 when on *Titanic*; third-class passenger (ticket price: £7 16s 7d); rescued in lifeboat unknown

Newsom, Miss Helen Monypeny – 30 December 1892 – aged 19 when on *Titanic*; first-class passenger (ticket price: £26 5s 8d); rescued in lifeboat 5

Vanderplancke, Miss Augusta Maria – 30 December 1893 – aged 18 when on *Titanic*; third-class passenger (ticket price: £18); *Titanic* victim

The following people died on this day:

Trout, Mrs Jessie L. – 30 December 1930 – aged 27 when on *Titanic*; second-class passenger (ticket price: £12 13s); rescued in lifeboat 9

Nakid, Mr Sahid – 30 December 1926 – aged 20 when on *Titanic*; third-class passenger (ticket price: £15 14s 10d); rescued in lifeboat unknown

Captain Arthur H. Rostron (captain of the *Carpathia*)

> 'That was the last message we got. It was either "Engine room nearly full" or "Engine room full," or "Engine room-filling." The exact words I could not give you. The impression was quite enough for me, as to the condition the ship was in.'

Titanic survivor quote

Mr Berk Pickard

> 'The steerage passengers, so far as I could see, were not prevented from getting up to the upper decks by anybody, or by closed doors, or anything else. While I was on the ship no one realized the real danger, not even the stewards. If the stewards knew, they were calm. It was their duty to try to make us believe there was nothing serious. Nobody was prevented from going up. They tried to keep us quiet. They said, "Nothing serious is the matter." Perhaps they did not know themselves. I did not realize it, the whole time, even to the last moment. Of course, I would never believe such a thing could happen.'

Day 366

31 December | On this day in *Titanic* history

The following people were born on this day:

Jukes, Mr Henry James – 31 December 1873 – aged 38 when on *Titanic*; greaser; *Titanic* victim

Cumings, Mrs Florence Briggs – 31 December 1876 – aged 35 when on *Titanic*; first-class passenger (ticket price: £71 5s 8d); rescued in lifeboat 4

Beesley, Mr Lawrence – 31 December 1877 – aged 34 when on *Titanic*; second-class passenger (ticket price: £13s); rescued in lifeboat 13

Hopgood, Mr Roland John C. – 31 December 1882 – aged 29 when on *Titanic*; fireman, *Titanic* victim

Wright, Mr Frederick – 31 December 1887 – aged 24 when on *Titanic*; squash court steward; *Titanic* victim

Savage, Mr Charles Joseph – 31 December 1888 – aged 23 when on *Titanic*; steward; rescued in lifeboat 11

Taylor, Mr Bernard Cuthbert – 31 December 1889 – aged 22 when on *Titanic*; steward in third class; rescued in lifeboat 15

Cohen, Mr Gurshon 'Gus' – 31 December 1893 – aged 19 when on *Titanic*; third-class passenger (ticket price: £8 1s); rescued in lifeboat 12

The following people died on this day:

Couper, Mr Robert Frederick William – 31 December 1941 – aged 30 when on *Titanic*; fireman; rescued in lifeboat 3

Douglas, Mrs Mary Hélène – 31 December 1954 – aged 27 when on *Titanic*; first-class passenger (ticket price: £247 10s 5d); rescued in lifeboat 6

Titanic survivor quote

Mr Lawrence Beesley, quoted in *Toronto Daily Star*, Friday, 19 April 1912

> 'Presently, about 2 am, as near as I can remember, we observed her settling very rapidly, with the bows and the bridge completely under water. She slowly tilted straight on end with the stern vertically upwards, and as she did so the lights in the cabins and saloons, which had not flickered for a moment since we left, died out, came on again for a single flash, and finally went altogether. At the same time the machinery roared down through the vessel with a rattle and a groaning that could be heard for miles. But this was not quite the end. To our amazement she remained in that upright position for a time, which I estimate as five minutes, others in the boat say less, but it was certainly some minutes while we watched at least 150 feet of the *Titanic* towering up above the level of the sea and looming black against the sky.'

Bibliography

Bäbler, Günter, *Guide to the Crew of Titanic: The Structure of Working Aboard the Legendary Liner* (The History Press, 2017)

Behe, George, *On Board RMS Titanic: Memories of the Maiden Voyage* (The History Press, 2012)

Behe, George, *Voices from the Carpathia* (The History Press, 2015)

Beveridge, Bruce; Andrews, Scott; Hall, Steve; Kilstorner, Daniel and Braunschweiger, Art, *Titanic the Ship Magnificent: Volumes One and Two*; (The History Press, 2008)

Bryceson, Dave, *The Titanic Disaster* (Patrick Stephens Limited, 1997)

Caren, Eric and Goldman, Steve, *Extra Titanic: The Story of the Disaster in the Newspapers of the Day* (Bradley Books, 1998)

Fitch, Tad; Layton, J. Kent and Wormstedt, Bill, *On a Sea of Glass: The Life & Loss of the RMS Titanic* (Amberley, 2012)

Halpern, Samuel, *Report into the Loss of the SS Titanic: a Centennial Reappraisal* (The History Press, 2011)

Lord, Walter, *A Night to Remember* (Longman, 1956)

Lord, Walter, *The Night Lives On* (Avon Books, 1986)

Maltin, Tim, *101 Things You Thought You Knew About the Titanic – But Didn't!* (Malt House Books, 2011)

Maltin, Tim, *Titanic First Accounts* (Penguin Classics, 2012)

Merideth, Lee W., *Titanic Names: A Complete List of Passengers and Crew on the Fateful Voyage* (Rocklin Press, 2007)

Nesmeyanov, Eugene, *The Titanic Expeditions* (The History Press, 2018)

Streater, Les, *Titanic and the Impact of Lost Time* (Maritime Publishing Concepts, 2019)

Websites

www.encyclopedia-titanica.org
www.titanicinquiry.org
www.britishtitanicsociety.com

Appendix

Further information on the seven other Chinese passengers aboard *Titanic*
Incomplete details of the Chinese passengers (credit to *The Six*: the critically-acclaimed documentary from director Arthur Jones, historian Steven Schwankert and executive producer James Cameron, telling the untold story of RMS *Titanic*'s Chinese survivors)

Bing, Mr Lee – born 1887 in Hengtang, Guangdong, China – aged 25 when on *Titanic*; died 1943
Chip, Mr Chang – born 1880 – aged 32 when on *Titanic*; died 1914
Foo, Mr Cheong – born 1880 – aged 32 when on *Titanic*; rescued in lifeboat 13
Lam, Mr Ah – born 1874 – aged 38 when on *Titanic*
Hee, Mr Ling – born 1880 – aged 32 when on *Titanic*
Ling, Mr Lee did not survive the sinking of *Titanic*; died 15 April 1912
Lam, Mr Len did not survive the sinking of *Titanic*; died 15 April 1912